Air Quality Monitoring and Forecasting

Special Issue Editors

Pius Lee
Rick Saylor
Jeff McQueen

MDPI • Basel • Beijing • Wuhan • Barcelona • Belgrade

MDPI

Special Issue Editors
Pius Lee, Rick Saylor and Jeff McQueen
NOAA
USA

Editorial Office
MDPI
St. Alban-Anlage 66
Basel, Switzerland

This edition is a reprint of the Special Issue published online in the open access journal *Atmosphere* (ISSN 2073-4433) from 2017–2018 (available at: http://www.mdpi.com/journal/atmosphere/special_issues/air_monitoring).

For citation purposes, cite each article independently as indicated on the article page online and as indicated below:

Lastname, F.M.; Lastname, F.M. Article title. *Journal Name* **Year**, *Article number*, page range.

First Edition 2018

ISBN 978-3-03842-839-8 (Pbk)
ISBN 978-3-03842-840-4 (PDF)

Table of Contents

About the Special Issue Editors

Pius Lee received his Ph.D. in Atmospheric and Ecological Science in 1994 under the supervision of Professor Hiroshiro Kitada of the Toyohashi Technical University, Japan. Dr. Lee is an adjunct professor in the Center for Spatial Information Science and Systems in George Mason University, VA USA. Dr. Lee is leading the NOAA Air Resources Laboratory project supporting the research and operational eligibility testing of the US National Air Quality Forecasting Capability (NAQFC). Dr. Lee and Co-Editor McQueen started the NAQFC team in 2003 and has since been instrumental in advancing the science, scope of service and accuracy of NAQFC.

Rick Saylor is a Physical Scientist with the National Oceanic and Atmospheric Administration (NOAA) Air Resources Laboratory, Atmospheric Turbulence and Diffusion Division, in Oak Ridge, Tennessee, and an Associate Research Professor at the University of Tennessee in the Department of Civil and Environmental Engineering in Knoxville, Tennessee. He obtained a B.S., M.S. and Ph.D. in Chemical Engineering at the University of Kentucky and has over 28 years of experience as a researcher in atmospheric chemistry and air quality. His expertise includes analysis and interpretation of field measurement data of trace gases and particulate matter and modeling of atmospheric physics and chemistry across all scales from small-scale process models to regional- and global-scale 3-D models.

Jeff McQueen has been NOAA/NWS National Center for Environmental Prediction Air Quality and dispersion modeling program leader since 2003. Mr. McQueen leads efforts to develop, test, implement and maintain the NOAA air quality and dispersion modeling systems. Mr. McQueen and his staff have established a national capability for ozone and particulate matter prediction by driving EPA's CMAQ model with atmospheric fields from the NCEP North American Model (NAM) fields. Mr. McQueen has also chaired the Weather Research and Forecasting (WRF) Model Science Board and is a member of the WRF Model Atmospheric Chemistry and Ensemble Working Groups and OFCM Atmospheric Transport and Dispersion Joint Action Group. He currently co-chairs the Aerosols and Atmospheric Composition Unified Modeling Strategic Implementation Plan Working Group for NWS. Mr. McQueen's research interests include atmospheric modeling and use of physical parameterizations and coupled models (e.g., boundary layer, land-surface) to quantify atmospheric dispersion uncertainties.

atmosphere

MDPI

Editorial

Air Quality Monitoring and Forecasting

Pius Lee [1,2,*], Rick Saylor [1] and Jeff McQueen [3]

[1] NOAA/Air Resources Laboratory, NOAA Center for Weather and Climate Prediction,
College Park, MD 20740, USA; rick.saylor@noaa.gov
[2] Center for Spatial Information Science and Systems, George Mason University, Fairfax, VA 22030, USA
[3] NOAA/NCEP/Environmental Modeling Center, NOAA Center for Weather and Climate Prediction,
College Park, MD 20740, USA; jeff.mcqueen@noaa.gov
* Correspondence: pius.lee@noaa.gov

Received: 27 February 2018; Accepted: 28 February 2018; Published: 1 March 2018

Air quality forecasting is a vital tool for local health and air managers to make informed decisions on mitigation measures to reduce public exposure risk. Given a forecast of impending poor air quality, air quality managers may issue car-pooling advisories, authorize free public transportation or impose other mitigation and warning measures. Air composition monitoring and exposure records can inform long-term trends of major air pollutants and their health impacts. Epidemiologists use long term composition data to understand air pollution related diseases and mortality rates to support public health policies. This Special Issue highlights the interplay and co-benefit of air quality monitoring and forecasting.

Public health is under a constant threat by air pollution across the world in various degrees and manifestations. In China, rapid economic growth has resulted in increased occurrences of poor air quality. In this special issue, Lu et al. [1] of Wuhan University and Zhou et al. [2] of Chengdu University respectively studied urban haze and the distribution of multiple pollutants in China. Lyu et al. [3] of Tsinghua University and Georgia Institute of technology advanced particulate matter forecasts in China. Zhao et al. [4] of Nanjing University studied the strong response of emission controls during a recent Youth Olympics event in Nanjing, China.

Ground truth of air constituent concentrations is determined by measurements. Woodall et al. [5] of the US EPA conducted an intriguing study about hand held air composition measurement devices. Constantin et al. [6] of the University of Galati, Romania, used an ultralight trike and flux calculations to measure nitrogen dioxide vertical column density. Bray et al. [7] of North Carolina State University characterized pollutants emitted from coal-fired power plants in Eastern USA. Baker and Pan [8] of NOAA's Air Resources Laboratory, developed a software tool which utilized many in-situ and surface monitoring network measurements to evaluate forecast model performance. Lightstone et al. [9] of City College of New York explored neural networks as a means for air quality forecasts. Environmental and Climate Change Canada's Munoz-Alpizar et al. [10] studied the impact wildfire pollution on public health, and Ménard and Deshaies-Jacques [11,12] analyzed chemical data evaluations by cross-validation statistical analysis.

It is clear that air pollution remains a global problem and that air quality monitoring, forecasting and mitigation begins as a local effort conducted in concert with global partners. The articles selected in this Special Issue speak volumes to this fact many times over across the globe. We thank the editing office for their excellent support to realize this herculean achievement to collect and publish the cutting edge articles in this issue.

Conflicts of Interest: The authors declare no conflict of interest.

Atmosphere **2018**, *9*, 89

References

1. Lu, W.; Ai, T.; Zhang, X.; He, Y. An Interactive Web Mapping Visualization of Urban Air Quality Monitoring Data of China. *Atmosphere* **2017**, *8*, 148. [CrossRef]
2. Zhou, T.; Sun, J.; Yu, H. Temporal and Spatial Patterns of China's Main Air Pollutants: Years 2014 and 2015. *Atmosphere* **2017**, *8*, 137. [CrossRef]
3. Lyu, B.; Zhang, Y.; Hu, Y. Improving $PM_{2.5}$ Air Quality Model Forecasts in China Using a Bias-Correction Framework. *Atmosphere* **2017**, *8*, 147. [CrossRef]
4. Zhao, H.; Zheng, Y.; Li, T. Air Quality and Control Measures Evaluation during the 2014 Youth Olympic Games in Nanjing and its Surrounding Cities. *Atmosphere* **2017**, *8*, 100. [CrossRef]
5. Woodall, G.M.; Hoover, M.D.; Williams, R.; Benedict, K.; Harper, M.; Soo, J.-C.; Jarabek, A.M.; Stewart, M.J.; Brown, J.S.; Hulla, J.E.; et al. Interpreting Mobile and Handheld Air Sensor Readings in Relation to Air Quality Standards and Health Effect Reference Values: Tackling the Challenges. *Atmosphere* **2017**, *8*, 182. [CrossRef] [PubMed]
6. Constantin, D.-E.; Merlaud, A.; Voiculescu, M.; Dragomir, C.; Georgescu, L.; Hendrick, F.; Pinardi, G.; Van Roozendael, M. Mobile DOAS Observations of Tropospheric NO_2 Using an UltraLight Trike and Flux Calculation. *Atmosphere* **2017**, *8*, 78. [CrossRef]
7. Bray, C.D.; Battye, W.; Uttamang, P.; Pillai, P.; Aneja, V.P. Characterization of Particulate Matter ($PM_{2.5}$ and PM_{10}) Relating to a Coal Power Plant in the Boroughs of Springdale and Cheswick, PA. *Atmosphere* **2017**, *8*, 186. [CrossRef]
8. Baker, B.; Pan, L. Overview of the Model and Observation Evaluation Toolkit (MONET) Version 1.0 for Evaluating Atmospheric Transport Models. *Atmosphere* **2017**, *8*, 210. [CrossRef]
9. Lightstone, S.D.; Moshary, F.; Gross, B. Comparing CMAQ Forecasts with a Neural Network Forecast Model for $PM_{2.5}$ in New York. *Atmosphere* **2017**, *8*, 161. [CrossRef]
10. Munoz-Alpizar, R.; Pavlovic, R.; Moran, M.D.; Chen, J.; Gravel, S.; Henderson, S.B.; Ménard, S.; Racine, J.; Duhamel, A.; Gilbert, S.; et al. Multi-Year (2013–2016) $PM_{2.5}$ Wildfire Pollution Exposure over North America as Determined from Operational Air Quality Forecasts. *Atmosphere* **2017**, *8*, 179. [CrossRef]
11. Ménard, R.; Deshaies-Jacques, M. Evaluation of analysis by cross-validation. Part I: Using verification metrices. *Atmosphere* **2017**, *8*, 86.
12. Ménard, R.; Deshaies-Jacques, M. Evaluation of analysis by cross-validation. Part II: Diagnostics and optimization of analysis error covariance. *Atmosphere* **2017**, *8*, 70.

atmosphere

MDPI

Article

An Interactive Web Mapping Visualization of Urban Air Quality Monitoring Data of China

Wei Lu , Tinghua Ai *, Xiang Zhang and Yakun He

School of Resource and Environmental Sciences, Wuhan University, Wuhan 430079, China;
whuluwei@whu.edu.cn (W.L.); xiangzhang@whu.edu.cn (X.Z.); hyk1990@whu.edu.cn (Y.H.)
* Correspondence: tinghua_ai@whu.edu.cn; Tel.: +86-139-0863-9199

Received: 7 July 2017; Accepted: 10 August 2017; Published: 13 August 2017

Abstract: In recent years, main cities in China have been suffering from hazy weather, which is gaining great attention among the public, government managers and researchers in different areas. Many studies have been conducted on the topic of urban air quality to reveal different aspects of the air quality problem in China. This paper focuses on the visualization problem of the big air quality monitoring data of all main cities on a nationwide scale. To achieve the intuitive visualization of this dataset, this study develops two novel visualization tools for multi-granularity time series visualization (timezoom.js) and a dynamic symbol declutter map mashup layer for thematic mapping (symadpative.js). With the two invented tools, we develops an interactive web map visualization application of urban air quality data of all main cities in China. This application shows us significant air pollution findings at the nationwide scale. These results give us clues for further studies on air pollutant characteristics, forecasting and control in China. As the tools are invented for general visualization purposes of geo-referenced time series data, they can be applied to other environmental monitoring data (temperature, precipitation, etc.) through some configurations.

Keywords: air quality; environmental data visualization; spatial-temporal visualization; visual analytics

1. Introduction

Recent years, in China, frequent occurrences of hazy weather in big cities have aroused great attention on urban air quality issues among the public, government managers and academic researchers. Discussions about air quality of big cities (like Beijing, Wuhan, etc.) frequently appear in social media. Since 2012, the government has adopted two new air quality monitoring technical standards [1,2], then has built a real-time air quality reporting platform and an air quality forecasting system. Because air pollution has detrimental effects on human health, vegetation, crops, etc., it has great political, societal and economic impacts [3–5]. Therefore, with the increasing availability of urban air quality data and the great environmental challenges we are facing, many studies have been conducted to explore new approaches for understanding this big environmental monitoring dataset.

Environmental monitoring data can be described by multivariate time series observations generated from geo-located monitoring stations. For our research topic, urban air quality monitoring data consist of many air pollutant concentration values (such as fine particles, carbon monoxide, sulfur dioxide, nitrogen oxides zone, etc.), which are reported hourly from monitoring stations fixed at specific positions in a city. These geo-referenced time series data are an important study subject in the areas of geovisualization and environmental science.

Multi-dimensional data visualization considering spatial distributions, temporal granularities and multivariate thematic attributes has been an interesting question in geovisualization, as well as in the environmental science area. In geovisualization, many studies have focused on systematic theories of multivariate spatio-temporal data visualization [6]. Multiple static maps series, animation maps, space-time cubes and self-organizing maps (SOM) are used to deal with spatio-temporal data mapping [7,8]. In parallel, basic graph charts, such as parallel coordinate plots (PCL), are used to show multivariate thematic attributes. To achieve effective visualization of multivariate spatio-temporal data, combinations of these methods are used; such as the visualization system for space-time and multivariate patterns (VIS-STAMP) [8], which proposed a systematic way of joining those visualization strategies in multiple visual views interactively. However, expertise is needed to use these visualization systems, which are limited to domain experts. Thus, it requires some easy-to-use visualization applications to face a wider range of users.

Visualization of environmental data is important in the area of environmental science [9]. Conventional technologies in the Geographic Information System (GIS) and statistical map making skills are widely used to manage, analyze and visualize environmental data [10]. With the development of web technologies, open source GIS standards and web mapping tools are adopted in the analysis and presentation of environmental data-related studies not only for experts and managers, but also for the public in general [11–13]. These studies mainly focus on the architecture of environmental data visualization system building, and the visualization techniques are limited to conventional methods. Therefore, more effective visualization tools need to be developed for visualization of the increasing accumulated multi-dimensional environmental monitoring data. Currently, as urban air quality is an urgent societal problem for many policy-makers and a study topic for scholars, air quality data are gaining more attention.

Many studies have been conducted on mapping urban air pollutions mainly from two directions. Firstly, for the distribution of monitoring stations is rarely homogeneous, and spatial interpolations are used for pollution exposure maps [14], concentration maps of air pollution [15] and mapping spatio-temporal trends of air pollution [16]. Secondly, conventional charts and modern visualization tools are also being used for air quality data exploration [17] and visual analytics [18,19]. Our research follows the second direction and aims at providing a web mapping application for visualizing a whole year of air quality monitoring data of all cities in China with one dynamic and interactive map. This visualization application provides freedom of configuration for users to explore this multi-dimensional dataset from different angles. The major contributions of this paper are in three aspects: (1) we develop two novel web client tools for interactive visualization of spatio-temporal data; (2) we propose a mashup strategy for web mapping by combining and extending the function of different visualization tools, which can be generalized to visualize other kinds of spatio-temporal data, such as temperature, precipitation, etc.; (3) we implement an on-line interactive urban air quality data visualization application that helps to explore and analyze a big air quality dataset and that puts forward clues for further studies on air pollution.

2. Method and Data

Mapping spatio-temporal datasets is mapping changes of geographical features over time and space [7]. Because the information load is huge, the interaction and dynamics [20] of spatio-temporal visualization should be well designed. Multiple levels of spatial and temporal details are important considerations in spatio-temporal visualization. Mostly, the methods that are proposed in the literature for visualizing spatio-temporal data in a multi-scale perspective have focused on either the spatial or temporal aspect, rather than integrating both views over multiple scales. Qiang [21] and Van de Weghe [22] presented a continuous spatio-temporal model for space-time analysis. However, in the real visualization domain, space and time are recognized or recorded by discrete intervals, such as the tiled map service and the periodic timekeeping system. The tiled map uses a quad tree to represent multi-scale feature of space, while the timekeeping system uses the year, season, month, day and hour structure to record time-based events. In this section, we propose a space and time zooming method conforming to the tiled map service and timekeeping system. This method also conforms to the "overview first, zoom and filter, then details on demand" process [23].

Mapping the time component onto an axis on 2D space or 3D space is conventional method for time series visualization. The space-time cube [24] maps time to a 3D axis vertically, resulting in a 3D trajectory for time series datasets. As the storygraph [25], it maps x and y coordinates of space to two vertical axes and time to the horizontal axis, which can show trends in time series datasets. This paper provides a cartographic method to encode the time component of spatio-temporal data. The method separates the time component from the map space and uses the glyph map symbol to encode and interact with time, which gives more freedom for time representation and interaction. Furthermore, all symbols on the map are controlled by the dynamic map layer for displaying appropriate symbols at different map zoom levels and view extents.

Environmental monitoring data, i.e., the air quality monitoring data have multidimensional features, which can be structured as data cubes with a hierarchical structure [26]. Figures 1 and 2 illustrate the hierarchies of time and geographical information respectively. In this paper, the time structure in the dashed polygon part is used for the visualization, as the instance on the right side of Figure 1. For the geographical part, this paper focuses on the air quality of city points, each of which has a semantic importance illustrated in the table on the right side of Figure 2. The importance level of the city points influences their weight in dynamic map symbol selection when zooming and panning the map.

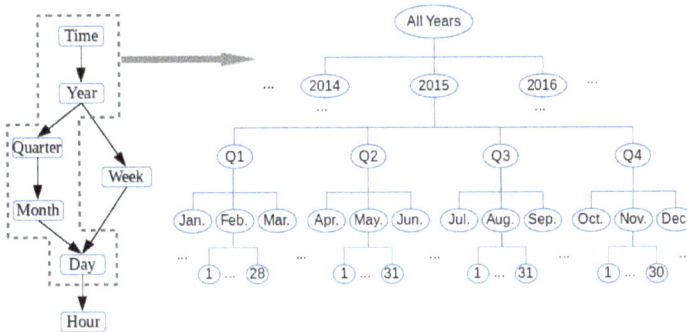

Figure 1. Hierarchical structure of the time dimension. The dashed polygon part is handled in this paper. An instance of this structure is on the right side.

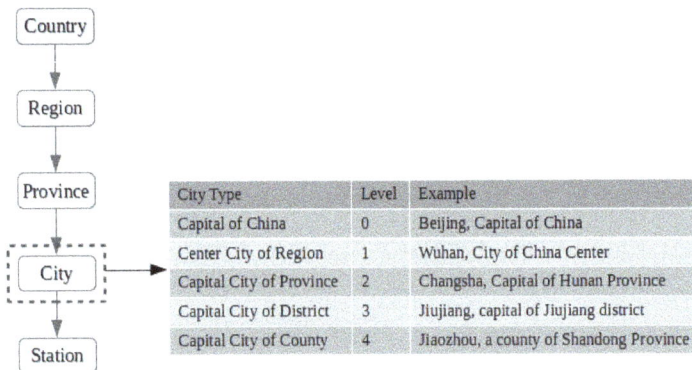

City Type	Level	Example
Capital of China	0	Beijing, Capital of China
Center City of Region	1	Wuhan, City of China Center
Capital City of Province	2	Changsha, Capital of Hunan Province
Capital City of District	3	Jiujiang, capital of Jiujiang district
Capital City of County	4	Jiaozhou, a county of Shandong Province

Figure 2. Hierarchical structure of the geographical dimension. In this paper, we focus on the city level, and the importance weight is in the table on the right side.

In the following subsections, the technologies and framework of the mapping application and the design and development of two JavaScript visualization tools, timezoom.js and symadaptive.js, are discussed. The two visualization components are mainly based on the general purpose web mapping library leaflet.js and the data visualization library D3.js. Then, the data and their processing for this study are presented.

2.1. Air Quality Mapping Technologies and Framework

As mentioned above, environmental monitoring data, i.e., urban air quality data are geo-referenced, and a map is very suitable for visualizing this dataset. To achieve the purpose of mapping such a big dataset on one dynamic web map application, we resort to some popular visualization tools in the GIS and information visualization area. With the booming web mapping tools, cartographers have more and more choices for their mapping works. Nevertheless, it is also a challenge for cartographers to know the characteristics of all of these tools and to maintain their own mapping frameworks. Roth [27] mentions this problem and provides a comprehensive analysis of current web mapping technologies, which gives us some advice to cope with the continued evolution of these technologies. To be more practical, cartographers need only three categories of tools for a web mapping work: application frameworks, mapping libraries and visualization libraries. Application frameworks work as the engine to drive the whole mapping application; mapping libraries as the map engine to glue all kinds of map services and geospatial data services; visualization libraries as the symbolization engine to visualize the non-spatial information within geographic entities. This is a kind of software mashup framework, with which the application engine, the map engine and the symbol engine can be selected by the map makers according to their visualization topics and development habits.

Among all of these mapping technologies, this study adopts web.py [28], leaflet.js [29] and D3.js [30] as the tool framework for the air mapping application. web.py works as the web application framework to drive the whole mapping application; leaflet.js works as the map engine to glue map services and data services; D3.js acts as a map symbolization engine to render the time series air quality data of the multivariate index. These three technologies work together for the air quality mapping application (Figure 3). This is a kind of technology mashup [31] working with the mapping mashup application. Though this study focuses on air quality monitoring data visualization, the framework and invented visualization tools can be applied to the visualization work of other similar spatio-temporal environmental monitoring data, such as meteorological monitoring data (temperature, precipitation, humidity, etc.).

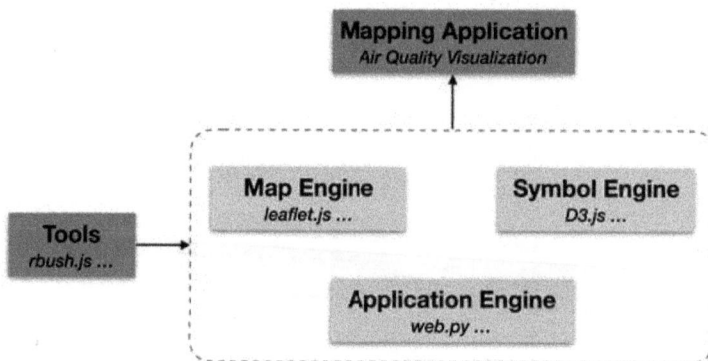

Figure 3. The air quality monitoring application framework. This paper proposes a mashup strategy to make use of multiple visualization and mapping technologies for web mapping applications.

2.2. Time Series Map Symbol Encoding: timezoom.js

Time series data visualization mostly has two kind of routines: the linear one and the cyclic one [32–34]. Linear ones take the perspective that time is a continuous concept represented by a time line. Cyclic ones [35] regard time as a periodic concept, such as a time keeping system. The timekeeping model is cyclic for the year, season, month and day, which originates from the astronomical observations of our ancestors. This paper constructs the time series air quality data with the hierarchical timekeeping system, and multiple temporal granularities are manipulated by map symbol interactions. In each node of this leveled structure, the values of each index of air quality are calculated as the mean value of lower level nodes. This hierarchic time structure is formulated in Formula (1). Figure 4 shows the timezoom.js symbol structure.

$$Time_H = G(Time_V, Time_S, Operation) \tag{1}$$

In Formula (1):

$Time_H$ is the time hierarchy structure that will be calculated;

$Time_V$ is the time series values' array, such as the 365 days of one year data value array, $\{t_1 : v_1, t_2 : v_2, ..., t_{365} : v_{365}\}$;

$Time_S$ is the time system for hierarchy construction, such as $\{Year, Season, Month, Day\}$. This can be decided by users and can be extended to hours, minutes and seconds levels, as shown in Figure 2;

$Operation$ is the operation for the value aggregation from the lower time level to the higher time level. This can be any statistical method for aggregation, such as average, median, quantile, etc.

The timezoom.js symbol is a radial tree map visualization tool based on D3.js. The root node of the hierarchy structure is at the center with leave nodes on the circumference. The values of all nodes are encoded with a defined color scheme. The initial state of this symbol shows the whole hierarchy. When clicking on any sector, the symbol will zoom to that sector and fold up other sectors; for example, when you click on the Q2 sector of the symbol, other sectors will be folded up, and the outside circle of the symbol will zoom to the days only in Q2 (Figure 4).

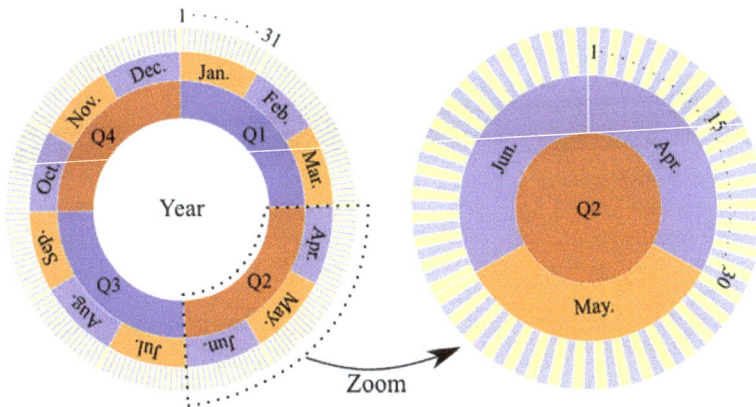

Figure 4. Multi-granularity time zooming interaction (timezoom.js): The inner circle shows the data value of the year, and Sectors Q1, Q2, Q3 and Q4 respectively indicate the four seasonal values and sectors January–December for the monthly data of a year; each month sector is surrounded by daily sectors. Clicking on each sector, the timezoom.js symbol will zoom to that time sector for detailed views of its value distributions.

2.3. Adaptive Map Symbol Control: symadaptive.js

Zooming control means adaptively selecting appropriate symbols at different zoom scales and view extents. Yang [36] proposes a strategy for multi-scale visualization of massive point data, which selects points with a heavy algorithm on server-side and makes light-weight symbol displacement on the client-side. Jari Korpi [37] evaluated point symbol clutter reduction methods for map mashups. According to these works, in this paper, the map symbol clutter reduction is achieved by symbol conflict detection with Rbush.js [38], which is a tool library based on the spatial index algorithm, Rtree [39]. At each map zoom level, symbols of points in the current map view extent are being tested for conflicts. When the symbol has conflicts with other symbols, the point that has a lower semantic weight will be dropped (Figure 5). The semantic level of each city is decided by the level of its administration division; see Figure 2.

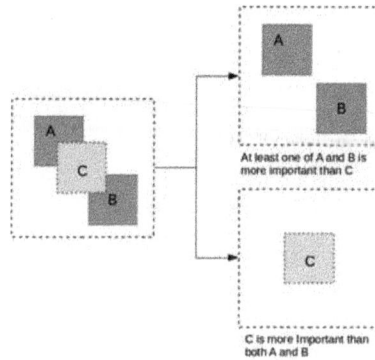

Figure 5. Multi-scale space zooming control (symadaptive.js): When the symbol C is add to the map, C will conflict with A and B. If C has a higher importance level than A and B, then C will be kept on the map, and A and B will be removed. For the other situation, if A or B has a higher importance level than C, C will be ignored.

In this paper, the mechanism of semantic importance-driven map symbol selection is proposed. In practice, one can have their own semantic level fields in their data, and this symadaptive.js layer tool can help to dynamically select appropriate symbols according to the assigned level weight field.

2.4. Data Sources and Processing

The Air Quality Index (AQI) is a number without any unit used to indicate how polluted the air is. It is adopted by the newly-published national environment protection standard Technical Regulation on Ambient Air Quality Index (on trial) in China. With this standard, different air pollutants have their own concentration level ranges, and the index value of each pollutant can be calculated by Formulas (2) and (3). The AQI is assigned the maximum value of the individual index value of all of the reported pollutants (Table 1). There are six ranges of the AQI values, and each range is assigned a descriptor and a visual color code, as shown in Table 2.

$$IAQI_P = \frac{IAQI_{Hi} - IAQI_{Lo}}{BP_{Hi} - BP_{Lo}}(C_P - BP_{Lo}) + IAQI_{Lo} \qquad (2)$$

$$AQI = max\{IAQI_1, IAQI_2, IAQI_3, ..., IAQI_n\} \qquad (3)$$

where:

$IAQI_P$ is the individual Air Quality Index of pollutant P,
C_P is the concentration of pollutant P,

BP_{Hi} is the the concentration division point of pollution P that is $\geq C_P$,
BP_{Lo} is the concentration breakpoint of pollution P that is $\leq C_P$,
$IAQI_{Hi}$ is the index division point (Table 3) corresponding to BP_{Hi},
$IAQI_{Lo}$ is the index division point (Table 3) corresponding to BP_{Lo}.

Table 1. Reported air index and pollutants and their descriptions.

Name	Description and Unit
AQI	Air Quality Index, value without unit, range 0–500
SO_2	Sulfur Dioxide, $\mu g/m^3$
NO_2	Nitrogen Dioxide, $\mu g/m^3$
CO	Carbon Monoxide, mg/m^3
O_3	Ozone, $\mu g/m^3$
O_3_8H	8-h average concentration of Ozone, $\mu g/m^3$
PM2.5	Particulate Matter, diameter less than 2.5 μm, $\mu g/m^3$
PM10	Particulate Matter, diameter less than 10 μm, $\mu g/m^3$

Table 2. Air Quality Index level divisions, descriptors and colors.

Divisions	Health Influences	Color Coding
0–50	Good: Satisfactory	Green (0,228,0)
51–100	Moderate: Acceptable, but influential for very sensitive groups.	Yellow (255,255,0)
101–150	Slightly Unhealthy: Influential for sensitive groups.	Orange (255,126,0)
151–200	Unhealthy	Red (255,0,0)
200–300	Very Unhealthy	Purple (153,76,0)
>300	Hazardous	Maroon (126,0,35)

We collect the hourly reported air quality data from the national real-time air quality reporting system. By 2017, there were 367 cities with 1497 monitoring sites in China reporting their Air Quality Index hourly, as shown in Figure 6. Daily air quality of cities is the object of this study. Therefore, the hourly reported data are aggregated into the daily reported data according to the standard [2]. First, the average hourly data of a city are calculated as the mean pollutant concentration of all monitoring stations. Then, the daily value is calculated as the mean concentration values of 24 h if it has 16 h of validated values in a day; otherwise, the daily value is invalidated. The final data are daily records of all cities with 24-h average concentrations of 5 pollutants (SO_2, NO_2, CO, PM10, PM2.5) and 24-h maximum concentrations of 2 pollutants (O_3, O_3-8H). All of the concentration values are converted to an Individual Air Quality Index(IAQI) value ranging from 0–500 according to Table 3 and Formulas (2) and (3). Thus, we can get the Air Quality Index (AQI) value and the primary pollutants of each city each day.

Table 3. Individual Air Quality Index and corresponding pollutants' concentration limits.

IAQI Ranges	SO_2 $\mu g/m^3$	NO_2 $\mu g/m^3$	PM10 $\mu g/m^3$	PM2_5 $\mu g/m^3$	CO mg/m^3	O_3 $\mu g/m^3$	O_3_8H $\mu g/m^3$
0	0	0	0	0	0	0	0
50	50	40	50	35	2	160	100
100	150	80	150	75	4	200	160
150	475	180	250	115	14	300	215
200	800	280	350	150	24	400	265
300	1600	565	420	250	36	800	800
400	2100	750	500	350	48	1000	-
500	2620	940	600	500	60	1200	-

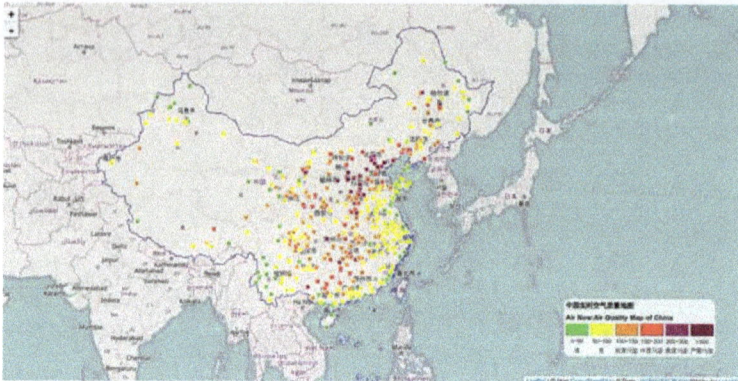

Figure 6. Air Quality Monitoring Network in China. This image shows a snapshot air quality map of monitoring stations in China, at 8:00, 12 December 2016. (refer to: http://mapviz.xyz:8080/).

3. Results and Discussion

Based on the discourse above, this study designs the air mapping application in terms of the navigation of space, time and theme [40]. The whole application is built on the framework illustrated in Figure 3, with the two invented JavaScript visualization tools: timezoom.js and symadaptive.js.

3.1. Air Quality Data Map Visualization Design

3.1.1. Spatial Navigation

Air quality data of all cities are loaded by symadaptive.js, and the time series data are used to build the timezoom.js symbol for each city point. These geo-located glyph symbols are overlaid on an OpenStreetMap tiled map background. The display of all symbols is under the control of symadaptive.js. Meanwhile, all of the symbols can be hovered over and clicked to invoke the display of detailed air quality information about the city point interacted with. The weight of each city is decided by its administrative level: central city of region, province capitol, district capitol and county capitol (see Figure 2). Figure 7 shows the interaction effects of zooming the map to different scales.

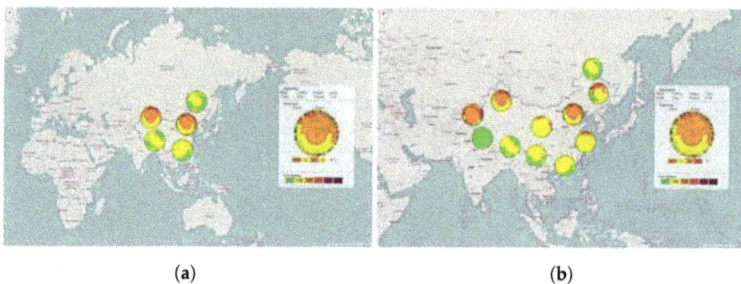

(a)

(b)

Figure 7. *Cont.*

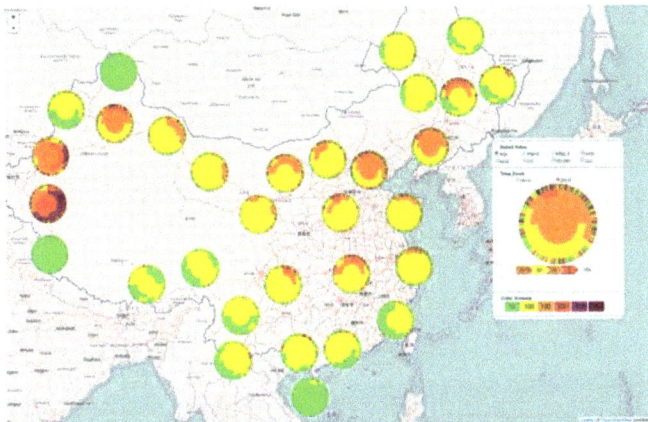

(c)

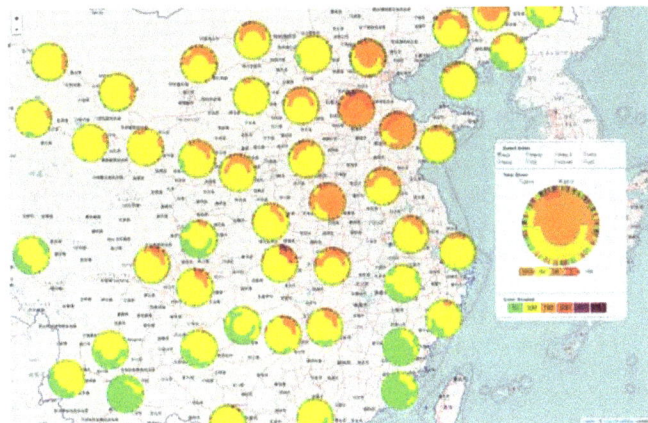

(d)

Figure 7. Spatial navigation. When zooming the map, the symbols are selected dynamically at each zoom level with the function provided by symadaptive.js. (**a**)–(**d**) illustrate map symbols at Zoom Levels 3–6 on the OpenStreetMap tiled map background. (**a**) Map Zoom Level 3; (**b**) Map Zoom Level 4; (**c**) Map Zoom Level 5; and (**d**) Map Zoom Level 6.

3.1.2. Temporal Navigation

The temporal component of air quality data is presented by the timezoom.js component. The timezoom.js symbol is driven by the date hierarchy structure of air quality data. All symbols are event connected. Therefore, when clicking on one map symbol at a specific time section, all symbols on the map will zoom and change their appearance simultaneously to show air quality data according to the chosen time section. For a better and efficient query, a bigger time wheel is designed for temporal navigation on the map side panel. Figure 8 shows the interactions with the time series data by the timezoom.js symbol.

(a)

(b)

(c)

(d)

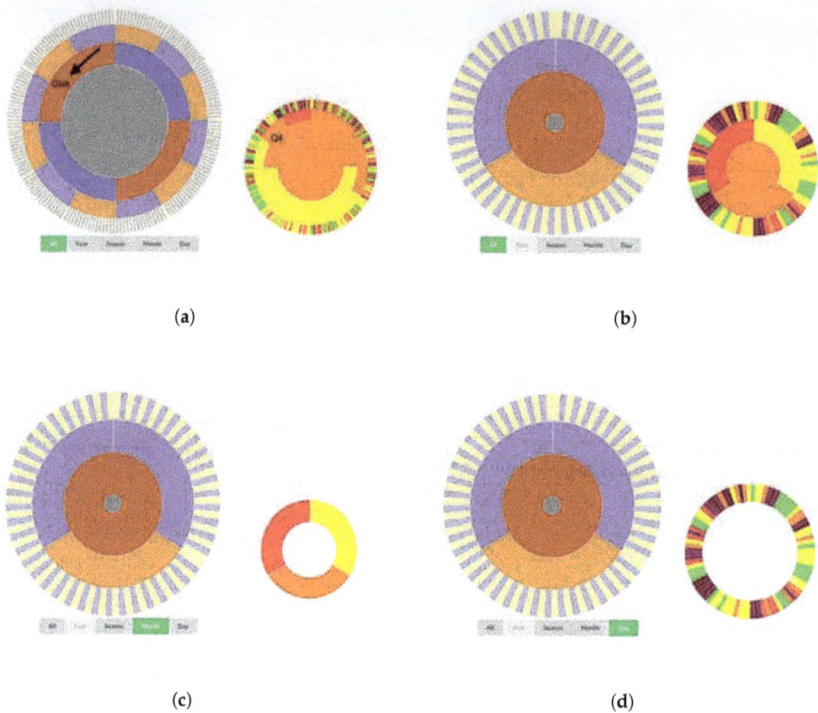

Figure 8. Temporal navigation. The left circle symbol shows the structure of the time symbol, and the right is the map symbol that will be displayed on the map. The toggle buttons below are for the time filter to focus on the time granularities of interest. (**a**) shows the full year data in a time symbol, and (**b**) shows the zoomed symbol to Quarter 4 (October, November and December). (**c**,**d**) show the symbols that are filtered by the month and day value.

3.1.3. Thematic Navigation

Air quality has a total description as AQI, which contains several important air quality sub-indexes, PM10, PM2.5, SO_2, NO_2, O_3_8h, O_3, CO. We design a side panel on the map for thematic attribution selection. When the thematic attribute changes, the map symbol of all points will change accordingly. Thus, it is easy to switch the map view among the whole air quality map and individual pollutant concentration maps.

To sum up, this study presents an urban air quality mapping application based on invented tools: symadaptive.js and timezoom.js. The map of air quality starts with visualizing the data of central cities in six major parts of China (see (Figure 9)).

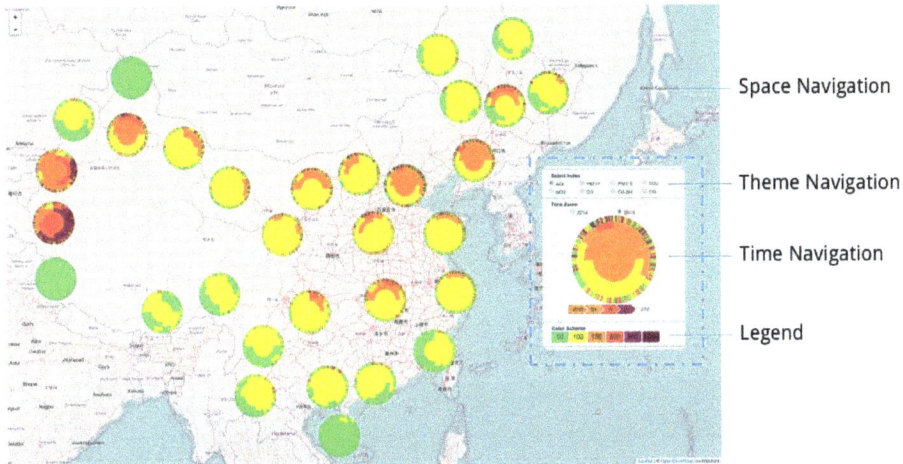

Figure 9. Initial view of the online air quality mapping application.

3.2. Results and Analysis

In the air quality application, the concentrations of all of the pollutants are illustrated with graded color hues under equal interval classification of all values of the year. The AQI map is rendered with the commonly-used standard color scheme (Table 2). With the thematic navigation radio button, we can render all air maps of different pollutants. In Figure 10 is presented a series of maps of the AQI value and seven important pollutants' concentrations at a nationwide scale, and the dataset can be switched between two years, 2014 and 2015 (Figure 9). From this series of maps, several significant findings are shown clearly. These findings provide some clues for in-depth research on the air pollutant causality and relationships among air pollutants from a spatio-temporal view.

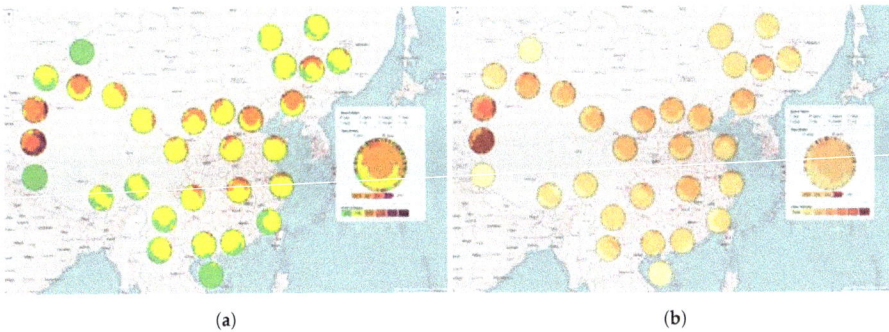

(a)   (b)

Figure 10. *Cont.*

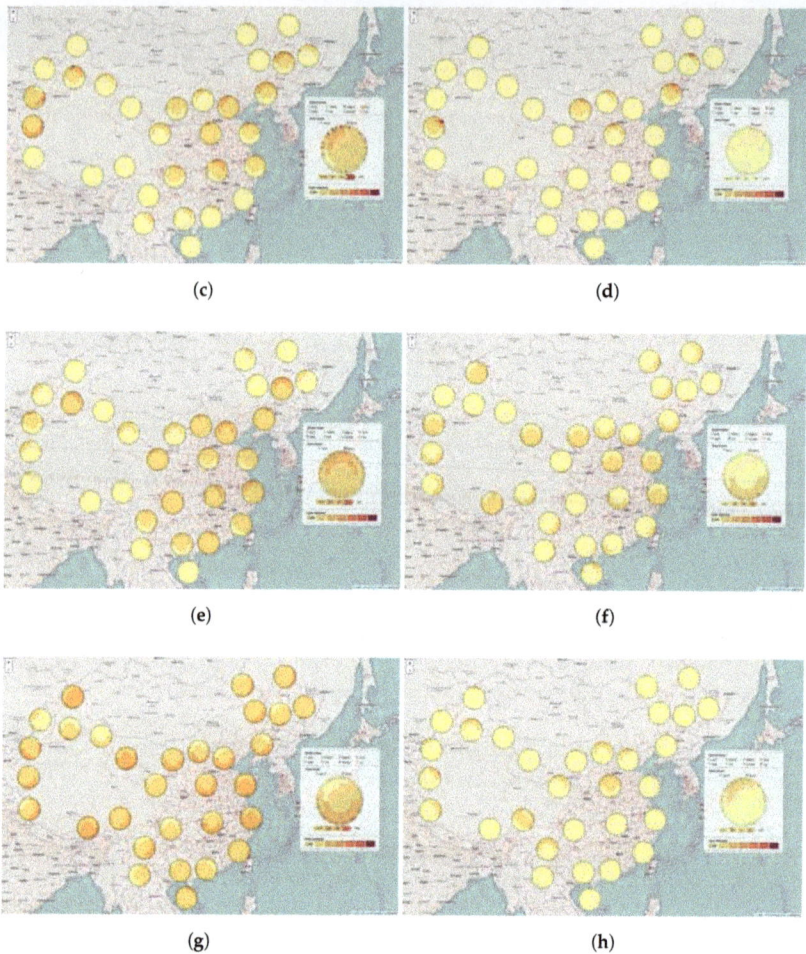

Figure 10. Air quality maps 2015 at OpenStreetMap Zoom Level 5. (**a**) is the AQI map of 2015; (**b**)–(**g**) are the air pollutants' concentration map in 2015 for PM10, PM2.5, SO_2, NO_2, O_3 and CO respectively. (**a**) AQI map of 2015; (**b**) PM10 concentration map of 2015; (**c**) PM2.5 concentration map of 2015; (**d**) SO_2 concentration map of 2015; (**e**) NO_2 concentration map of 2015; (**f**) O_3 concentration map of 2015; (**g**) O_3 8H concentration map of 2015; (**h**) CO concentration map of 2015.

3.3. Nationwide Air Quality Condition of China

The interactive mapping application gives us an overview of the air quality condition of China at a nationwide scale throughout a whole year's time (Figure 10a). First, for most of the cities in China, it is more likely to have a good air condition during summer days than winter days. Some special places are in the northwest part of China, where several cities have a terrible air condition throughout the year. This situation should be given more attention in further research. Second, air quality maps of a different time granularities show a clear ribbon pattern along the coastline and southwest part of China where cities have better air conditions than other part of China; meanwhile, the cities in the north part of China have the worst air condition. The possible reasons for such patterns reside in two aspects. On the one hand, the coastline cities have better air circulation conditions for air purification than hinterland areas of China. On the other hand, the cities in the southwest are less developed than the hinterland; thus, the pollutant emission is lower. Nevertheless, these are hypotheses that need further study. Furthermore, the patterns can provide auxiliary information for policy-makers to adopt different measures in different cities.

3.4. Spatio-Temporal Pattern of Air Pollutants

With the thematic navigation radio button, one can have air maps of different pollutants. In Figure 10, from this series of maps, several significant patterns are shown clearly. First, we can easily indicate that PM10 and PM2.5 more likely contribute to the AQI value due to their similar pattern in terms of space and time (see Figure 10b,c, PM10, PM2.5 concentration maps). Actually, the public cares more about the PM10 and PM2.5 index in China, and they are the critical impacts of smog air conditions. Second, NO_2 is mainly caused by automobile exhaust, which is more likely to be worse in big cities, as shown in the NO_2 concentration maps (Figure 10e). Third, from the map of SO_2, one can find that the concentration of SO_2 is more likely to be serious in winter months in the north of China (see SO_2 concentration maps in Figure 10d). This situation can be connected to the burning of coal for heating in winter of cities in the north of China, which emits a great amount of SO_2. Fourth, there is an obvious situation that the value of CO is hardly serious enough to impact public health, as CO is a deadly poisonous gas that has critical control in China (see the CO concentration maps in Figure 10h).

Unlike other pollutants, which are emitted directly into the air by some specific sources, ozone (O_3) is created by sunlight acting on NOx and VOC in the air. Thus, the index value of O_3 is higher on sunny days throughout the year or in areas that have a longer sunlight duration and stronger sunlight intensity. In the concentration map of O_3 (Figure 10f,g), we can find that most cities in China will have higher O_3 values in Seasons 2 and 3, when sunshine is greater and the hours longer during the daytime. In Lhasa City, Tibet, the O_3 value is high across the year because of its high elevation and thin air, which causes the high intensity of sunlight. Some cities in the south of China and along the coastline will show different O_3 value patterns (see the O_3 concentration maps in Figure 10f,g). This kind of situation may be caused by unstable weather conditions in the south of China; for example, in summer there are many rainy days in which sunlight intensity is mild.

All of the findings mentioned above are hypotheses. Through these clues, we can design further studies on these topics and make more reliable conclusions. Moreover, these illustrations can help environmental regulation governors and the general public to have intuitive images of the air condition of China.

4. Conclusions

The research presented here describes a novel combination of modern mapping technologies, with which this study develops an online mapping application of air quality of China. In this study, an open web platform is fully used to collect the time series air quality data consistently. Then, data visualization tools (D3.js) and web mapping tools (leaflet.js, rbush.js) are well combined to produce a fine interactive mapping application of spatio-temporal data. From the application, we can get a whole view of the air quality condition of China at a nationwide scale and a year time span at multiple spatio-temporal granularities. This interactive map application clearly presents several significant findings of air quality in China, which provide good assistance for visual air quality analysis and clues for in-depth studies on air pollution.

The lessons we learned from this study reside in three aspects. First, there are more and more open data about our living environment, into which we can delve and find important results for making our living environment better. Second, as cartographers, we should make full use of new technologies for data visualization and web mapping. A good combination of these excellent tools can give us greater power for environmental data visualization. Third, the visual form of data is more expressive than the raw data table, and it would give deeper insights into the data analysis. In other

Atmosphere **2017**, *8*, 148

words, nowadays, with more and more open data, fine and flexible visualization tools and the crowd wisdom of the public, we can have a clear vision of the environment around us.

The future work is to enhance the efficiency of the timezoom.js symbol and to extend it to hour granularities for a more detailed time level. What is more, as time goes on and the air quality data are accumulated, a year selection mechanism should be added to the timezoom.js symbol, and the comparison function should be enhanced. At the same time, some air quality study problems can be defined from the previous discussions of the findings with the visualization, then our future work will collect evidence to test our hypotheses.

Acknowledgments: This research was supported by the National Key Research and Development Program of China (Grant No. 2017YFB0503500), and the National Natural Science Foundation of China (Grant No. 41531180).

Author Contributions: Wei Lu proposed the idea and did the visualization tools development, as well as the paper writing. Tinghua Ai gave important experiment suggestions and editing of the paper. Xiang Zhang and Yakun He provided suggestions on paper writing.

Conflicts of Interest: The authors declare no conflict of interest.

Appendix A. Mapping and Visualization Technologies Used

leaflet.js: leaflet.js is a lightweight, extensible, open-source JavaScript library for interactive maps, which was developed by Vladimir Agafonkin with a group of dedicated contributors. Its simplicity and extensibility provide the freedom to customize desirable plugins for specific use. For the map symbol, the Icon class component of leaflet.js can be extended as the HTML(Hypertext Markup Language) <div> element ,which provides the freedom of combination with other data visualization library.

D3.js: D3.js is short for Data-Driven Document, which is a JavaScript library for manipulate DOM(Document Object Model) elements based on data. It also provides many powerful visualization components based on HTML, SVG(Scalable Vector Graphics) and CSS(Cascading Stylesheets). Thus, we can utilize these visualization facilities to represent data as SVG elements, which can be embedded in a <div> element. Therefore, we can freely extend the Icon component of leaflet.js by using D3's visualization components.

rbush.js: rbush.js is a high-performance JavaScript R-Tree-based 2D spatial index library for points and rectangles by Vladimir Agafonkin. It can be used for conflict detection for map symbols when zooming. We can extend the LayerGroup class of leaflet.jsusing the collision detection function of RBush to realize a self-declutter map layer.

web.py: web.py is a light-weight web application framework in the Python language. It is simple and powerful, and it is an open sources software tool that can be used for whatever purpose with absolutely no restrictions.

References

1. MEP China. *Ambient Air Quality Standards. GB 3095-2012*; China Environmental Science Press: Beijing, China, 2012.
2. MEP China. *Technical Regulation on Ambient Air Quality Index (on Trial) (HJ633-2012)*; China Environmental Science Press: Beijing, China, 2012.
3. Chen, C.; Arjomandi, M.; Balmes, J.; Tager, I.; Holland, N. Effects of Chronic and Acute Ozone Exposure on Lipid Peroxidation and Antioxidant Capacity in Healthy Young Adults. *Environ. Health Perspect.* **2007**, *115*, 1732–1737.
4. Sicard, P.; Lesne, O.; Alexandre, N.; Mangin, A.; Collomp, R. Air quality trends and potential health effects—Development of an aggregate risk index. *Atmos. Environ.* **2011**, *45*, 1145–1153.
5. Ochoa-Hueso, R.; Munzi, S.; Alonso, R.; Arróniz-Crespo, M.; Avila, A.; Bermejo, V.; Bobbink, R.; Branquinho, C.; Concostrina-Zubiri, L.; Cruz, C.; et al. Ecological impacts of atmospheric pollution and interactions with climate change in terrestrial ecosystems of the Mediterranean Basin: Current research and future directions. *Environ. Pollut.* **2017**, *227*, 194–206.

6. Andrienko, N.; Andrienko, G.; Gatalsky, P. Exploratory spatio-temporal visualization: An analytical review. *J. Vis. Lang. Comput.* **2003**, *14*, 503–541.
7. Kraak, M.J.; Ormeling, F. *Cartography: Visualization of Spatial Data*; Guilford Press: New York, NY, USA, 2011.
8. Guo, D.; Chen, J.; MacEachren, A.M.; Liao, K. A visualization system for space-time and multivariate patterns (vis-stamp). *IEEE Trans. Vis. Comput. Graph.* **2006**, *12*, 1461–1474.
9. Rink, K.; Scheuermann, G.; Kolditz, O. Visualisation in environmental sciences. *Environ. Earth Sci.* **2014**, *72*, 3749–3751.
10. Grünfeld, K. Integrating spatio-temporal information in environmental monitoring data-a visualization approach applied to moss data. *Sci. Total Environ.* **2005**, *347*, 1–20.
11. Huang, B. Web-based dynamic and interactive environmental visualization. *Comput. Environ. Urban Syst.* **2003**, *27*, 623–636.
12. Hugentobler, M.; Hurni, L. Web cartography with open standards-a solution to cartographic challenges of environmental management. *Environ. Model. Softw.* **2010**, *25*, 988–999.
13. Kulawiak, M.; Prospathopoulos, A.; Perivoliotis, L.; Kioroglou, S.; Stepnowski, A.; łuba, M. Interactive visualization of marine pollution monitoring and forecasting data via a Web-based GIS. *Comput. Geosci.* **2010**, *36*, 1069–1080.
14. Long, Y.; Wang, J.; Wu, K.; Zhang, J. Population Exposure to Ambient PM 2.5 at the Subdistrict Level in China. Available online: https://papers.ssrn.com/sol3/papers.cfm?abstract_id=2486602 (accessed on 27 August 2014).
15. Rohde, R.A.; Muller, R.A. Air pollution in China: Mapping of concentrations and sources. *PLoS ONE* **2015**, *10*, e0135749.
16. Sicard, P.; Serra, R.; Rossello, P. Spatiotemporal trends in ground-level ozone concentrations and metrics in France over the time period 1999–2012. *Environ. Res.* **2016**, *149*, 122–144.
17. Huan, L.; Hong, F.; Feiyue, M. A Visualization Approach to Air Pollution Data Exploration—A Case Study of Air Quality Index (PM2.5) in Beijing, China. *Atmosphere* **2016**, *7*, 35.
18. Chung, K.L.; Qu, H.; Chan, W.Y.; Guo, P.; Xu, A.; Lau, K.H. Visual Analysis of the Air Pollution Problem in Hong Kong. *IEEE Trans. Vis. Comput. Graph.* **2007**, *13*, 1408–1415.
19. Zhang, Y.L.; Cao, F. Fine particulate matter (PM2.5) in China at a city level. *Sci. Rep.* **2015**, *5*. doi:10.1038/srep14884.
20. MacEachren, A.M.; Kraak, M.J. Research challenges in geovisualization. *Cartogr. Geogr. Inf. Sci.* **2001**, *28*, 3–12.
21. Qiang, Y.; Chavoshi, S.H.; Logghe, S.; De Maeyer, P.; Van de Weghe, N. Multi-scale analysis of linear data in a two-dimensional space. *Inf. Vis.* **2013**, *3*, 248–265.
22. Van de Weghe, N.; De Roo, B.; Qiang, Y.; Versichele, M.; Neutens, T.; De Maeyer, P. The continuous spatio-temporal model (CSTM) as an exhaustive framework for multi-scale spatio-temporal analysis. *Int. J. Geogr. Inf. Sci.* **2014**, *28*, 1047–1060.
23. Shneiderman, B. The eyes have it: A task by data type taxonomy for information visualizations. In Proceedings of the IEEE Symposium on Visual Languages, Boulder, CO, USA, 3–6 September 1996; pp. 336–343.
24. Bach, B.; Dragicevic, P.; Archambault, D.; Hurter, C.; Carpendale, S. A review of temporal data visualizations based on space-time cube operations. In Proceedings of the Eurographics Conference on Visualization, Wales, UK, 9–13 June 2014.
25. Shrestha, A.; Zhu, Y.; Miller, B.; Zhao, Y. Storygraph: Telling stories from spatio-temporal data. In *International Symposium on Visual Computing*; Springer: Berlin, Germany, 2013; pp. 693–702.
26. Stolte, C.; Tang, D.; Hanrahan, P. Polaris: A system for query, analysis, and visualization of multidimensional relational databases. *IEEE Trans. Vis. Comput. Graph.* **2002**, *8*, 52–65.
27. Roth, R.E.; Donohue, R.G.; Sack, C.M.; Wallace, T.R.; Buckingham, T. A process for keeping pace with evolving web mapping technologies. *Cartogr. Perspect.* **2015**, 25–52, doi:10.14714/CP78.1273.
28. Swartz, A. Web.py Home Page. Available online: http://webpy.org/ (accessed on 4 December 2016).
29. Agafonkin, V. Leaflet-a JavaScript library for interactive maps. Available online: http://leafletjs.com/ (accessed on 4 December 2016).
30. Bostock, M. D3.js-Data-Driven Documents. Available online: http://d3js.org/ (accessed on 4 December 2016).

31. Wood, J.; Dykes, J.; Slingsby, A.; Clarke, K. Interactive visual exploration of a large spatio-temporal dataset: Reflections on a geovisualization mashup. *IEEE Trans. Vis. Comput. Graph.* **2007**, *13*, 1176–1183.

32. Peuquet, D.J. It's about time: A conceptual framework for the representation of temporal dynamics in geographic information systems. *Ann. Assoc. Am. Geogr.* **1994**, *84*, 441–461.

33. Aigner, W.; Miksch, S.; Schumann, H.; Tominski, C. *Visualization of Time-Oriented Data*; Springer: London, UK, 2011; p. 232.

34. Munzner, T. *Visualization Analysis and Design*; CRC Press: Boca Raton, FL, USA, 2014.

35. Carlis, J.V.; Konstan, J.A. Interactive visualization of serial periodic data. In Proceedings of the 11th Annual ACM Symposium on User Interface Software and Technology, New York, NY, USA, 1–4 November 1998; pp. 29–38.

36. Min, Y.; Tinghua, A.; Wei, L.; Xiaoqiang, C.; Qi, Z. A Real-time Generalization and Multi-scale Visualization Method for POI Data in Volunteered Geographic Information. *Acta Geodaetica Cartogr. Sin.* **2015**, *44*, 228.

37. Korpi, J.; Ahonen-Rainio, P. Clutter reduction methods for point symbols in map mashups. *Cartogr. J.* **2013**, *50*, 257–265.

38. Agafonkin, V. RBush-a high-performance JavaScript R-tree-based 2D spatial index for points and rectangles. Available online: https://github.com/mourner/rbush (accessed on 4 December 2016).

39. Guttman, A. *R-Trees: A Dynamic Index Structure for Spatial Searching*; Springer: Berlin, Germany, 1984; Volume 14.

40. Neumann, A. Spatial, Temporal and Thematic Navigation: Visualizing Biographies of European Artists Using a Task Oriented User Interface Design Approach. Ph.D. Thesis, ETH, Zürich, Switzerland, 2010.

atmosphere

MDPI

Letter

Temporal and Spatial Patterns of China's Main Air Pollutants: Years 2014 and 2015

Tiancai Zhou [1,2] , Jian Sun [2,*] and Huan Yu [1,*]

[1]   College of Earth Sciences, Chengdu University of Technology, Chengdu 610059, China;
      ztc18108279610@163.com
[2]   Institute of Geographic Sciences and Natural Resources Research, Chinese Academy of Sciences,
      Beijing 100101, China
*   Correspondence: sunjian@igsnrr.ac.cn (J.S.); yuhuan0622@126.com (H.Y.)

Received: 20 June 2017; Accepted: 25 July 2017; Published: 27 July 2017

Abstract: China faces unprecedented air pollution today. In this study, a database (SO_2, NO_2, CO, O_3, $PM_{2.5}$ (particulate matter with aerodynamic diameter less than 2.5 μm), and PM_{10} (particulate matter with aerodynamic diameter less than 10 μm) was developed from recordings in 188 cities across China in 2014 and 2015 to explore the spatial-temporal characteristics, relationships among atmospheric contaminations, and variations in these contaminants. Across China, the results indicated that the average monthly concentrations of air pollutants were higher from November to February than in other months. Further, the spatial patterns of air pollutants showed that the most polluted areas were located in Shandong, Henan, and Shanxi provinces, and the Beijing-Tianjin-Hebei region. In addition, the average daily concentrations of air pollutants were also higher in spring and winter, and significant relationships between the principal air pollutants (negative for O_3 and positive for the others) were found. Finally, the results of a generalized additive model (GAM) indicated that the concentrations of PM_{10} and O_3 fluctuate dynamically; there was a consistent increase in CO and NO_2, and $PM_{2.5}$ and SO_2 showed a sharply decreasing trend. To minimize air pollution, open biomass burning should be prohibited, the energy efficiency of coal should be improved, and the full use of clean fuels (nuclear, wind, and solar energy) for municipal heating should be encouraged from November to February. Consequently, an optimized program of urban development should be highlighted.

Keywords: air pollutants; dynamics; spatial patterns; generalized additive model; policy recommendations

1. Introduction

Haze is an atmospheric effect that has become a serious global issue [1,2], as it affects species diversity, global climate, and human health [3], social, and economic [4]. In general, the chemical aspects of haze have been studied—specifically its physical and chemical properties, including the elements Cd, Cr, Cu, Fe, and Mn, gaseous pollutants (O_3, NO_x, SO_2, CO_x), and inorganic aerosols (SO_4^{2-}, NO_3^-, and NH_4^+ [5–7]. Source apportioning has shown that haze is attributable primarily to dust storms, biomass burning [8], coal consumption, and vehicle exhaust [9–11]. Further, secondary inorganic and organic aerosols should not be neglected [12]. Some studies have focused on the long-range transport mechanism of haze, which is controlled by meteorological conditions [13,14]; thus, affluent moisture, warm advection in the lower troposphere, and stable atmospheric stratification favor the concentration of haze [15]. Other studies have investigated the side effects of haze, including reduced visibility [16], adverse effects on health [17], and so on.

China's air quality has deteriorated in recent years, and haze affects increasing areas [18]. Moreover, numerous studies have found that the regions of Beijing [19], North China [20], Wuhan [21], and Guangzhou [22] have excessively high concentrations of haze. Other studies have examined the formation and evolution of haze [23,24], and the role of meteorological conditions in haze transportation (e.g., wind and relative humidity) [25–27]. The urbanization is another driving force in the development of haze [28]. The varied characteristics of haze in different seasons have also been reported: for example, haze in summer in East China [29] and Beijing [30], autumn in Shanghai [31], and winter in Beijing [32,33], Shanghai [34], the China Loess Plateau [35], and the North China Plain [36], all of which have relatively higher concentrations of haze.

Meanwhile, the high concentration of organic carbon in PM (particulate matter) indicates that the main source of haze is biomass burning [37], and that the high concentration of NO_2 is closely related to vehicle emissions [34]. Moreover, high concentrations of SO_2 and NO_2 contribute significantly in the formation of secondary aerosols [38]. Thus, secondary aerosols (sulfate, nitrate, and organic matter) are produced when combined with gaseous pollutants (NO_2, SO_2, O_3, and formaldehyde) and main atmospheric particles [39]. These constitute the main chemical compositions of particulate matter with aerodynamic diameter less than 2.5 µm ($PM_{2.5}$) in Hangzhou [40], and the heterogeneous chemical processes promote and sustain the growth of haze [41,42]. Further, the spatial patterns and temporal variation of $PM_{2.5}$ have been simulated [43], and the simulated values of meteorological conditions (temperature, and wind speed and direction) agree with the values observed. The diurnal cycle of land-sea breezes among the Pearl River Delta is the primary influence in the transport of particulate matter with aerodynamic diameter less than 10 µm (PM_{10}) [44].

Many studies have indeed investigated the spatiotemporal patterns of haze across China. However, the dynamics and relationships of air pollutants during long time series (2 years) have been insufficiently discussed. Hence, a special study was performed to more fully reveal the spatiotemporal distribution of the principal air pollutants at present and their future variations in China. Within this context, the objectives of this study were to: (1) analyze the dynamic and spatiotemporal distribution of the principal air pollutants for 2014 and 2015, respectively; (2) reveal the dynamics and the relationships between the principal air pollutants in the most polluted cities, respectively; and (3) identify the principal variable trends in air pollutants over major Chinese cities based on a generalized additive model (GAM). Finally, we present strategic policy recommendation for reducing the levels of air pollutants according to specific haze characteristics in China.

2. Experiments

The main air pollutants (SO_2, NO_2, CO, O_3, $PM_{2.5}$, and PM_{10}) were monitored by the continuous air-monitoring stations (CAMS, Figure 1) covering 188 cities of China. The stations were established based on the standard "Technical regulation for selection of ambient air quality monitoring stations (HJ 664-2013)", and the data were collected through the Ministry of Environmental Protection of the People's Republic of China (Available online: http://datacenter.mep.gov.cn/). In addition, atmospheric contamination was compiled daily and monthly for 2014 and 2015.

In this study, we used the Inverse Distance Weighted (IDW) in ArcGIS 10.2 (ESRI, Inc., Redlands, CA, USA) to acquire the spatial graphs, SigmaPlot 10.0 (Systat Software, Inc., Chicago, IL, USA) was used to obtain the rose diagram, and Microsoft Office Visio (2007) (Microsoft corporation, Redmond, WA, USA) for fishbone diagrams. The Performance Analytics module of R software (R Core Development Team, R Foundation for Statistical Computing, Vienna, Austria) was employed to obtain the scatterplot matrices and conduct correlation as well as regression analyses. The two-tailed Pearson's correlations at $p = 0.05$ were employed to determine the correlations between different variables. The GAM (the non-linear relationship between the pollutants and day-of-year trends) was modeled flexibly by the log link function, and we used the *mgcv* module in R to predict air pollutants in China. The package *mgcv* (the gam fit function, gam function, family function were contained) of R was used to fit GAMs specified by presenting a symbolic description of the additive predictor as well as a description of the error distribution.

Atmosphere **2017**, *8*, 137

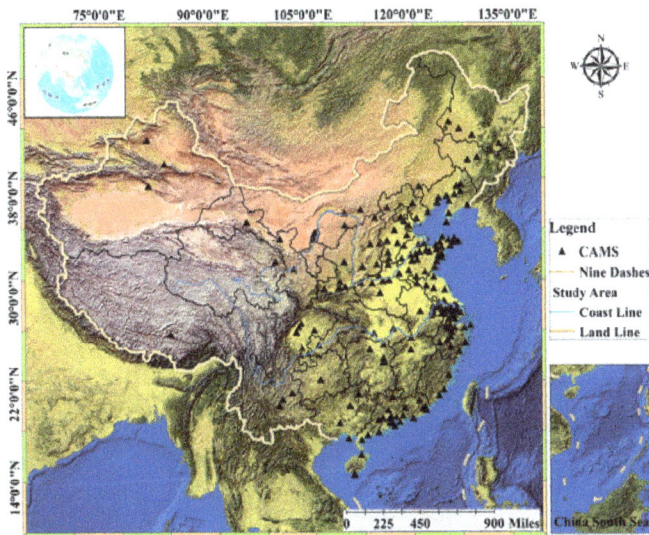

Figure 1. The study area.

3. Results

3.1. Dynamics of the Principal Air Pollutants

The average monthly concentrations of the principal air pollutants over China in 2014 are shown in Figure 2 and Table S1. The highest levels of $PM_{2.5}$, PM_{10}, CO, SO_2, and NO_2 were 106.5 $\mu g/m^{-3}$, 154.1 $\mu g/m^{-3}$, 1.69 mg/m^{-3}, 62.0 $\mu g/m^{-3}$, and 50.4 $\mu g/m^{-3}$ in January. The lowest $PM_{2.5}$ and PM_{10} levels were 39.8 $\mu g/m^{-3}$ and 69.9 $\mu g/m^{-3}$ in September, and the lowest values of CO, SO_2, and NO_2 were 0.93 mg/m^{-3}, 19.7 $\mu g/m^{-3}$, and 25.86 $\mu g/m^{-3}$ in July. In contrast, the highest and lowest levels of O_3 were 131.6 $\mu g/m^{-3}$ in July and 61.3 $\mu g/m^{-3}$ in December.

A comparison of the average monthly concentrations of $PM_{2.5}$, PM_{10}, CO, SO_2, and NO_2 indicated that the values were higher in January, February, November, and December, followed by March, April, May, and October, with lower concentrations from June to September. However, the opposite trend was observed with O_3, which had the highest average monthly levels from May to August, followed by March, April, September, and October, with lower values in January, February, November, and December. Further, the highest $PM_{2.5}$, PM_{10}, CO, SO_2, and NO_2 levels were higher than the lowest ones, respectively, indicating that the principal air pollutants were highest in January.

Similar trends were observed in 2015 (Figure 2). The highest levels of $PM_{2.5}$, PM_{10}, CO, and NO_2 were 86.8 $\mu g/m^{-3}$, 129.3 $\mu g/m^{-3}$, 1.60 mg/m^{-3}, and 48.0 $\mu g/m^{-3}$ in December, and the highest levels of SO_2 and O_3 were 53.3 $\mu g/m^{-3}$ in January and 130.8 $\mu g/m^{-3}$ in August (Table S1). The lowest $PM_{2.5}$ and PM_{10} levels were 36.4 $\mu g/m^{-3}$ and 65.4 $\mu g/m^{-3}$ in September, and the lowest values of CO, SO_2, and NO_2 were 0.83 mg/m^{-3}, 15.7 $\mu g/m^{-3}$, and 24.4 $\mu g/m^{-3}$ in July. The lowest level of O_3 was 55.7 $\mu g/m^{-3}$ in August (Table S1).

The average monthly $PM_{2.5}$, PM_{10}, CO, SO_2, and NO_2 concentrations in January and December were higher than in other months. Thus, the highest $PM_{2.5}$, PM_{10}, CO, SO_2, and NO_2 levels were 2.38, 1.98, 1.91, 3.39, 1.97, and 2.35 times greater than the lowest ones. However, the average monthly concentration of O_3 was higher from May to August, followed by March, April, September, and October, with lower values found from January to December.

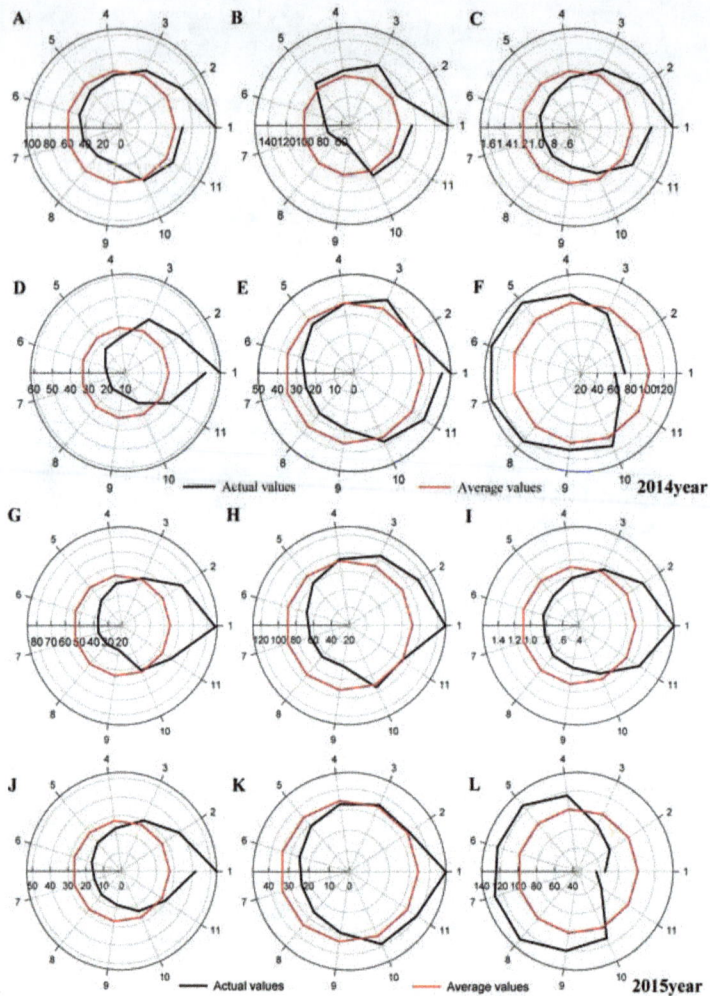

Figure 2. The monthly dynamics of the principal air pollutants over China (2014 and 2015). Graphs (**A–F**) represent the monthly concentration of $PM_{2.5}$ ($\mu g/m^{-3}$), PM_{10} ($\mu g/m^{-3}$), CO (mg/m^{-3}), SO_2 ($\mu g/m^{-3}$), NO_2 ($\mu g/m^{-3}$), and O_3 ($\mu g/m^{-3}$) in 2014, respectively; Graphs (**G–L**) represent the monthly concentration of $PM_{2.5}$ ($\mu g/m^{-3}$), PM_{10} ($\mu g/m^{-3}$), CO (mg/m^{-3}), SO_2 ($\mu g/m^{-3}$), NO_2 ($\mu g/m^{-3}$), and O_3 ($\mu g/m^{-3}$) in 2015, respectively.

3.2. Spatial Analysis of the Principal Air Pollutants

We obtained the average concentrations of the principal air pollutants during the four months (January, February, November, and December), except for O_3, the average value of which was calculated from May to August. Following the processes described above, we measured the spatial distribution of the principal air pollutants in 2014 (Figure 3). The highest and lowest levels of $PM_{2.5}$ and PM_{10} were 191.4 $\mu g/m^{-3}$ and 296.9 $\mu g/m^{-3}$ in Baoding, and 26.4 $\mu g/m^{-3}$ and 44.5 $\mu g/m^{-3}$ in Sanya (Table S2). The provinces of Shanxi, Shandong, Henan, and Shanxi, and the Beijing-Tianjin-Hebei region were the most polluted. It should be noted that most areas of China were affected by $PM_{2.5}$ and PM_{10}, including central and northern China. With respect to CO, SO_2, NO_2, and O_3, the distribution narrowed, and was more centralized in the Beijing-Tianjin-Hebei region, and Shandong and Henan provinces. The highest values of CO, SO_2, NO_2, and O_3 were 3.62 mg/m^{-3} in Baoding, 151.5 µg/m^{-3} in Yangquan, 85.2 µg/m^{-3} in Baoding, and 225.2 µg/m^{-3} in Wuhan (Table S2), respectively.

The air quality improved in 2015 in comparison to 2014 (Figure 3). The most polluted areas were primarily in the provinces of Hebei, Shandong, and Henan, with the highest levels of $PM_{2.5}$, PM_{10}, and CO (170.4 µg/m^{-3}, 254.2 µg/m^{-3}, and 3.57 mg/m^{-3}) in Baoding (Table S2). The highest values of SO_2, NO_2, and O_3 were 137.0 µg/m^{-3} in Shenyang, 77 µg/m^{-3} in Xingtai, and 184.6 µg/m^{-3} in Dezhou (Table S2), respectively. The lowest values of the principal air pollutants were observed in Hainan province.

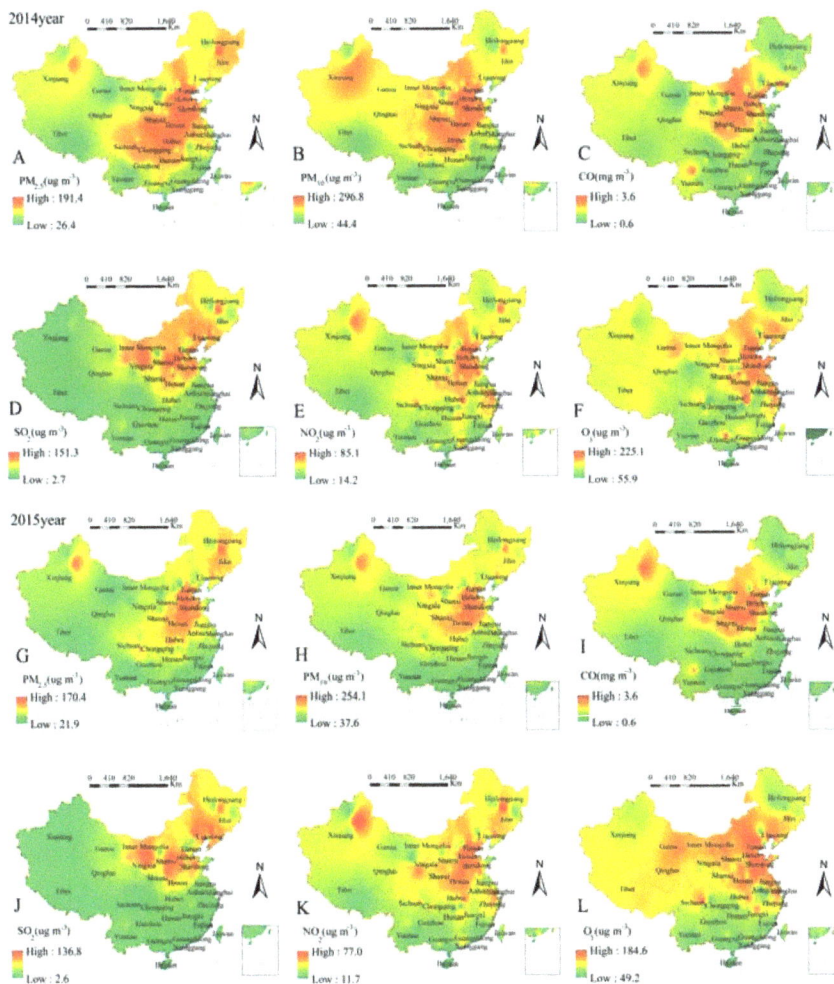

Figure 3. The spatial pattern of the mean concentrations (during January, February, November, and December) of the principal air pollutants over China (2014 and 2015). Graphs (**A–F**) represent the spatial pattern of $PM_{2.5}$, PM_{10}, CO, SO_2, NO_2, and O_3 in 2014, respectively; Graphs (**G–L**) represent the spatial pattern of $PM_{2.5}$, PM_{10}, CO, SO_2, NO_2, and O_3 in 2015, respectively.

The dynamic and spatiotemporal distribution of the principal air pollutants indicated that the cities Baoding, Xingtai, Handan, Shijiazhuang, Hengshui, Dezhou, Tangshan, Heze, Langfang, Liaocheng, Zibo, Laiwu, Anyang, Linyi, Yichang, Zhengzhou, and Pingdingshan were the most polluted regions; hence, the average daily concentrations of the principal air pollutants were analyzed in 2014 and 2015 (Figure 4). The average daily concentrations of air pollutants (SO_2, NO_2, CO, $PM_{2.5}$, and PM_{10}) were also higher in the spring and winter, except for O_3, which had the higher average daily levels in the summer. Besides, the average daily concentrations of SO_2, NO_2, CO, $PM_{2.5}$, and PM_{10} decreased at a rate of 16.2 $(\mu g/m^{-3})$/year, 2.8 $(\mu g/m^{-3})$/year, 0.1 (mg/m^{-3})/year, 11.8 $(\mu g/m^{-3})$/year, and 17.8 $(\mu g/m^{-3})$/year from 2014 to 2015, respectively. On the contrary, the average daily concentrations of O_3 increased from 102.2 $(\mu g/m^{-3})$/year to 105.4 $(\mu g/m^{-3})$/year between 2014 and 2015.

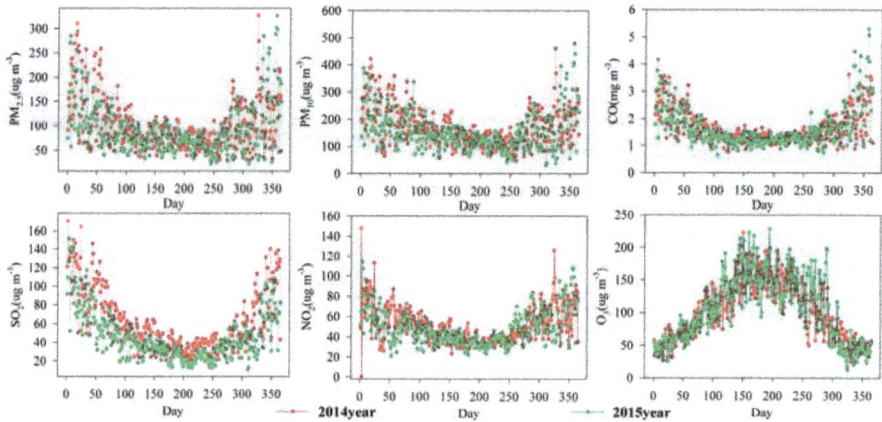

Figure 4. The daily dynamics of the principal air pollutants in the 17 most polluted cities (2014 and 2015).

3.3. The Dynamics of the Principal Air Pollutants in Beijing-Tianjin-Hebei during January, February, November, and December

As for the mosted polluteted region (Beijing-Tianjin-Hebei), Figure 5 exhibits the dynamic of the principal air pollutants during January, February, November, and December in 2014 and 2015. In Beijing, the concentrations of $PM_{2.5}$, CO, and NO_2 slightly increased; while the concentrations of PM_{10}, SO_2, and O_3 decreased at a rate of 3.5 $(\mu g/m^{-3})$/year, 14.7 $(\mu g/m^{-3})$/year, and 3.0 $(\mu g/m^{-3})$/year, respectively. All the principal air pollutants in Tianjin showed a decreasing trend, especially for the SO_2, which decreased at a rate of 30.9 $(\mu g/m^{-3})$/year. Similarly, all the principal air pollutants in Shijiazhuang also decreased, except for the CO, which increased at a rate of 0.1 (mg/m^{-3})/year. In addition, the concentration of PM_{10} obviously decreased at a rate of 83.1 $(\mu g/m^{-3})$/year in Shijiazhuang.

3.4. The Dynamics of the Principal Air Pollutants in Shandong, Henan, and Hebei during Autumn

In terms of the major agricultural provinces Shandong, Henan, and Hebei, the dynamics of the principal air pollutants in Jinan and Zhengzhou during July and September are presented in Figure 6. In Jinan, all the principal air pollutants showed a decreasing trend, especially for SO_2, which decreased at a rate of 18.0 $(\mu g/m^{-3})$/year. However, the average concentrations of $PM_{2.5}$, PM_{10}, NO_2, and O_3 in Zhengzhou increased at a rate of 7.0 $(\mu g/m^{-3})$/year, 14.7 $(\mu g/m^{-3})$/year, 9.4 $(\mu g/m^{-3})$/year, and 38.6 $(\mu g/m^{-3})$/year, respectively. Fortunately, the concentration of SO_2 and CO presented a decreasing trend in Zhengzhou during autumn. In addition, all the principal air pollutants in Shijiazhuang also decreased, except for the NO_2, which slightly increased at a rate of 0.4 ($\mu g/m^{-3}$)/year.

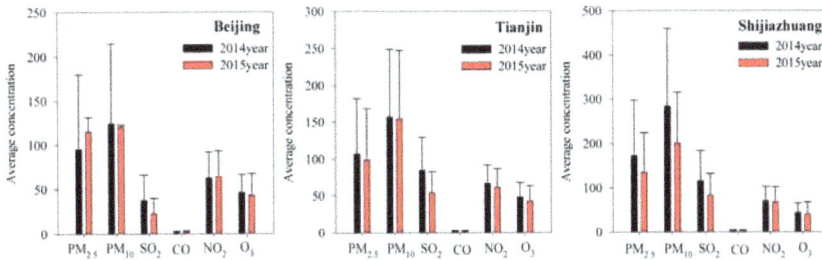

Figure 5. The average concentration of the principal air pollutants in Beijing-Tianjin-Hebei during January, February, November, and December (2014 and 2015). Units: $PM_{2.5}$ ($\mu g/m^{-3}$), PM_{10} ($\mu g/m^{-3}$), CO (mg/m^{-3}), SO_2 ($\mu g/m^{-3}$), NO_2 ($\mu g/m^{-3}$), and O_3 ($\mu g/m^{-3}$).

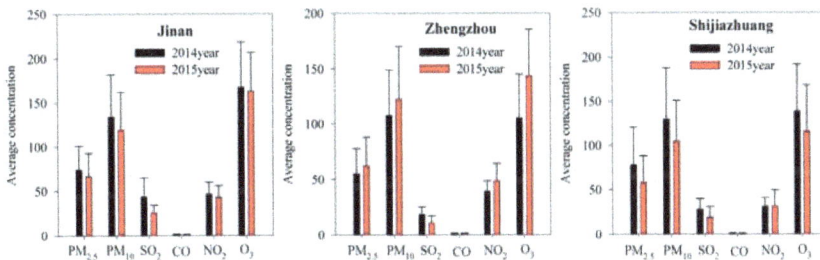

Figure 6. The average concentration of the principal air pollutants in Jinan, Zhengzhou, and Shijiazhuang in autumn (2014 and 2015). Units: $PM_{2.5}$ ($\mu g/m^{-3}$), PM_{10} ($\mu g/m^{-3}$), CO (mg/m^{-3}), SO_2 ($\mu g/m^{-3}$), NO_2 ($\mu g/m^{-3}$), and O_3 ($\mu g/m^{-3}$).

3.5. The Relationships among the Principal Air Pollutants

As Figure 7 shows, the correlation coefficients among air pollutants were significant at $p < 0.05$ in the most polluted 17 cities (2014). Our results indicated that there were positive correlations among $PM_{2.5}$, PM_{10}, CO, SO_2, and NO_2, and negative correlations for O_3. As for $PM_{2.5}$, and the correlation coefficients between PM_{10}, CO, SO_2, NO_2, and O_3 were 0.94, 0.89, 0.74, 0.82, and -0.42, respectively. High correlation coefficients among CO, SO_2, and NO_2 (0.88 for CO and SO_2, 0.89 for CO and NO_2, 0.87 for SO_2 and NO_2) were observed. We also found significant negative correlation coefficients among O_3, CO, SO_2, and NO_2 (-0.57 for O_3 and CO, -0.60 for O_3 and SO_2, and -0.56 for O_3 and NO_2).

Figure 8 shows that there were significant relationships among the principal air pollutants in 2015. The correlation coefficients between $PM_{2.5}$ and PM_{10}, CO, SO_2, NO_2, and O_3 were 0.94, 0.93, 0.73, 0.84, and -0.38. In addition, there was a slight decrease in the correlation coefficients among CO, SO_2, and NO_2 compared to those in 2014, with values of 0.77 for CO and SO_2, 0.87 for CO and NO_2, and 0.82 for SO_2 and NO_2. A similar decrease in the correlation coefficients among O_3, CO, SO_2, and NO_2 was observed (-0.54 for O_3 and CO, -0.52 for O_3 and SO_2, and -0.44 for O_3 and NO_2).

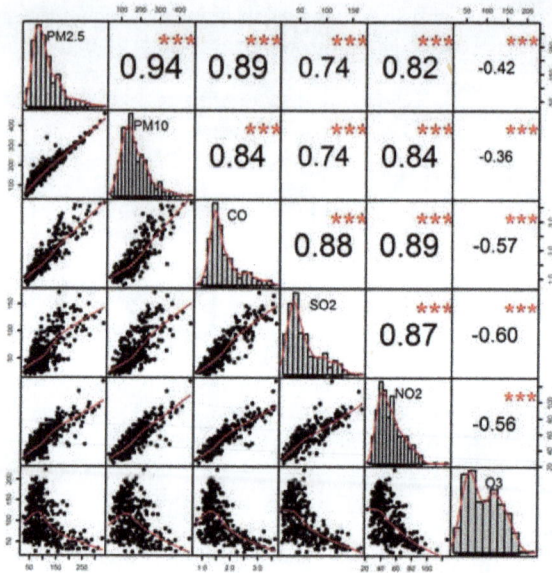

Figure 7. The relationships among the principal air pollutants in the 17 most polluted cities (2014). All the numbers and red stars represent the correlation coefficients and significant relationships among the principal air pollutants.

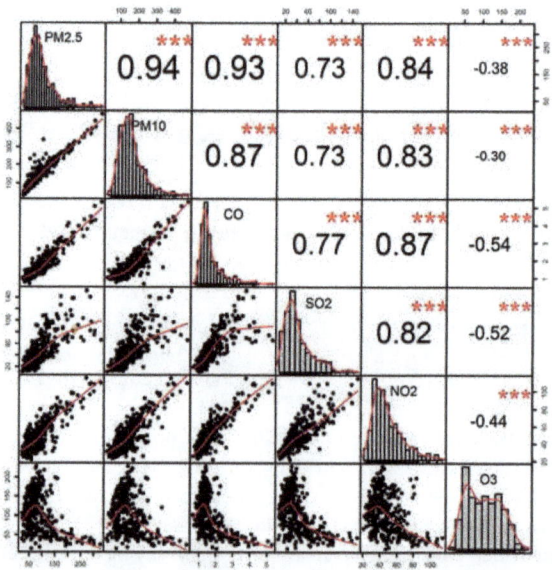

Figure 8. The relationships among the principal air pollutants in the most polluted cities (2015). All the numbers and red stars represent the correlation coefficients and significant relationships among the principal air pollutants.

3.6. GAM Prediction of the Trend for Principal Air Pollutants

The GAM was employed to reveal the future dynamics of atmospheric pollutants (Figure 9), and the non-linear relationship between the pollutants and day-of-year trends was modeled flexibly by the log link function. The results showed that although $PM_{2.5}$ and SO_2 will decrease sharply in the future, further observation is imperative. Unfortunately, a consistent increase in NO_2 and CO was predicted. Furthermore, a slight increase of PM_{10} was observed until it peaked between $100 < x < 150$, and then decreased slightly over the remaining days. Meanwhile, the estimated O_3 showed a slight increase before $x = 50$ and then decreased until approximately $x = 125$, followed by a slight upward trend during the remaining days. In summary, a small variation in PM_{10} and O_3 indicated that the simulated results were reliable and desirable.

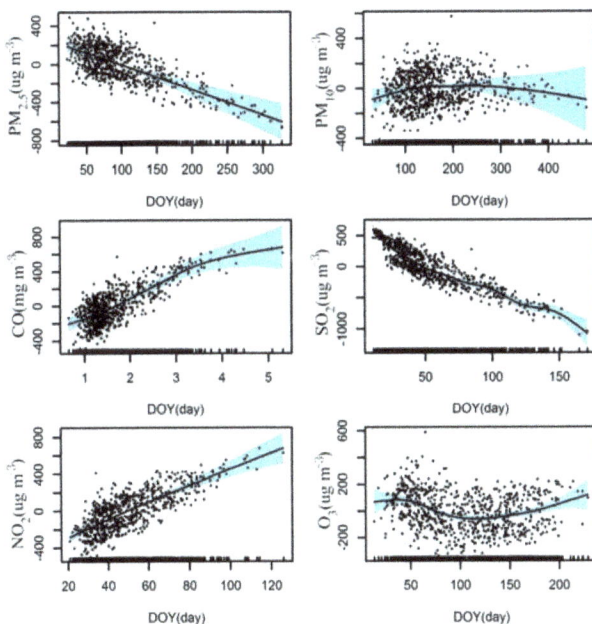

Figure 9. Predicted changes in the principal air pollutants together with temporal dynamics by generalized additive model (GAM) analysis. Rugplot on the x-axis represents the DOY (day of year), and the light blue belts indicate the credible intervals.

4. Discussion

4.1. Temporal and Spatial Patterns of China's Principal Air Pollutants

The concentration of the principal air pollutants was higher in January and December—a phenomenon also observed in Fuzhou [45] and Shanghai [46,47]. In January 2013, two severe air pollution events happened in Beijing, during which the hourly concentration of $PM_{2.5}$ rose to 680 μg/m^{-3} and 530 μg/m^{-3}, respectively [38]. In Shanghai, the highest and lowest levels of PM_{10}, NO_2, and SO_2 were found in winter and autumn, with vehicle emissions and meteorological conditions the most probable causes [36]. In the North China Plain, high emissions of atmospheric contamination, biomass burning, and stable weather conditions contributed to the haze events in winter [48], and higher energy consumption and motor vehicle emissions occurred in winter in Guangzhou [49]. If natural gas is supplied for municipal heating in Beijing rather than coal, air pollutant emissions ($PM_{2.5}$, SO_2, NO_X, etc.) would decrease by 52% in winter [50] because the inter-transport of pollution and the secondary aerosols O_3, OC (organic carbon) and VOC (volatile organic compounds) promote pollution. Thus, reducing the levels of air pollutants during the cold season is imperative for public health [51].

The spatial patterns of the average monthly simulation values indicated that in 2014 the most polluted regions were the Beijing-Tianjin-Hebei region and the provinces of Shanxi, Shandong, Henan, Hubei, and Shanxi. Industrial and domestic sources of pollution and agricultural emissions are the chief regional contributors to pollution in these regions [52]. Compared to 2014, the distribution patterns of air pollutants show a slight shift towards a smaller area in 2015, becoming more centralized in Hebei, Shandong, and Henan provinces, where haze episodes were also more frequent. A similar situation prevailed in the North China Plain [53], Guangzhou [22], and the Yangtze River Delta [26,54]. The stricter laws protecting air quality by government may account for this decreasing trend between 2014 and 2015 to some extent.

The significant relationships among air pollutants illustrated that there could be similar or interacting sources for the pollutions. Many studies have been undertaken to explore the source of large areas of haze. Biomass open burning contributes 47% of the $PM_{2.5}$ in the Yangtze River Delta [55], is an important precursor for O_3 [56], and is a stressor on marine ecosystems [57]. The fact that the inter-transport of pollution accompanies high humidity is the dominant reason for the haze in East China [29,46] and Northern Taiwan [27]. Dust was a major source of pollution in eastern Inner Mongolia [58], local emissions and regional transport accounted for pollution in Nanjing [59], coal consumption and industry increased pollution in Beijing in winter [32,50], and industrial pollution and vehicle emissions were the dominant local contributors to the levels of NO_2 and $PM_{2.5}$ in Shanghai [31,34].

Specifically, for the most polluted region—the northeast of China—the mineral dust from the deserts of western China contributes significantly to the concentration of PM [60,61]. The long-range transport dust plumes mixed with regional pollutants aid the formation of haze episodes [53,60–62]. Thus, the local pollutants also have significant contributions to the widespread haze pollution [48]. Meanwhile, during haze episodes, the secondary inorganic pollutants evident increased in PM [36], suggesting a joint effect among them [63]. Under unfavorable meteorological conditions, the interaction of PM and the secondary inorganic pollutants produced a large amount of aerosols with the characteristic of low visibility [36].

In general, the dust, municipal heating, and vehicle and local emissions are the dominant contributors to haze in winter [31], and meteorological conditions contribute significantly in the distribution, formation, and evolution of haze: for example, higher relative humidity and weaker wind speed contribute to haze [23,25,30]. In addition, there is a positive relationship between visibility and wind speed, and a negative relationship with relative humidity [45,49,59]. Secondary inorganic ions were also positively correlated with stable weather conditions (higher relative humidity), which determined the specific chemical composition of haze [64]. In summary, the median or highest values and the distribution area of air pollutants showed a decreasing trend from 2014 to 2015, which may be explained in part by the improved energy efficiency and stricter laws protecting air quality.

4.2. Policy Recommendations

The estimated values of $PM_{2.5}$ and SO_2 showed a sharp decreasing trend. However, we observed a consistent increasing trend in CO and NO_2, while the estimated values of PM_{10} and O_3 remained stable overall. Thus, the concentration of CO and NO_2 in China will continue to increase. As we know already, vehicle emissions are a key factor in the high concentration of NO_2, which has been caused by the rapid increase in the number of vehicles and industrial parks [31,34]. Therefore, technological innovation is imperative in coal gasification, liquefaction, and storage to improve the energy efficiency of coal [65]. Indeed, replacing fossil fuels with clean fuel (natural gas, nuclear energy, etc.) for municipal heating in winter would be an effective long-term measure to mitigate the dense haze north of the Yellow River [50]. In addition, stringent measures to control particle emissions (biomass burning) should be implemented during specific periods based on meteorological conditions [29,66], while mass transit should be encouraged to avoid energy waste. Further, the central and local government should subsidize the manufacture of hybrid, flex-fuel, and electric automobiles [67,68], and remove market-entry barriers to promote competition among companies [65]. Finally, we need to improve the proportion of the tertiary industry to optimize the economic structure that uses less energy and produces fewer particle emissions [69,70].

As Figure 10 shows, this study highlighted the measures needed to decrease current sources of air pollutants. Indeed, replacing fossil fuels with clean fuels (natural gas, nuclear, wind, and solar energy) will help to address air pollution's root causes. To transform and improve air quality, restricting biomass burning during specific periods (according to meteorological conditions) is imperative, and it may be necessary to implement strict environmental policies forbidding open biomass burning when municipal heating is highest. We should also consider recycling biological feed. Equally important, the issue of vehicle emissions requires the cooperation of government, manufacturers, retailers, and consumers—particularly in the development and promotion of flex-fuel (hybrid) and electric automobiles. Further, ecosystem restoration projects such as the Three-North Shelterbelt System can be effective in defending against sandstorms while simultaneously improving the regional environment. Remote sensing should also be employed to monitor the formation and development of haze, to record, measure, and evaluate the sources of pollution, and monitor the trajectory of haze over the long-term. Indeed, predicting the values and levels of air pollutants and communicating this information by mobile messages will cement trust between individuals and governments. Further, analyzing the structure and function of cities based on local conditions to facilitate scientific urban planning and formulate reasonable population distribution policies should help to mitigate pollution. All of the aforementioned underscore the essential need to share information, provide support, and increase cooperation among different departments to achieve effective haze control. Of course, trade-offs between environmental objectives, economic growth, energy consumption, living standards, and funding are inevitable, but we believe that these can be balanced for the benefit and welfare of all.

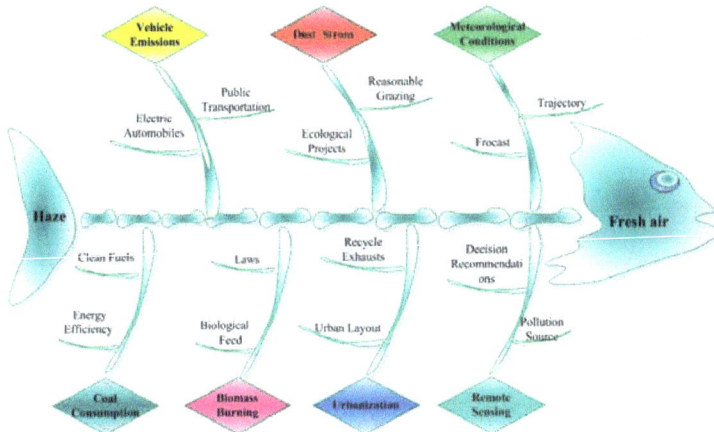

Figure 10. The mind map for decision-makers.

5. Conclusions

Because particulate haze episodes over China have increased in recent years, it is imperative that we understand pollutant pathways and propose recommendations for a policy framework. In general, November to February demonstrated the highest concentrations of air pollutants ($PM_{2.5}$, PM_{10}, CO, SO_2, and NO_2), excluding O_3 levels, which were highest from May to August. Further, the most highly polluted areas were in the provinces of Shandong, Henan, and Shanxi, and in the Beijing-Tianjin-Hebei region. Fortunately, most of the principal air pollutants presented a decreasing or relatively stable trend in Beijing-Tianjin-Hebei during January, February, November, and December. Meanwhile, most of the principal air pollutants also presented a decreasing trend in Jinan and Shijiazhuang in autumn. However, in Zhengzhou, the haze events during autumn were unoptimistic. Although the conflict between clean air and economic growth will continue, measures can be taken to mitigate the sources of air pollutants and to use resources to control, reduce, and manage air pollution.

To improve our understanding of the formation and frequency of haze, satellite observations and monitoring of high-risk areas are important subjects for future research. We need to understand the specific effects of meteorological conditions on the transport mechanism of air pollutants, and the role they play in the secondary formation process.

Supplementary Materials: The following are available online at www.mdpi.com/2073-4433/8/8/137/s1: Table S1: The average monthly concentration of air pollutants over China during 2014 and 2015; Table S2: The mean concentration of air pollutants over China among four months (January, February, November, and December).

Acknowledgments: This work was supported by the Ministry of Environmental Protection of the People's Republic of China. The research was funded as well by the West Light Foundation of The Chinese Academy of Sciences, and the Open Fund of the Key Laboratory of Mountain Surface Processes and Eco-regulation.

Author Contributions: J.S. contributed to the study design, J.S., T.Z. and H.Y. were involved in drafting the manuscript, approving the final draft, and agree to be accountable for the work. All authors read and approved the final manuscript.

Conflicts of Interest: The authors declare no conflict of interest.

References

1. He, X.Y.; Hu, J.B.; Chen, W.; Li, X.Y. Haze removal based on advanced haze-optimized transformation (AHOT) for multispectral imagery. *Int. J. Remote Sens.* **2010**, *31*, 5331–5348. [CrossRef]
2. Han, L.J.; Zhou, W.Q.; Pickett, S.T.A.; Li, W.F.; Li, L. An optimum city size? The scaling relationship for urban population and fine particulate ($PM_{2.5}$) concentration. *Environ. Pollut.* **2016**, *208*, 96–101. [CrossRef] [PubMed]
3. Carrasco, L.R. Silver lining of Singapore's haze. *Science* **2013**, *341*, 342–343. [CrossRef] [PubMed]
4. Hao, Y.; Liu, Y.-M. The influential factors of urban $PM_{2.5}$ concentrations in China: A spatial econometric analysis. *J. Clean. Prod.* **2016**, *112*, 1443–1453. [CrossRef]
5. Malandrino, M.; Casazza, M.; Abollino, O.; Minero, C.; Maurino, V. Size resolved metal distribution in the PM matter of the city of Turin (Italy). *Chemosphere* **2016**, *147*, 477–489. [CrossRef] [PubMed]
6. Salinas, S.V.; Chew, B.N.; Miettinen, J.; Campbell, J.R.; Welton, E.J.; Reid, J.S.; Yu, L.Y.E.; Liew, S.C. Physical and optical characteristics of the October 2010 haze event over Singapore: A photometric and lidar analysis. *Atmos. Res.* **2013**, *122*, 555–570. [CrossRef]
7. Kim, K.W.; Kim, Y.J.; Bang, S.Y. Summer time haze characteristics of the urban atmosphere of Gwangju and the rural atmosphere of Anmyon, Korea. *Environ. Monit. Assess.* **2008**, *141*, 189–199. [CrossRef] [PubMed]
8. Zhao, H.M.; Zhang, X.L.; Zhang, S.C.; Chen, W.W.; Tong, D.Q.; Xiu, A.J. Effects of Agricultural Biomass Burning on Regional Haze in China: A Review. *Atmosphere* **2017**, *8*, 88. [CrossRef]
9. Szidat, S. Atmosphere. Sources of Asian Haze. *Science* **2009**, *323*, 470–471. [CrossRef] [PubMed]
10. Begum, B.A.; Hopke, P.K. Identification of haze-creating sources from fine particulate matter in Dhaka aerosol using carbon fractions. *J. Air Waste Manag.* **2013**, *63*, 1046–1057. [CrossRef]
11. Pernigotti, D.; Belis, C.A.; Spano, L. Specieurope: The European data base for PM source profiles. *Atmos. Pollut. Res.* **2016**, *7*, 307–314. [CrossRef]
12. Huang, R.J.; Zhang, Y.; Bozzetti, C.; Ho, K.F.; Cao, J.J.; Han, Y.; Daellenbach, K.R.; Slowik, J.G.; Platt, S.M.; Canonaco, F. High secondary aerosol contribution to particulate pollution during haze events in China. *Nature* **2014**, *514*, 218–222. [CrossRef] [PubMed]
13. Soleiman, A.; Ohtman, M.; Samah, A.A.; Sulaiman, N.M.; Radojevic, M. The occurrence of haze in Malaysia: A case study in an urban industrial area. *Pure Appl. Geophys.* **2003**, *160*, 221–238. [CrossRef]
14. Abdullaev, S.F.; Maslov, V.A.; Nazarov, B.I. Study of dust haze in arid zone. *Izv. Atmos. Ocean. Phys.* **2013**, *49*, 276–284. [CrossRef]

15. Zhang, R.H.; Li, Q.; Zhang, R.N. Meteorological conditions for the persistent severe fog and haze event over eastern China in January 2013. *Sci. China Earth Sci.* **2014**, *57*, 26–35.

16. Elias, T.; Haeffelin, M.; Drobinski, P.; Gomes, L.; Rangognio, J.; Bergot, T.; Chazette, P.; Raut, J.C.; Colomb, M. Particulate contribution to extinction of visible radiation: Pollution, haze, and fog. *Atmos. Res.* **2009**, *92*, 443–454. [CrossRef]

17. Sun, J.; Zhou, T.C. Health risk assessment of China's main air pollutants. *BMC Public Health* **2017**, *17*, 212. [CrossRef] [PubMed]

18. Zhang, X.Y.; Wang, Y.Q.; Niu, T.; Zhang, X.C.; Gong, S.L.; Zhang, Y.M.; Sun, J.Y. Atmospheric aerosol compositions in China: Spatial/temporal variability, chemical signature, regional haze distribution and comparisons with global aerosols. *Atmos. Chem. Phys.* **2012**, *12*, 799.

19. Wang, Y.; Li, Z.Q.; Zhang, Y.; Wang, Q.; Ma, J.Z. Impact of aerosols on radiation during a heavy haze event in Beijing. *IOP Conf. Ser. Earth Environ. Sci.* **2014**, *17*, 012012.

20. Sun, Y.L.; Jiang, Q.; Wang, Z.F.; Fu, P.Q.; Li, J.; Yang, T.; Yin, Y. Investigation of the sources and evolution processes of severe haze pollution in Beijing in January 2013. *J. Geophys. Res. Atmos.* **2014**, *119*, 4380–4398. [CrossRef]

21. Zhang, M.; Ma, Y.Y.; Gong, W.; Zhu, Z.M. Aerosol optical properties of a haze episode in Wuhan based on ground-based and satellite observations. *Atmosphere* **2014**, *5*, 699–719. [CrossRef]

22. Zhang, Z.L.; Wang, J.; Chen, L.H.; Chen, X.Y.; Sun, G.Y.; Zhong, N.S.; Kan, H.D.; Lu, W.J. Impact of haze and air pollution-related hazards on hospital admissions in Guangzhou, China. *Environ. Sci. Pollut. Res.* **2014**, *21*, 4236–4244. [CrossRef] [PubMed]

23. Liu, X.G.; Li, J.; Qu, Y.; Han, T.; Hou, L.; Gu, J.; Chen, C.; Yang, Y.; Liu, X.; Yang, T.; et al. Formation and evolution mechanism of regional haze: A case study in the megacity Beijing, China. *Atmos. Chem. Phys.* **2013**, *13*, 4501–4514. [CrossRef]

24. Quan, J.; Zhang, Q.; He, H.; Liu, J.; Huang, M.; Jin, H. Analysis of the formation of fog and haze in North China Plain (NCP). *Atmos. Chem. Phys.* **2011**, *11*, 8205–8214. [CrossRef]

25. Bei, N.F.; Xiao, B.; Meng, N.; Feng, T. Critical role of meteorological conditions in a persistent haze episode in the Guanzhong basin, China. *Sci. Total Environ.* **2016**, *550*, 273–284. [CrossRef] [PubMed]

26. Tang, L.L.; Yu, H.X.; Ding, A.J.; Zhang, Y.J.; Qin, W.; Wang, Z.; Chen, W.T.; Hua, Y.; Yang, X.X. Regional contribution to PM_1 pollution during winter haze in Yangtze River Delta, China. *Sci. Total Environ.* **2016**, *541*, 161–166. [CrossRef] [PubMed]

27. Wang, S.H.; Hung, W.T.; Chang, S.C.; Yen, M.C. Transport characteristics of Chinese haze over Northern Taiwan in winter, 2005–2014. *Atmos. Environ.* **2016**, *126*, 76–86. [CrossRef]

28. Gabrielli, M.; Fracassetti, D.; Tirelli, A. Release of phenolic compounds from cork stoppers and its effect on protein-haze. *Food Control* **2016**, *62*, 330–336. [CrossRef]

29. Tao, M.H.; Chen, L.F.; Wang, Z.F.; Tao, J.H.; Su, L. Satellite observation of abnormal yellow haze clouds over East China during summer agricultural burning season. *Atmos. Environ.* **2013**, *79*, 632–640. [CrossRef]

30. Sun, Y.; Song, T.; Tang, G.Q.; Wang, Y.S. The vertical distribution of $PM_{2.5}$ and boundary-layer structure during summer haze in Beijing. *Atmos. Environ.* **2013**, *74*, 413–421. [CrossRef]

31. Wang, Y.J.; Li, L.; Chen, C.H.; Huang, C.; Huang, H.Y.; Feng, J.L.; Wang, S.X.; Wang, H.L.; Zhang, G.; Zhou, M.; et al. Source apportionment of fine particulate matter during autumn haze episodes in Shanghai, China. *J. Geophys. Res. Atmos.* **2014**, *119*, 1903–1914. [CrossRef]

32. Lin, Y.C.; Hsu, S.C.; Chou, C.C.K.; Zhang, R.J.; Wu, Y.F.; Kao, S.J.; Luo, L.; Huang, C.H.; Lin, S.H.; Huang, Y.T. Wintertime haze deterioration in Beijing by industrial pollution deduced from trace metal fingerprints and enhanced health risk by heavy metals. *Environ. Pollut.* **2016**, *208*, 284–293. [CrossRef] [PubMed]

33. Gui, K.; Che, H.Z.; Chen, Q.L.; An, L.C.; Zeng, Z.L.; Guo, Z.Y.; Zheng, Y.; Wang, H.; Wang, Y.Q.; Yu, J.; et al. Aerosol optical properties based on ground and satellite retrievals during a serious haze episode in December 2015 over Beijing. *Atmosphere* **2016**, *7*, 70. [CrossRef]

34. Zhao, W.C.; Cheng, J.P.; Li, D.L.; Duan, Y.S.; Wei, H.P.; Ji, R.X.; Wang, W.H. Urban ambient air quality investigation and health risk assessment during haze and non-haze periods in Shanghai, China. *Atmos. Pollut. Res.* **2013**, *4*, 275–281. [CrossRef]

35. Li, W.J.; Shi, Z.B.; Zhang, D.Z.; Zhang, X.Y.; Li, P.R.; Feng, Q.J.; Yuan, Q.; Wang, W.X. Haze particles over a coal-burning region in the China Loess Plateau in winter: Three flight missions in December 2010. *J. Geophys. Res. Atmos.* **2012**, *117*. [CrossRef]

36. Zhao, X.J.; Zhao, P.S.; Xu, J.; Meng, W.; Pu, W.W.; Dong, F.; He, D.; Shi, Q.F. Analysis of a winter regional haze event and its formation mechanism in the North China Plain. *Atmos. Chem. Phys.* **2013**, *13*, 5685–5696. [CrossRef]

37. Zhu, C.S.; Cao, J.J.; Tsai, C.J.; Shen, Z.X.; Liu, S.X.; Huang, R.J.; Zhang, N.N.; Wang, P. The rural carbonaceous aerosols in coarse, fine, and ultrafine particles during haze pollution in northwestern China. *Environ. Sci. Pollut. Res.* **2016**, *23*, 4569–4575. [CrossRef] [PubMed]

38. Wang, Y.S.; Yao, L.; Wang, L.L.; Liu, Z.R.; Ji, D.S.; Tang, G.Q.; Zhang, J.K.; Sun, Y.; Hu, B.; Xin, J.Y. Mechanism for the formation of the January 2013 heavy haze pollution episode over central and eastern China. *Sci. China Earth Sci.* **2014**, *57*, 14–25. [CrossRef]

39. Zhu, T.; Shang, J.; Zhao, D.F. The roles of heterogeneous chemical processes in the formation of an air pollution complex and gray haze. *Sci. China Chem.* **2011**, *54*, 145–153. [CrossRef]

40. Jansen, R.C.; Shi, Y.; Chen, J.M.; Hu, Y.J.; Xu, C.; Hong, S.M.; Li, J.; Zhang, M. Using hourly measurements to explore the role of secondary inorganic aerosol in $PM_{2.5}$ during haze and fog in Hangzhou, China. *Adv. Atmos. Sci.* **2014**, *31*, 1427–1434. [CrossRef]

41. Shen, X.J.; Sun, J.Y.; Zhang, X.Y.; Zhang, Y.M.; Zhang, L.; Che, H.C.; Ma, Q.L.; Yu, X.M.; Yue, Y.; Zhang, Y.W. Characterization of submicron aerosols and effect on visibility during a severe haze-fog episode in Yangtze River Delta, China. *Atmos. Environ.* **2015**, *120*, 307–316. [CrossRef]

42. Lin, Y.F.; Huang, K.; Zhuang, G.S.; Fu, J.S.; Wang, Q.Z.; Liu, T.N.; Deng, C.R.; Fu, Q.Y. A multi-year evolution of aerosol chemistry impacting visibility and haze formation over an Eastern Asia megacity, Shanghai. *Atmos. Environ.* **2014**, *92*, 76–86. [CrossRef]

43. Catalano, M.; Galatioto, F.; Bell, M.; Namdeo, A.; Bergantino, A.S. Improving the prediction of air pollution peak episodes generated by urban transport networks. *Environ. Sci. Policy* **2016**, *60*, 69–83. [CrossRef]

44. Chen, X.L.; Feng, Y.R.; Li, J.N.; Lin, W.S.; Fan, S.J.; Wang, A.Y.; Fong, S.K.; Lin, H. Numerical simulations on the effect of sea-land breezes on atmospheric haze over the Pearl River Delta Region. *Environ. Model. Assess.* **2009**, *14*, 351–363.

45. Zhang, F.W.; Xu, L.L.; Chen, J.S.; Chen, X.Q.; Niu, Z.C.; Lei, T.; Li, C.M.; Zhao, J.P. Chemical characteristics of $PM_{2.5}$ during haze episodes in the urban of Fuzhou, China. *Particuology* **2013**, *11*, 264–272. [CrossRef]

46. Wang, X.M.; Chen, J.M.; Cheng, T.T.; Zhang, R.Y.; Wang, X.M. Particle number concentration, size distribution and chemical composition during haze and photochemical smog episodes in Shanghai. *J. Environ. Sci.* **2014**, *26*, 1894–1902. [CrossRef] [PubMed]

47. Wang, H.L.; Qiao, L.P.; Lou, S.R.; Zhou, M.; Ding, A.J.; Huang, H.Y.; Chen, J.M.; Wang, Q.; Tao, S.K.; Chen, C.H.; et al. Chemical composition of $PM_{2.5}$ and meteorological impact among three years in urban Shanghai, China. *J. Clean. Prod.* **2016**, *112*, 1302–1311. [CrossRef]

48. Gao, M.; Carmichael, G.R.; Wang, Y.; Saide, P.E.; Yu, M.; Xin, J.; Liu, Z.; Wang, Z. Modeling study of the 2010 regional haze event in the North China Plain. *Atmos. Chem. Phys.* **2016**, *16*, 1673–1691. [CrossRef]

49. Tan, J.H.; Duan, J.C.; Chen, D.H.; Wang, X.H.; Guo, S.J.; Bi, X.H.; Sheng, G.Y.; He, K.B.; Fu, J.M. Chemical characteristics of haze during summer and winter in Guangzhou. *Atmos. Res.* **2009**, *94*, 238–245. [CrossRef]

50. Li, H.C.; Yang, S.Y.; Zhang, J.; Qian, Y. Coal-based synthetic natural gas (SNG) for municipal heating in China: Analysis of haze pollutants and greenhouse gases (GHGs) emissions. *J. Clean. Prod.* **2016**, *112*, 1350–1359. [CrossRef]

51. Fang, Z.Q.; Zhu, H.L.; Bao, W.Z.; Preston, C.; Liu, Z.; Dai, J.Q.; Li, Y.Y.; Hu, L.B. Highly transparent paper with tunable haze for green electronics. *Energ. Environ. Sci.* **2014**, *7*, 3313–3319. [CrossRef]

52. Wang, L.T.; Wei, Z.; Yang, J.; Zhang, Y.; Zhang, F.F.; Su, J.; Meng, C.C.; Zhang, Q. The 2013 severe haze over southern Hebei, China: Model evaluation, source apportionment, and policy implications. *Atmos. Chem. Phys.* **2014**, *14*, 3151–3173. [CrossRef]

53. Zhang, X.L.; Huang, Y.B.; Zhu, W.Y.; Rao, R.Z. Aerosol characteristics during summer haze episodes from different source regions over the coast city of North China Plain. *J. Quant. Spectrosc. Radiat. Transf.* **2013**, *122*, 180–193. [CrossRef]

54. Wang, S.; Zhou, C.; Wang, Z.; Feng, K.; Hubacek, K. The characteristics and drivers of fine particulate matter ($PM_{2.5}$) distribution in China. *J. Clean. Prod.* **2017**, *142*, 1800–1809. [CrossRef]

55. Cheng, Z.; Wang, S.; Fu, X.; Watson, J.G.; Jiang, J.; Fu, Q.; Chen, C.; Xu, B.; Yu, J.; Chow, J.C.; et al. Impact of biomass burning on haze pollution in the Yangtze River delta, China: A case study in summer 2011. *Atmos. Chem. Phys.* **2014**, *14*, 4573–4585. [CrossRef]

56. Zhang, Y.W.; Zhang, X.Y.; Zhang, Y.M.; Shen, X.J.; Sun, J.Y.; Ma, Q.L.; Yu, X.M.; Zhu, J.L.; Zhang, L.; Che, H.C. Significant concentration changes of chemical components of PM_1 in the Yangtze River Delta area of China and the implications for the formation mechanism of heavy haze-fog pollution. *Sci. Total Environ.* **2015**, *538*, 7–15. [CrossRef] [PubMed]

57. Jaafar, Z.; Loh, T.L. Linking land, air and sea: Potential impacts of biomass burning and the resultant haze on marine ecosystems of Southeast Asia. *Glob. Chang. Biol.* **2014**, *20*, 2701–2707. [CrossRef] [PubMed]

58. Guo, J.P.; Zhang, X.Y.; Cao, C.X.; Che, H.Z.; Liu, H.L.; Gupta, P.; Zhang, H.; Xu, M.; Li, X.W. Monitoring haze episodes over the Yellow Sea by combining multisensor measurements. *Int. J. Remote Sens.* **2010**, *31*, 4743–4755. [CrossRef]

59. Kang, H.Q.; Zhu, B.; Su, J.F.; Wang, H.L.; Zhang, Q.C.; Wang, F. Analysis of a long-lasting haze episode in Nanjing, China. *Atmos. Res.* **2013**, *120*, 78–87. [CrossRef]

60. Tao, M.H.; Chen, L.F.; Su, L.; Tao, J.H. Satellite observation of regional haze pollution over the North China Plain. *J. Geophys. Res. Atmos.* **2012**, *117*, D12. [CrossRef]

61. Yang, T.; Wang, X.Q.; Wang, Z.F.; Sun, Y.L.; Zhang, W.; Zhang, B.; Du, Y.M. Gravity-current driven transport of haze from North China Plain to Northeast China in Winter 2010-Part I: Observations. *Sola* **2012**, *8*, 13–16. [CrossRef]

62. Ye, X.X.; Song, Y.; Cai, X.H.; Zhang, H.S. Study on the synoptic flow patterns and boundary layer process of the severe haze events over the North China Plain in January 2013. *Atmos. Environ.* **2016**, *124*, 129–145. [CrossRef]

63. Kirrane, E.; Svendsgaard, D.; Ross, M.; Buckley, B.; Davis, A.; Johns, D.; Kotchmar, D.; Long, T.C.; Luben, T.J.; Smith, G.; et al. A Comparison of risk estimates for the effect of short-term exposure to PM, NO_2 and CO on cardiovascular hospitalizations and emergency department visits: Effect size modeling of study findings. *Atmosphere* **2011**, *2*, 688–701. [CrossRef]

64. Han, B.; Zhang, R.; Yang, W.; Bai, Z.P.; Ma, Z.Q.; Zhang, W.J. Heavy haze episodes in Beijing during January 2013: Inorganic ion chemistry and source analysis using highly time-resolved measurements from an urban site. *Sci. Total Environ.* **2016**, *544*, 319–329. [CrossRef] [PubMed]

65. Zhang, X.H.; Wu, L.Q.; Zhang, R.; Deng, S.H.; Zhang, Y.Z.; Wu, J.; Li, Y.W.; Lin, L.L.; Li, L.; Wang, Y.J.; et al. Evaluating the relationships among economic growth, energy consumption, air emissions and air environmental protection investment in China. *Renew. Sustain. Energy Rev.* **2013**, *18*, 259–270. [CrossRef]

66. Mu, M.; Zhang, R.H. Addressing the issue of fog and haze: A promising perspective from meteorological science and technology. *Sci. China Earth Sci.* **2014**, *57*, 1–2. [CrossRef]

67. Sun, J.; Wang, J.N.; Wei, Y.Q.; Li, Y.R.; Liu, M. The Haze Nightmare Following the Economic Boom in China: Dilemma and Tradeoffs. *Int. J. Environ. Res. Public Health* **2016**, *13*, 402. [CrossRef] [PubMed]

68. Wu, X.; Tan, L.; Guo, J.; Wang, Y.; Liu, H.; Zhu, W. A study of allocative efficiency of $PM_{2.5}$ emission rights based on a zero sum gains data envelopment model. *J. Clean. Prod.* **2016**, *113*, 1024–1031. [CrossRef]

69. Wang, S.S.; Zhou, D.Q.; Zhou, P.; Wang, Q.W. CO_2 emissions, energy consumption and economic growth in China: A panel data analysis. *Energy Policy* **2011**, *39*, 4870–4875. [CrossRef]

70. Yuan, X.L.; Mu, R.M.; Zuo, J.; Wang, Q.S. Economic development, energy consumption, and air pollution: A critical assessment in China. *Hum. Ecol. Risk Assess.* **2015**, *21*, 781–798. [CrossRef]

![atmosphere logo] *atmosphere*

MDPI

Article

Improving PM$_{2.5}$ Air Quality Model Forecasts in China Using a Bias-Correction Framework

Baolei Lyu [1,†], Yuzhong Zhang [2,‡] and Yongtao Hu [3,*]

1   Department for Earth System Science, Tsinghua University, Beijing 100084, China; baoleilv@foxmail.com
2   School of Earth and Atmospheric Sciences, Georgia Institute of Technology, Atlanta, GA 30332, USA;
    zhangyz.pku@gmail.com
3   School of Civil and Environmental Engineering, Georgia Institute of Technology, Atlanta, GA 30332, USA
*   Correspondence: yh29@mail.gatech.edu; Tel.: +01-404-385-4558
†   Current address: Huayun Sounding Meteorological Technology Corporation, Beijing 102299, China.
‡   Current address: School of Engineering and Applied Sciences, Harvard University, Cambridge,
    MA 02138, USA.

Received: 11 July 2017; Accepted: 9 August 2017; Published: 13 August 2017

Abstract: Chinese cities are experiencing severe air pollution in particular, with extremely high PM$_{2.5}$ levels observed in cold seasons. Accurate forecasting of occurrence of such air pollution events in advance can help the community to take action to abate emissions and would ultimately benefit the citizens. To improve the PM$_{2.5}$ air quality model forecasts in China, we proposed a bias-correction framework that utilized the historic relationship between the model biases and forecasted and observational variables to post-process the current forecasts. The framework consists of four components: (1) a feature selector that chooses the variables that are informative to model forecast bias based on historic data; (2) a classifier trained to efficiently determine the forecast analogs (clusters) based on clustering analysis, such as the distance-based method and the classification tree, etc.; (3) an error estimator, such as the Kalman filter, to predict model forecast errors at monitoring sites based on forecast analogs; and (4) a spatial interpolator to estimate the bias correction over the entire modeling domain. One or more methods were tested for each step. We applied five combinations of these methods to PM$_{2.5}$ forecasts in 2014–2016 over China from the operational AiMa air quality forecasting system using the Community Multiscale Air Quality (CMAQ) model. All five methods were able to improve forecast performance in terms of normalized mean error (NME) and root mean square error (RMSE), though to a relatively limited degree due to the rapid changing of emission rates in China. Among the five methods, the CART-LM-KF-AN (a Classification And Regression Trees-Linear Model-Kalman Filter-Analog combination) method appears to have the best overall performance for varied lead times. While the details of our study are specific to the forecast system, the bias-correction framework is likely applicable to the other air quality model forecast as well.

Keywords: PM$_{2.5}$; forecast; post-processing; CMAQ

1. Introduction

The fine particulate matter with an aerodynamic diameter of less than 2.5 µm (PM$_{2.5}$) is the dominant air pollutant in most Chinese cities in recent years. In 2016, the nationwide annual mean-observed PM$_{2.5}$ concentration was 47 µg/m^3, which exceeded 35 µg/m^3, the level II national ambient air quality standards (NAQQS) of China (GB3095-2012), by more than 34%. The pollution levels are much higher in densely populated regions [?]. For example, in the Beijing-Tianjin-Hebei (BTH) area the average annual mean of PM$_{2.5}$ concentration was 71 µg/m^3 in 2016, twice of the standard, which would cause severe adverse health effects [? ?]. It is critical to provide the public and administrative agencies air pollution alerts in advance, to help citizens take timely protective actions (e.g., wearing masks, staying indoors), and to help governments control emissions through dynamic management actions [? ?].

Empirical models and deterministic models are the two major approaches to forecast air quality in the near future. The empirical approaches usually employ statistical models that relate predictor and explanatory variables [? ?]. These methods can be easily implemented as long as there are sufficient observations available for training the statistical model. However, they have difficulty in forecasting air pollution in longer-term and larger-scales, and have no means to predict pollutant compositions or to provide emission controlling implications [? ?]. The deterministic approaches overcome the above inabilities of statistical models by adopting chemical transport models (CTMs), e.g., the Community Multi-scale Air Quality- model (CMAQ) [?], and especially, can produce forecasts in a much longer lead time, with comparable accuracy, to meet the requirements of the dynamic management practice. CTMs start with a detailed emissions inventory and forecasted meteorological fields, and solve a series of mathematical equations in space and time to simulate air pollutants' fates with the evolution of physical and chemical processes in the atmosphere. Regional air quality forecasting systems that are based upon CTMs have been widely established in the world to provide operational air quality forecasts in real-time [? ? ? ?]. However, there could be significant prediction errors in these forecasts due to emission inventory uncertainties, meteorology forecast uncertainties, and the missing physical and chemical mechanisms in the CTMs [?].

One way to improve the deterministic model forecasting performance is to utilize the statistical methods-based, post-processing techniques to adjust the current forecasts. Fundamentally, these post-processing techniques are bias correction techniques that utilize the deterministic model's historic errors to correct the current model forecasts (from now on, the word "model" exclusively refers to the deterministic model). The simplest bias correction technique is the moving mean method, which directly applies the averaged forecast bias of the previous time period to the current model forecast [?]. The Kalman filter has also been used to derive future bias from model's historical performance [? ? ? ?]. A more advanced bias correction technique is the analog method which first clusters the historic forecasts into resembling analogs, and then derives future bias from the historic analog members [? ? ?]. The analog method considers the distinction of performance levels between different analogs, which might be related to the variations of pollution levels (which are due to various air masses and emission events). The analog method can be further combined with the Kalman filter and other methods to determine the ensemble bias within the same group of analog members [?]. These bias correction techniques have been demonstrated to decrease model forecast errors in weather forecasts [? ?], and O_3 [? ? ? ?] and $PM_{2.5}$ forecasts [? ?]. It is noteworthy that the previous bias correction studies on $PM_{2.5}$ forecasts were all conducted for areas with much cleaner air than in China (e.g., US, UK, Italy, and Portugal) [? ? ?], with their annual mean $PM_{2.5}$ concentrations around 10 $\mu g/m^3$ or below, and with relatively small day-to-day variations. Also, most of these bias correction studies were conducted for 1–2 day lead time model forecasts.

In this study, we propose a bias-correction framework that explores and utilizes the relationships between model biases and predictor variables to improve nationwide $PM_{2.5}$ model forecasts in China. Within the same framework, we tested and compared five bias-correction techniques to post-process the 1-day, 3-day, and 5-day lead-time model forecasts from a nationwide air quality forecasting system in China that has been in operation for over three years. The purpose of the study is to test the existing bias-correction techniques in China, which includes very heavily polluted areas with very large variations in $PM_{2.5}$ concentration. For example, the $PM_{2.5}$ concentration in Beijing can vary from over 250 $\mu g/m^3$ to below 20 $\mu g/m^3$ within a couple of days. We utilized the framework to test the Classification and Regression Trees (CART) method to determine analogs with better distinctions. The study was also aimed to test the effectiveness of the bias-correction techniques for longer lead-time forecasts, such as the 3-day and 5-day in advance, which can better support the dynamic air quality management practice.

2. Experiments

2.1. PM$_{2.5}$ Model Forecasts

The daily PM$_{2.5}$ model forecasts from 2014 to 2016 were obtained from an operational air quality forecasting system (Available online: www.aimayubao.com) built upon the CMAQ model (version 5.0.2), along with the Weather Research and Forecasting Model (WRF, version 3.4.1) [?] and an emissions processing component. The emissions inventory used (called AiMa inventory) was compiled and projected from various inventories and information sources and was further adjusted by utilizing inverse modeling techniques. The CMAQ modeling is configured with the saprc07tc gas chemistry mechanism and the aero6 aerosol module. The WRF simulation is driven by NCEP's GFS 0.5-degree global weather forecast products. The forecasting system produces 144 h forecasts (called AiMa forecasts) at each cycle that covers 5 days local time. The operation of the forecasting system started on 4 February 2014. The model grids have a spatial resolution of 12 km covering the entirety of China. The configurations of the forecasting system have not been changed since operation, ensuring the consistency of the forecasting error distribution. The daily forecasts for the lead times of 1, 3, and 5 days were used in this study. The relative longer forecasting lead time (compared to 1–2 days lead time in previous studies [? ?]) would enable us to evaluate whether the post-correction method is useful for dynamic management practices [?], which usually requires about a 3–5 day lead time to take actions.

2.2. PM$_{2.5}$ Observations

The air quality observational data from the Chinese air quality monitoring network was obtained from the official real-time air quality monitoring publishing platform [?]. The monitoring network had 945 monitors in operation in 2014 and expanded to 1496 monitors in 2015. The network measured PM$_{2.5}$ concentrations with the widely used TEOM (Tapered Element Oscillating Microbalance) instruments. The 24 hourly observations within a day from 0:00 through 24:00 (Beijing local time) were averaged to calculate daily mean PM$_{2.5}$ concentrations. In case that two or more monitors are within the same grid cell, the averages of the observations from these monitors are used. To distinguish from the original observational data from monitors, we refer to this dataset as observations at "avg-monitors". As a result, the 945 monitors operational since 2014 were assigned to 545 different grid cells and all the 1496 monitors operational since 2015 were assigned to 840 grid cells (Figure ??a). Figure ??b illustrates the spatial patterns of the observed PM$_{2.5}$ pollution levels in China in year 2016.

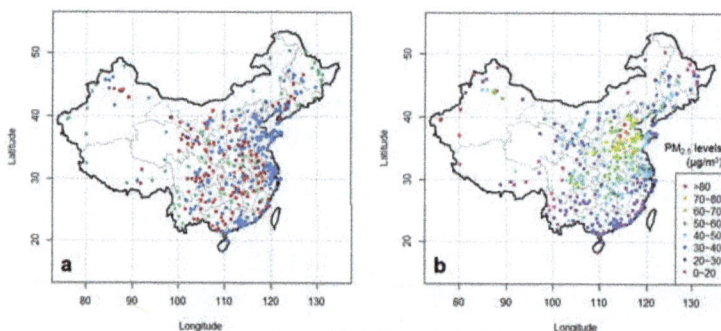

Figure 1. (**a**) Avg-monitors used in this study. The monitors located in the same grid cell have been averaged to derive avg-monitors. The blue dots denote the monitors operational since 2014. The red and green dots both denote monitors operational since 2015, but the red dots represent monitors that are used for evaluating the "final product". (**b**) Annual mean PM$_{2.5}$ concentrations observed at avg-monitors in 2016.

2.3. Bias-Correction Framework

To improve forecast performance, we propose a bias-correction framework as a post-processing procedure for model forecasts. The framework utilizes historic data to establish relationships between the model forecast biases and a variety of model-simulated or observed variables, and then uses these relationships to correct the current forecast. The framework is conducted in four steps: (1) feature selection; (2) forecast analog determination; (3) local correction estimation; and (4) correction spread. Figure **??** illustrates the framework and the methods we tested in each step.

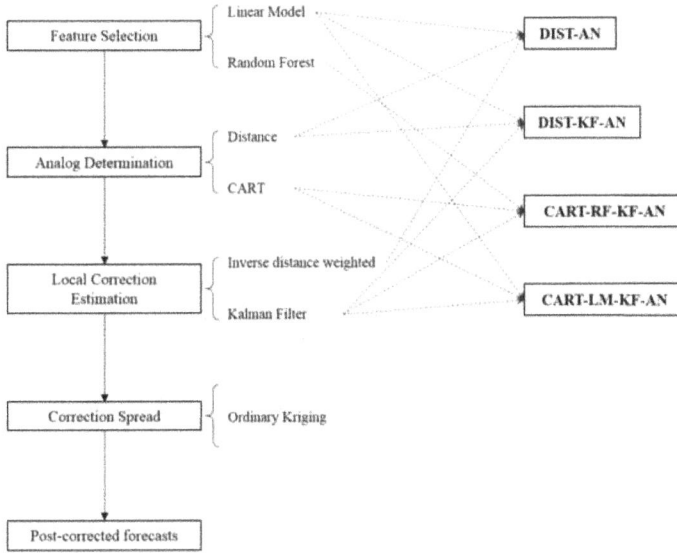

Figure 2. Bias-correction framework with its four steps and the four method combinations that are tested in this study. The fifth method tested is the 7-day Moving Average (7-Day-MA) method, which can be regarded as a special case within the framework but is not explicitly shown here.

2.3.1. Feature Selection

In the "feature selection" step, we choose a group of variables containing information about model biases from a pool of modeled or observed variables. By eliminating non-informative variables, we can improve the predictability of model biases by analogs. In this study, data from 545 avg-monitors (blue dots in Figure **??**) in 2014 and 2015 are used as historic data for selecting informative variables, or features, at each of these avg-monitors. Only the variables that are selected in this step will be used in the following "analog determination" step.

Here, we consider in total 19 candidate variables. Five of them are observation variables (i.e., the mean observed concentrations of the five criteria air pollutants, CO, NO_2, O_3, $PM_{2.5}$, and SO_2) on the day prior to the forecasting cycle, and the rest of the candidate variables are model output, including model forecasted concentrations of the five criteria air pollutants, forecasted daily mean $PM_{2.5}$ composition concentrations (SO_4, NO_3, NH_4, EC and OC), and four meteorological variables (i.e., wind speed, temperature, planetary boundary layer heights, and relative humidity). We tested two methods for feature selection: (1) a linear regression model (denoted as LM); and (2) the random forest algorithm (denoted as RF). The LM method constructed a linear regression model with the $PM_{2.5}$ forecast biases being the explained variable and the candidate variables being the explanatory variables. Those explanatory variables with p-value < 0.05 [**? ?**] were retained as informative variables for analog determination. With the RF method [**?**], the variables with large importance indicators are selected as informative variables. We implemented these feature selection algorithm with the R software (the lm function for LM and the Boruta function for RF) [?].

Figure ?? shows the number counts of the avg-monitors that selected each candidate variable by the LM and RF algorithms. On average, the LM algorithm selected about six informative variables for each avg-monitors. In contrast, the RF algorithm selected about 15 on average. This difference is likely an indication that RF is less effective than LM at distinguishing the information content among variables. As we will find out later in our analysis (Table ??), a method involving LM (e.g., the CART-LM-KF-AN method, see definition in Figure ??) often outperformed a method involving RF (e.g., CART-RF-KF-AN). This comparison highlights the importance of a proper feature selection procedure.

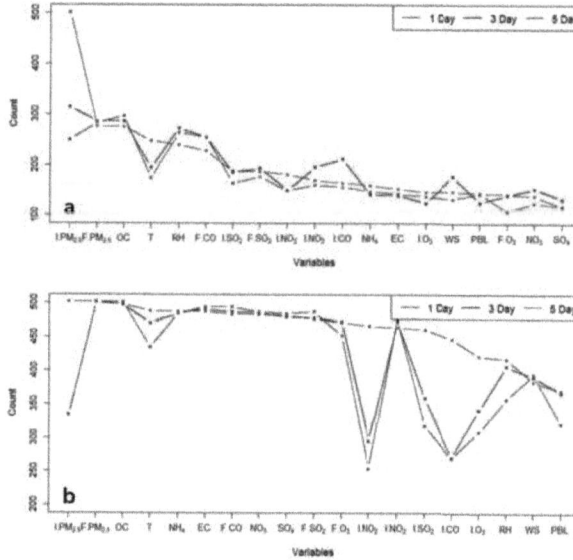

Figure 3. Number counts of the avg-monitors that used each of the 19 features respectively by the linear regression (**a**) and the random forest (**b**) based feature selection method.

Table 1. Performance statistics, R^2, NME, and RMSE (in ug/m^3) after the "local correction estimation" step.

Lead Time	Metrics	Raw	7-Day-MA	DIST-AN	DIST-KF-AN	CART-RF-KF-AN	CART-LM-KF-AN
1 day	R^2	0.46	0.45	0.46	0.48	0.48	0.49
	NME	0.49	0.39	0.40	0.40	0.45	0.41
	RMSE	32.2	27.0	27.0	27.2	29.7	27.5
3 day	R^2	0.38	0.35	0.33	0.37	0.38	0.39
	NME	0.50	0.43	0.46	0.45	0.44	0.43
	RMSE	31.4	27.8	29.0	28.0	28.2	27.5
5 day	R^2	0.34	0.31	0.28	0.31	0.33	0.34
	NME	0.51	0.46	0.49	0.47	0.46	0.45
	RMSE	32.4	29.1	30.3	28.8	29.5	28.7

Among the 19 variables, PM$_{2.5}$ observations on the initial day of the forecast cycle (I.PM$_{2.5}$) and CMAQ PM$_{2.5}$ forecast (F.PM$_{2.5}$) were the top two most selected variables for the 1-day lead time, indicating that they contain the most information about the short-term model forecast errors. As expected, the I.PM$_{2.5}$ is less informative for longer lead times, resulting in a decrease in selection counts by the LM algorithm for the 3 and 5 day lead times (Figure ??). For all lead times in question, model-forecast air pollutant variables, such as F.PM$_{2.5}$, OC, and F.CO, and meteorological variables, such as RH, were also frequently selected by the LM algorithm at many avg-monitors. The forecasted OC and F.CO were frequently selected, likely because they are indicative of model biases in emissions and transport.

2.3.2. Analog Determination

In the "analog determination" step, we search for a "forecast analog", an ensemble of previous model forecasts that are similar to the current forecast to be corrected [? ?]. The bias information in the forecast analog is then used to estimate the correction to be applied in the current forecast in the following "local correction estimation" step. The similarity between current and historic forecasts is measured in terms of the informative variables we selected in the "feature selection" step, using two different methods: (1) Euclidean distance between two forecasts in the feature space (denoted as DIST) and (2) classification predicted by the CART algorithm [?] (denoted as CART). The CART algorithm generates a decision tree which minimizes the total deviations within the branches of the tree. The algorithm is widely used in remote sensing image processing for land cover classification [? ?], ecology modeling [? ?] and pattern recognition studies [?]. We implemented the CART calculation with the rpart() function in the R software [?].

2.3.3. Local Correction Estimation

In the "local correction estimation" step, we estimated the forecast bias at individual avg-monitors based on the "forecast analog" determined in the previous "analog determination" step.

A straightforward way to estimate the bias is the Distance-based Analog (DIST-AN) method, which takes an inverse distance weighted average of the biases in the forecast analog.

$$PM_{cp,T+k} = PM_{p,T+k} - \delta_{T+k} = PM_{p,T+k} - \frac{\sum_{m=1}^{M} \frac{PM_{p,tm} - PM_{o,tm}}{d_{tm,T+k}}}{\sum_{m=1}^{M} \frac{1}{d_{tm,T+k}}} \tag{1}$$

where the $PM_{cp,T+k}$ and $PM_{p,T+k}$, respectively, refer to the corrected and raw model forecasts of the PM$_{2.5}$ concentrations. δ_{T+k} refers to the correction, which is calculated by the inverse distance weighted mean forecast biases of the M analogs.

In addition to the inverse distance weighted average, the Kalman filter (KF), known for its easy implementation, fast convergence speed, and effectiveness at eliminating data noises, is also tested in this study to estimate forecast errors. The Kalman filter works on an ordered set of inputs. Previous studies have used the input dataset ordered by time [? ?]. In this study, we implemented the Kalman filter on a set of analogs ordered by distances. The Kalman filter used a dynamic weighting method to fuse the observations and estimations at time t, as shown in the equation below:

$$\hat{x}_{t+1|t} = \hat{x}_{t|t-1} + K_t \left(y_t - \hat{x}_{t|t-1} \right) \tag{2}$$

where the $\hat{x}_{t+1|t}$ refers to the estimation of forecast error at the time $t+1$ using the information at time t. The $\hat{x}_{t|t-1}$ denotes the forecast error estimation at the time t. The y_t refers to the observed forecast error at time t. The weighting factor K_t was called Kalman gain, which was calculated through the optimization of estimation and observation noises. The detailed approach for K_t estimation can be found in Delle Monache, Nipen, Liu, Roux, and Stull [?]. Depending on the methods used in the "analog determination" step, the Kalman filter is applied in the Distance-based Kalman Filtering Analog (DIST-KF-AN) method, the CART linear model Kalman Filtering Analog (CART-LM-KF-AN) method, and the CART random forest Kalman Filtering Analog (CART-RF-KF-AN) method.

Additionally, we also tested the 7-day Moving Average (7-Day-MA). With this method, the correction is computed as the average of the forecast biases in a 7-day window prior to the forecasting cycle. This method is chosen for its fast computation and easy implementation. The 7-day window length has also been used in previous studies on model post-correction [? ?]. The 7-Day-MA method can be regarded as a special case of the analog method, in which the 7 days prior to the forecasting cycle are the forecast analog and each day is weighted equally.

2.3.4. Correction Spread

After the "local correction estimation" is done, we further spread the estimated corrections at avg-monitors to the entire domain including model grids containing no monitors. The corrections were then applied to the original gridded model forecast to obtain bias-corrected forecasts over the whole China domain. In this study, we used ordinary Kriging [?] to spatially interpolate the biases from monitors to the entire domain.

2.4. Evaluation

Following the post-process framework, we tested five combinations of methods (Figure ??). For example, the CART-LM-KF-AN method uses the LM method for "feature selection", the CART method for "analog determination", and the KF method for "local correction estimation". Readers can refer to Figure ?? for an illustration of how varied methods for each step are combined. To evaluate the performance in varied steps of the procedure, we reported separately the performance statistics for the "local correction estimation" and "correction spread" steps. The reported performance statistics include the coefficient of determination (R^2), the normalized mean error (NME), and the root mean square error (RMSE) [?].

The observation and model outputs from 4 February 2014 to 31 December 2015 are used as historic data, with which we selected informative variables and searched for forecast analogs. In the "local correction estimation" step, we used the model output for 2016 to estimate the bias-corrected forecasts at the 545 individual avg-monitors. We then used the corresponding 2016 observations to evaluate the performance after the "local correction estimation" step. Note that not all avg-monitors were used in the "local correction estimation". After the "correction spread" step, we then used the data from the remaining 211 avg-monitors to evaluate the performance of the "final product" at locations without observations. The 211 avg-monitors were so selected that they were adequately apart from each other and from the 545 avg-monitors that were used in the "local correction estimation". These 211 avg-monitors are marked as red dots in Figure ??. The performance evaluation was conducted separately for 1-day, 3-day, and 5-day lead forecasts.

3. Results

3.1. Performance in Estimating Local Corrections

Although up to the "local correction estimation" step we only computed the local corrections at locations with observations, these local corrections were crucial for the performance of the final product. Therefore, in this study, we applied five different methods (Figure ??) and evaluated them at avg-monitors. Figure ?? presents the performance of the raw $PM_{2.5}$ model forecasts in predicting observations at avg-monitors in 2016. The annual mean biases of model forecasts in 2016 were generally negative in the southern and northwestern part of China, while they were positive (mostly 0–10 μg/m^3) in the middle part of China, regardless of the forecast lead-times. According to the distributions of statistical metrics of R^2 and NME, the raw model forecast performed much better in North China and East China than in other regions, while it performed much worse in West China. Spatial patterns with geographical divisions in biases and other performance statistical metrics, i.e., R^2, NME, and RMSE here, indicate potential non-uniform uncertainties in the emission rates estimation among regions and varied prediction errors in meteorological forecast fields over different terrains. We do observe a significant degradation in the performance of predicting meteorological variables at the surface in regions with complex terrains (results not shown). However, as the lead-time increases, the performance of the raw PM$_{2.5}$ model forecasts only degrade slightly (Figure **??** and Table **??**), implying that the error does not grow much in predicting meteorology during the forecasting period of 144 h.

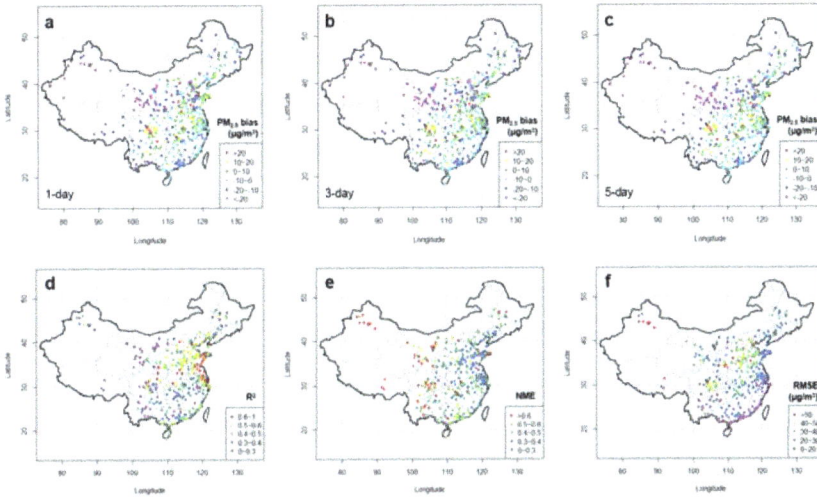

Figure 4. Performance of the raw PM$_{2.5}$ model forecasts in predicting observations at avg-monitors in 2016: (**a,b,c**) annual mean biases for the 1, 3, and 5-day lead times and (**d,e,f**) annual mean R^2, NME, and RMSE for the 1-day lead-time.

Table **??** summarizes the performance statistics of each method for different lead times and Table **??** summarizes the percentage changes of the performance statistics with respect to the raw model forecasts. Compared with the raw model forecasts, all the bias-correction methods are able to decrease the NME and the RMSE at all the three lead times. The reductions in the NME are 7.4–19.3%, 7.3–14.4%, and 4.5–12.2%, and the reductions in RMSE are 7.8–16.1%, 7.5–12.5%, and 6.3–11.3% for 1-day, 3-day, and 5-day lead times, respectively, showing that these post-processing techniques are effective to improve the PM$_{2.5}$ forecast at locations with observations. Figure **??** shows that all methods can improve NME and RMSE at the majority of avg-monitors. For example, the CART-LM-KF-AN method decreases NME and RMSE at about 70% to 80% avg-monitors for all three lead times.

Table 2. Percentage changes (%) of the performance statistics with respect to original model forecast by the five methods in the "correction estimation" step.

Lead Time	Metrics	7-Day-MA	DIST-AN	DIST-KF-AN	CART-RF-KF-AN	CART-LM-KF-AN
1 day	R^2	−2.7	0.3	3.1	3.5	6.1
	NME	−19.3	−18.3	−17.3	−7.4	−14.7
	RMSE	−16.0	−16.1	−15.6	−7.8	−14.6
3 day	R^2	−8.3	−13.3	−4.0	−1.5	0.7
	NME	−13.6	−7.3	−10.0	−12.6	−14.4
	RMSE	−11.6	−7.5	−10.9	−10.1	−12.5
5 day	R^2	−8.9	−17.7	−7.0	−2.0	−0.3
	NME	−11.2	−4.5	−8.9	−10.7	−12.2
	RMSE	−10.0	−6.3	−11.1	−9.0	−11.3

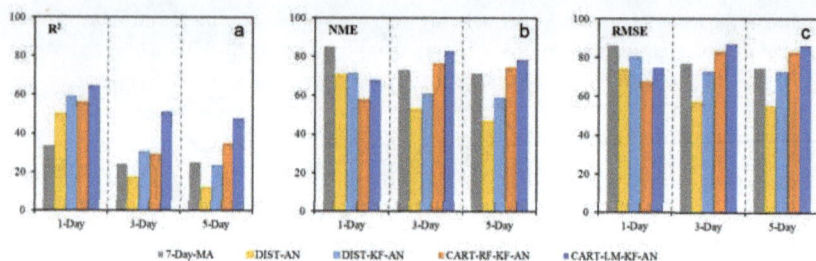

Figure 5. Percentage (%) of avg-monitors with increased R^2 values (**a**), and decreased NME (**b**) and RMSE values (**c**) by the five post-correction methods and for the 1, 3, and 5-day lead times.

For most of these methods, however, R^2 only increases marginally for the 1-day lead time and even decreases slightly in the 3-day and 5-day lead times. For example, DIST-AN decreases the R^2 by 17.7% for the 3-day lead time. The ineffectiveness in increasing R^2 may reflect that these analog-based methods, although good at reducing biases, do not improve the ability to capture the variability in the data, especially for longer lead times. Among the five methods in question, the CART-LM-KF-AN method has the largest R^2 for all the three lead times, with a 6% increase in R^2 for the 1-day lead and essentially no change for the 3-day and 5-day leads from the raw model forecast.

The results also show the impact of forecasting lead times on the performance of bias-correction techniques (????, Figure ??). In general, the enhancement in the forecast performance decreases with the longer lead time. The decreasing effectiveness of the post-processing procedure with lead times may partly result from the fact that the increasing uncertainties in the model forecasted meteorology and pollutant concentrations lead to larger uncertainties in the analog determination for the longer lead times. Among the five methods in this study, the performance of the CART-LM-KF-AN method is most insensitive to varied lead times (NME 0.41, 0.43, 0.45, and RMSE 27.5, 27.5, 28.7 ug/m^3 for 1-day, 3-day, and 5-day lead times, respectively), showing that the combination of the CART and LM methods constitutes a more robust analog determination algorithm for the PM$_{2.5}$ model forecast in China. In contrast, the performance of the DIST-AN method (NME 0.40, 0.46, 0.49, and RMSE 27.0, 29.0, 30.3 ug/m^3 for 1-day, 3-day, and 5-day lead times, respectively) and the 7-Day-MA method (NME 0.39, 0.43, 0.46 and RMSE 27.0, 27.8, 29.1 ug/m^3 for 1-day, 3-day, and 5-day lead time, respectively) degrade significantly with longer lead times. Although the performance of the CART-LM-KF-AN method is similar to or a little worse than the 7-Day-MA and DIST-AN for the 1-day lead time, CART-LM-KF-AN outperforms other methods for a longer lead time, which is a good property for the purpose of dynamic air quality management.

3.2. Performance of the Final Product

After we estimated the correction for the model forecasts at each of the individual avg-monitors, we spread the local corrections across the entire Chinese domain by spatially interpolating the local corrections with ordinary Kriging. Figure ?? shows an example of the "final product" of a 5-day lead forecast for 30 December 2016, using the CART-LM-KF-AN method.

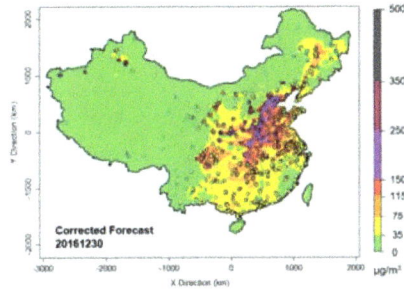

Figure 6. Bias-corrected CMAQ PM$_{2.5}$ forecast over China for 30 December 2016 (5-day lead time) by the CART-LM-KF-AN method. The dots represent observed PM$_{2.5}$ levels.

By comparing the "final product" with observations at 211 avg-monitors (whose data were not used in the "local correction estimation" step), we can evaluate the performance of the "final product" at locations without avg-monitors. Table **??** shows that the correction estimated through the spatial interpolation can also effectively reduce forecast errors, even at locations without PM$_{2.5}$ monitors. Depending on the methods and lead times, the fraction of avg-monitors that finds improvements in NME and RMSE varies from 50 to 80% (Figure **??**). The improvements, for most methods, are slightly less but still comparable to those at locations with observations (**????**), indicating that, compared to the forecast errors at locations with observations, the uncertainties induced by spatial interpolation are likely insignificant. In other words, the "local correction estimation" (including feature selection and analog determination) rather than "correction spread" is the "bottle-neck" in the post-correction framework. Efforts to further improve the performance should be directed to improve the estimation of local corrections.

Table 3. Performance statistics for R^2, NME, and RMSE (in ug/m^3) at locations without monitors after correction spread.

Lead Time	Metrics	Raw	7-Day-MA	DIST-AN	DIST-KF-AN	CART-RF-KF-AN	CART-LM-KF-AN
1 day	R^2	0.38	0.33	0.40	0.39	0.44	0.43
	NME	0.48	0.47	0.42	0.46	0.42	0.41
	RMSE	28.4	27.7	25.6	27.4	24.5	24.3
3 day	R^2	0.34	0.29	0.33	0.34	0.33	0.33
	NME	0.49	0.47	0.44	0.47	0.44	0.44
	RMSE	27.7	26.8	25.3	26.6	25.8	25.6
5 day	R^2	0.31	0.26	0.29	0.31	0.30	0.30
	NME	0.50	0.49	0.46	0.48	0.46	0.46
	RMSE	28.3	27.6	26.1	27.1	26.5	26.5

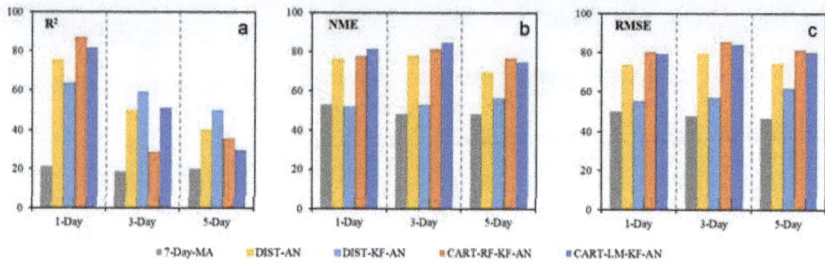

Figure 7. Percentage (%) of avg-monitors with increased R^2 values (**a**) and decreased NME (**b**) and RMSE values (**c**) by the five post-correction methods and for the 1, 3, and 5-day lead times by spatially interpolating the estimated biases.

3.3. Discussion

In comparison with previous studies conducted in the U.S., our results generally show less improvements (in terms of percentage) from the raw model forecasts. For example, Djalalova, Delle, Monache, and Wilczak [?] used KF-AN (similar to DIST-KF-AN in this study) to post-correct hourly CMAQ $PM_{2.5}$ forecasts for the 1-day lead time and reduced the MAE values by 65% from 8.5 $\mu g/m^3$ to 3 $\mu g/m^3$. Kang, et al. [?] also applied the KF-AN method to daily $PM_{2.5}$ forecasts and reduced RMSE by 33% from 7.5 $\mu g/m^3$ to 5 $\mu g/m^3$. Previous studies have found that the KF- and AN-based methods better perform in lower pollution level regions [?]. The differences in the performance between this and previous studies may result from the fact that $PM_{2.5}$ levels in China are much higher than in the U.S. Consistent with previous studies, our analysis also shows that the percentage improvements by the CART-LM-KF-AN method are generally larger in relatively cleaner regions (e.g., the Pearl River Delta in South china, Northeast China, and other remote regions) than in heavily polluted regions (e.g., the North China Plain and the Yangtze River Delta in East China) (Figure ??), suggesting that there might be important factors missing in the trained relationship between model biases and predictor variables over polluted regions. One such factor is the fast-changing emissions in both magnitude and distribution in regions such as the North China Plain and the Yangtze River Delta during the modeled three years [? ? ?], a result of increasingly more strict emission control enforcements and/or economic fluctuations. The significant change of emission rates in these regions between the training years (2014–2015) and the prediction year (2016) could confound the trained bias correction relationships. In contrast, the actual emission rates were likely to vary insignificantly in the cleaner regions over the same time period.

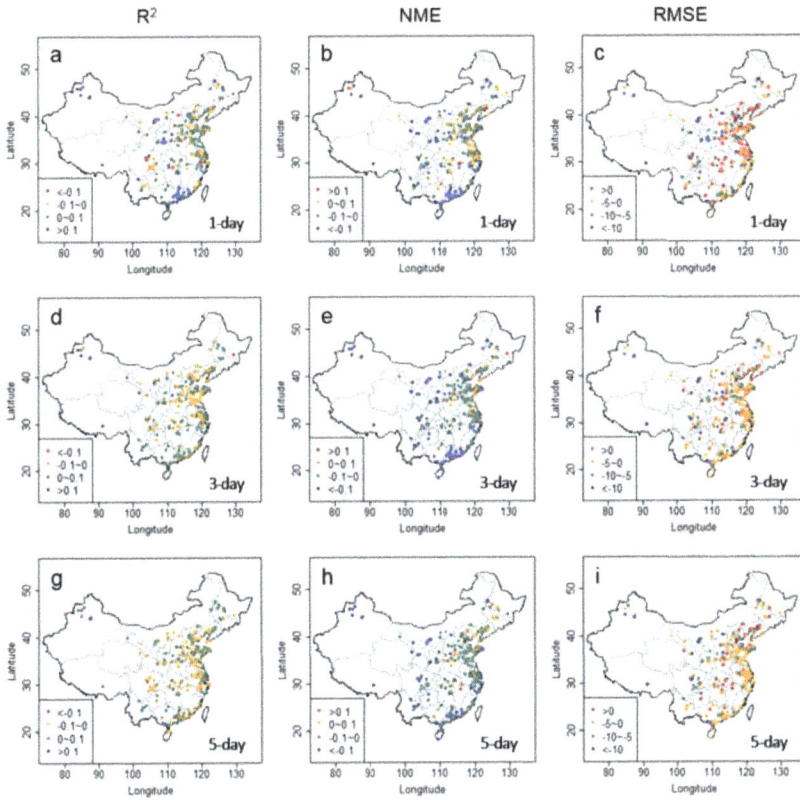

Figure 8. The difference in R^2 (**a,d,g**), NME (**b,e,h**), and RMSE (**c,f,i**) values between bias-corrected forecasts by the CART-LM-KF-AN method and raw CMAQ forecasts (Corrected forecast minus raw forecasts) at the 545 avg-monitors for the 1, 3, and 5-day lead times. The differences in RMSE is in μg/m^3.

4. Conclusions

To improve the PM$_{2.5}$ forecast, we proposed a bias-correction framework that utilized the relationships between biases and select forecasted and observational variables. The framework consists of four steps: feature selection, analog determination, local correction estimation, and correction spread. We applied this bias-correction framework to PM$_{2.5}$ forecasts in 2014–2016 over China from the operational AiMa air quality forecasting system using the CMAQ model. Five methods, differing in how to perform feature selection, analog determination, and local correction estimation, were tested in this study, and we found all the five methods were able to improve the overall forecast performance in terms of RMSE and NME, though to a relatively limited degree.

Based on our results, we recommend the CART-LM-KF-AN method. In most cases, the performance of this method is better or comparable to other methods. Particularly, the method shows consistent improvement for longer lead times (3–5 day) when other methods degrade in their performance. This is important for dynamic air quality management, as this type of practice often requires longer lead time.

In comparison with previous studies that were all conducted in areas outside of China, our results generally show fewer improvements (in terms of percentage) from the raw model forecast, especially in regions with much higher pollution levels. A major reason for this is that the fast-changing emissions in high pollution regions of China can confound the relationship between model biases and predictor variables. On the other side, the correction spread is not found to be a significant source of errors at locations without monitors. Future efforts should be directed to improve the performance in the local correction estimation, especially to explore methods that can build relationships that better represent the missing factors of changing emissions.

Acknowledgments: The authors thank the Hangzhou AiMa Technologies, Inc. for providing the archived AiMa 5-day air quality forecasting products for years 2014–2016.

Author Contributions: Baolei Lyu, Yuzhong Zhang, and Yongtao Hu conceived and designed the study; Baolei Lyu gathered data and implemented the algorithm; Baolei Lyu and Yuzhong Zhang performed the analysis; Baolei Lyu, Yuzhong Zhang, and Yongtao Hu wrote the paper.

Conflicts of Interest: The authors declare no conflict of interest.

References

1. Zhang, Y.L.; Cao, F. Fine particulate matter (PM$_{2.5}$) in China at a city level. *Sci. Rep.* **2015**, *5*, 14884. [CrossRef] [PubMed]
2. Fang, D.; Wang, Q.g.; Li, H.; Yu, Y.; Lu, Y.; Qian, X. Mortality effects assessment of ambient PM$_{2.5}$ pollution in the 74 leading cities of China. *Sci. Total Environ.* **2016**, *569–570*, 1545–1552. [CrossRef] [PubMed]
3. Gao, M.; Saide, P.E.; Xin, J.; Wang, Y.; Liu, Z.; Wang, Y.; Wang, Z.; Pagowski, M.; Guttikunda, S.K.; Carmichael, G.R. Estimates of health impacts and radiative forcing in winter haze in eastern China through constraints of surface PM$_{2.5}$ predictions. *Environ. Sci. Technol.* **2017**, *51*, 2178–2185. [CrossRef] [PubMed]
4. Hu, Y.; Odman, M.T.; Chang, M.E.; Russell, A.G. Operational forecasting of source impacts for dynamic air quality management. *Atmos. Environ.* **2015**, *116*, 320–322. [CrossRef]
5. Zhang, Y.; Bocquet, M.; Mallet, V.; Seigneur, C.; Baklanov, A. Real-time air quality forecasting, part I: History, techniques, and current status. *Atmos. Environ.* **2012**, *60*, 632–655. [CrossRef]
6. Lv, B.; Cobourn, W.G.; Bai, Y. Development of nonlinear empirical models to forecast daily PM$_{2.5}$ and ozone levels in three large Chinese cities. *Atmos. Environ.* **2016**, *147*, 209–223. [CrossRef]
7. Perez, P.; Salini, G. PM$_{2.5}$ forecasting in a large city: Comparison of three methods. *Atmos. Environ.* **2008**, *42*, 8219–8224. [CrossRef]
8. Eder, B.; Kang, D.; Mathur, R.; Yu, S.; Schere, K. An operational evaluation of the Eta–CMAQ air quality forecast model. *Atmos. Environ.* **2006**, *40*, 4894–4905. [CrossRef]
9. Zhang, Y.; Bocquet, M.; Mallet, V.; Seigneur, C.; Baklanov, A. Real-time air quality forecasting, part II: State of the science, current research needs, and future prospects. *Atmos. Environ.* **2012**, *60*, 656–676. [CrossRef]
10. Byun, D.; Schere, K.L. Review of the governing equations, computational algorithms, and other components of the models-3 community multiscale air quality (CMAQ) modeling system. *Appl. Mech. Rev.* **2006**, *59*, 51–77. [CrossRef]
11. Mckeen, S.; Grell, G.; Peckham, S.; Wilczak, J.; Djalalova, I.; Hsie, E.Y.; Frost, G.; Peischl, J.; Schwarz, J.; Spackman, R. An evaluation of real-time air quality forecasts and their urban emissions over eastern texas during the summer of 2006 second texas air quality study field study. *J. Geophys. Res. Atmos.* **2009**, *114*, D7. [CrossRef]
12. Vaughan, J.; Lamb, B.; Frei, C.; Wilson, R.; Bowman, C.; Figueroa-Kaminsky, C.; Otterson, S.; Boyer, M.; Mass, C.; Albright, M.; et al. A numerical daily air quality forecast system for the Pacific Northwest. *Bull. Am. Meteorol. Soc.* **2004**, *85*, 549–561. [CrossRef]
13. Otte, T.L.; Pouliot, G.; Pleim, J.E.; Young, J.O.; Schere, K.L.; Wong, D.C.; Lee, P.C.S.; Tsidulko, M.; McQueen, J.T.; Davidson, P.; et al. Linking the Eta model with the community multiscale air quality (CMAQ) modeling system to build a national air quality forecasting system. *Weather Forecast.* **2005**, *20*, 367–384. [CrossRef]
14. Zhou, G.Q.; Xu, J.M.; Xie, Y.; Chang, L.Y.; Gao, W.; Gu, Y.X.; Zhou, J. Numerical air quality forecasting over eastern China: An operational application of WRF-Chem. *Atmos. Environ.* **2017**, *153*, 94–108. [CrossRef]
15. Binkowski, F.S.; Roselle, S.J. Models-3 community multiscale air quality (CMAQ) model aerosol component 1. Model description. *J. Geophys. Res. Atmos.* **2003**, *108*, 335–346. [CrossRef]

16. Djalalova, I.; Delle Monache, L.; Wilczak, J. PM$_{2.5}$ analog forecast and kalman filter post-processing for the community multiscale air quality (CMAQ) model. *Atmos. Environ.* **2015**, *108*, 76–87. [CrossRef]

17. Delle Monache, L.; Nipen, T.; Deng, X.X.; Zhou, Y.M.; Stull, R. Ozone ensemble forecasts: 2. A kalman filter predictor bias correction. *J. Geophys. Res. Atmos.* **2006**, *111*, D05308. [CrossRef]

18. Kang, D.W.; Mathur, R.; Rao, S.T. Implementation of real-time bias-corrected o-3 and PM$_{2.5}$ air quality forecast and their performance evaluations during 2008 over the continental united states. In *Air Pollution Modeling and its Application XX*; Steyn, D.G., Rao, S.T., Eds.; Springer: Berlin/Heidelberg, Germany, 2010; p. 283.

19. Djalalova, I.; Wilczak, J.; McKeen, S.; Grell, G.; Peckham, S.; Pagowski, M.; DelleMonache, L.; McQueen, J.; Tang, Y.; Lee, P.; et al. Ensemble and bias-correction techniques for air quality model forecasts of surface o-3 and PM$_{2.5}$ during the texaqs-ii experiment of 2006. *Atmos. Environ.* **2010**, *44*, 455–467. [CrossRef]

20. De Ridder, K.; Kumar, U.; Lauwaet, D.; Blyth, L.; Lefebvre, W. Kalman filter-based air quality forecast adjustment. *Atmos. Environ.* **2012**, *50*, 381–384. [CrossRef]

21. Delle Monache, L.; Djalalova, I.; Wilczak, J. Analog-based postprocessing methods for air quality forecasting. In *Air Pollution Modeling and its Application XXIII*; Steyn, D.G., Rao, S.T., Eds.; Springer: Berlin/Heidelberg, Germany, 2014; pp. 237–239.

22. Huang, J.P.; McQueen, J.; Wilczak, J.; Djalalova, I.; Stajner, I.; Shafran, P.; Allured, D.; Lee, P.; Pan, L.; Tong, D.; et al. Improving NOAA NAQFC PM$_{2.5}$ predictions with a bias correction approach. *Weather Forecast.* **2017**, *32*, 407–421. [CrossRef]

23. Delle Monache, L.; Nipen, T.; Liu, Y.; Roux, G.; Stull, R. Kalman filter and analog schemes to postprocess numerical weather predictions. *Mon. Weather Rev.* **2011**, *139*, 3554–3570. [CrossRef]

24. Dolman, B.K.; Reid, I.M. Bias correction and overall performance of a VHF spaced antenna boundary layer profiler for operational weather forecasting. *J. Atmos. Sol. Terr. Phys.* **2014**, *118*, 16–24. [CrossRef]

25. Mckeen, S.; Wilczak, J.; Grell, G.; Djalalova, I.; Peckham, S.; Hsie, E.Y.; Gong, W.; Bouchet, V.; Menard, S.; Moffet, R. Assessment of an ensemble of seven real-time ozone forecasts over eastern north america during the summer of 2004. *J. Geophys. Res. Atmos.* **2005**, *110*, 3003–3013. [CrossRef]

26. Monache, L.D.; Grell, G.A.; Mckeen, S.; Wilczak, J.; Pagowski, M.O.; Peckham, S.; Stull, R.; Mchenry, J.; Mcqueen, J. A kalman-filter bias correction of ozone deterministic, ensemble-averaged, and probabilistic forecasts. *Tellus* **2006**, 60b.

27. Wilczak, J.; Mckeen, S.; Djalalova, I.; Grell, G.; Peckham, S.; Gong, W.; Bouchet, V.; Moffet, R.; Mchenry, J.; Mcqueen, J. Bias-corrected ensemble and probabilistic forecasts of surface ozone over eastern North America during the summer of 2004. *J. Geophys. Res. Atmos.* **2012**, *111*, 6443–6445. [CrossRef]

28. Kang, D.; Mathur, R.; Rao, S.T. Real-time bias-adjusted O$_3$ and PM$_{2.5}$ air quality index forecasts and their performance evaluations over the continental united states. *Atmos. Environ.* **2010**, *44*, 2203–2212. [CrossRef]

29. Crooks, J.L.; Özkaynak, H. Simultaneous statistical bias correction of multiple PM$_{2.5}$ species from a regional photochemical grid model. *Atmos. Environ.* **2014**, *95*, 126–141. [CrossRef]

30. Neal, L.S.; Agnew, P.; Moseley, S.; Ordonez, C.; Savage, N.H.; Tilbee, M. Application of a statistical post-processing technique to a gridded, operational, air quality forecast. *Atmos. Environ.* **2014**, *98*, 385–393. [CrossRef]

31. Silibello, C.; Bolignano, A.; Sozzi, R.; Gariazzo, C. Application of a chemical transport model and optimized data assimilation methods to improve air quality assessment. *Air. Qual. Atmos. Health* **2014**, *7*, 283–296. [CrossRef]

32. Skamarock, W.C.; Klemp, J.B.; Dudhia, J.; Gill, D.O.; Barker, D.M.; Wang, W.; Powers, J.G. *A Description of the Advanced Research WRF Version 2*; No. NCAR/TN-468+ STR; Mesoscale and Microscale Meteorology Division, National Center For Atmospheric Research: Boulder, CO, USA, 2005.

33. Zhou, Y.; Wu, Y.; Yang, L.; Fu, L.X.; He, K.B.; Wang, S.X.; Hao, J.M.; Chen, J.C.; Li, C.Y. The impact of transportation control measures on emission reductions during the 2008 olympic games in Beijing, China. *Atmos. Environ.* **2010**, *44*, 285–293. [CrossRef]

34. China National Urban Air Quality Real-Time Publishing Platform. Available online: http://106.37.208.233:20035 (accessed on 14 January 2017).

35. Malik, M.B. Applied linear regression. *Technometrics* **2005**, *47*, 371–372. [CrossRef]

36. Jones, P.W.; Quirk, F.H.; Baveystock, C.M.; Littlejohns, P. A self-complete measure of health status for chronic airflow limitation. The st. George's respiratory questionnaire. *Am. Rev. Respir. Dis.* **1992**, *145*, 1321–1327. [CrossRef] [PubMed]

37. Chen, Y.W.; Lin, C.J. Combining svms with various feature selection strategies. In *Feature Extraction*; Guyon, I., Nikravesh, M., Gunn, S., Zadeh, L.A., Eds.; Springer: Berlin/Heidelberg, Germany, 2006; Volume 3, pp. 315–324.

38. Team, R.C. *R: A Language and Environment for Statistical Computing*; The R Team: Vienna, Austria, 2016.

39. Breiman, L.; Friedman, J.H.; Olshen, R.; Stone, C.J. Classification and regression trees. *Biometrics* **1984**, *40*, 358.

40. Liu, J.; Sun, D.; He, F.; Zhang, W.; Guan, X. Land use/cover classification with classification and regression tree applied to MODIS imagery. *J. Appl. Sci.* **2013**, *13*, 3070–3773.

41. Youssef, A.M.; Pourghasemi, H.R.; Pourtaghi, Z.S.; Al-Katheeri, M.M. Landslide susceptibility mapping using random forest, boosted regression tree, classification and regression tree, and general linear models and comparison of their performance at Wadi Tayyah Basin, asir region, Saudi Arabia. *Landslides* **2016**, *13*, 839–856. [CrossRef]

42. Mertens, M.; Nestler, I.; Huwe, B. Gis-based regionalization of soil profiles with classification and regression trees (CART). *J. Plant. Nutr. Soil Sci.* **2002**, *165*, 39–43. [CrossRef]

43. De'Ath, G.; Fabricius, K.E. Classification and regression trees: A powerful yet simple technique for the analysis of complex ecological data. *Ecology* **2000**, *81*, 3178–3192. [CrossRef]

44. Jain, A.K.; Duin, R.P.W.; Mao, J. Statistical pattern recognition: A review. *IEEE Trans. Pattern Anal.* **2000**, *22*, 4–37. [CrossRef]

45. Kang, D.; Mathur, R.; Rao, S.T.; Yu, S. Bias adjustment techniques for improving ozone air quality forecasts. *J. Geophys. Res. Atmos.* **2007**, *113*, 2036–2044. [CrossRef]

46. Oliver, M.A.; Webster, R. Kriging: A method of interpolation for geographical information systems. *Int. J. Geogr. Inf. Syst.* **1990**, *4*, 313–332. [CrossRef]

47. Kang, D.; Mathur, R.; Rao, S.T. Assessment of bias-adjusted $PM_{2.5}$ air quality forecasts over the continental united states during 2007. *Geosci. Model. Dev.* **2009**, *2*, 309–320.

48. Duncan, B.N.; Lamsal, L.N.; Thompson, A.M.; Yoshida, Y.; Lu, Z.F.; Streets, D.G.; Hurwitz, M.M.; Pickering, K.E. A space-based, high-resolution view of notable changes in urban NOX pollution around the world (2005–2014). *J. Geophys. Res. Atmos.* **2016**, *121*, 976–996. [CrossRef]

49. Krotkov, N.A.; McLinden, C.A.; Li, C.; Lamsal, L.N.; Celarier, E.A.; Marchenko, S.V.; Swartz, W.H.; Bucsela, E.J.; Joiner, J.; Duncan, B.N.; et al. Aura omi observations of regional SO_2 and NO_2 pollution changes from 2005 to 2015. *Atmos. Chem. Phys.* **2016**, *16*, 4605–4629. [CrossRef]

50. Wu, Y.; Zhang, S.J.; Hao, J.M.; Liu, H.; Wu, X.M.; Hu, J.N.; Walsh, M.P.; Wallington, T.J.; Zhang, K.M.; Stevanovic, S. On-road vehicle emissions and their control in China: A review and outlook. *Sci. Total Environ.* **2017**, *574*, 332–349. [CrossRef] [PubMed]

atmosphere

MDPI

Article

Air Quality and Control Measures Evaluation during the 2014 Youth Olympic Games in Nanjing and its Surrounding Cities

Hui Zhao [1], Youfei Zheng [1,2,*] and Ting Li [2]

[1] Key Laboratory for Aerosol-Cloud-Precipitation of China Meteorological Administration,
    Nanjing University of Information Science and Technology, Nanjing 210044, China; zhaohui_nuist@163.com
[2] Key Laboratory of Atmospheric Environment Monitoring and Pollution Control,
    Collaborative Innovation Center of Atmospheric Environment and Equipment Technology,
    Nanjing University of Information Science and Technology, Nanjing 210044, China; c2015liting@163.com
*   Correspondence: zhengyf@nuist.edu.cn

Academic Editors: Pius Lee, Rick Saylor and Jeff McQueen
Received: 28 April 2017; Accepted: 3 June 2017; Published: 4 June 2017

Abstract: Air pollution had become a vital concern for the 2014 Youth Olympic Games in Nanjing. In order to control air pollutant emissions and ensure better air quality during the Games, the Nanjing municipal government took a series of aggressive control measures to reduce pollutant emissions in Nanjing and its surrounding cities during the Youth Olympic Games. The Air Quality Index (AQI) is an index of air quality which is used to inform the public about levels of air pollution and associated health risks. In this study, we use the AQI and air pollutant concentrations data to evaluate the effectiveness of the implementation of control measures. The results suggest that the emission reduction measures significantly improved air quality in Nanjing. In August 2014, the mean concentrations of $PM_{2.5}$, PM_{10}, SO_2, NO_2, CO and O_3 were 42.44 $\mu g \cdot m^{-3}$, 59.01 $\mu g \cdot m^{-3}$, 11.12 $\mu g \cdot m^{-3}$, 31.09 $\mu g \cdot m^{-3}$, 0.76 $mg \cdot m^{-3}$ and 38.39 $\mu g \cdot m^{-3}$, respectively, and fell by 35.92%, 36.75%, 20.40%, 15.05%, 8.54% and 47.15%, respectively, compared to the prophase mean before the emission reduction. After the emission reduction, the mean concentrations of $PM_{2.5}$, PM_{10}, SO_2, NO_2, and O_3 increased by 20.81%, 41.84%, 22.84%, 21.16% and 60.93%, respectively, which is due to the cancellation of temporary atmospheric pollution control measures. The air pollutants diurnal variation curve during the emission reduction was lower than the other two periods, except for CO. In addition, the AQI of Nanjing and its surrounding cities showed a downward trend, compared with July 2014. The most of effective method to control air pollution is to implement the measures of regional cooperation and joint defense and control, and reduce local emissions during the polluted period, such as airborne dust, coal-burning, vehicle emissions, mobile sources and industrial production.

Keywords: air pollutants; $PM_{2.5}$; Youth Olympic Games; meteorological conditions; emission reduction; air quality

---

1. Introduction

As the largest developing country in the world, China has suffered from serious air pollution due to the rapid economic development and industrial reconstruction in the past three decades. It has become one of the major environmental concerns in some Chinese cities, including Beijing, Shanghai, Guangzhou and Nanjing [1]. Some major air pollutants in the atmosphere, including particulate matter (PM), sulfur dioxide (SO_2), carbon monoxide (CO), nitrogen dioxide (NO_2) and ozone (O_3), have attracted increasing attention due to their impacts on air quality, visibility reduction, human health and global climate [2–6]. From June 2000, the China National Environmental Monitoring Center started reporting the status of ambient air quality by using an air pollution index (API), which is calculated based on the highest index of 24-h average concentrations of PM_{10}, SO_2 and NO_2. From March 2012, the Ministry of Environmental Protection (MEP) started using the air quality index (AQI), which is calculated based on the concentrations of the six major pollutants including $PM_{2.5}$, PM_{10}, SO_2, NO_2, CO and O_3. The AQI ranges from 0 to 500. The greater the value of the AQI, the higher the level of air pollution. Currently, the AQI is widely used to describe the quality of the air.

Nanjing, the capital city of Jiangsu Province, is located in the western Yangtze River Delta and has a population of over 8.2 million. As a highly industrialized and urbanized city, Nanjing suffers from severe air pollution. It hosted the 2nd Summer Youth Olympic Games (YOG) from 17 to 28 August 2014. In view of the significant air pollution in Nanjing, to ensure good air quality during the YOG, the Nanjing municipal government actively took temporary control measures to reduce pollutant emissions. These measures mainly included dust control, coal-burning control, vehicle emission control, industrial production halts, construction site shutdowns and regional joint prevention and control. In addition, other cities in the surrounding area cooperated with Nanjing to guarantee good air quality during the same period.

In recent years, China has hosted many major international events. The local government has implemented numerous stringent local emission standards, which significantly reduced the emissions and concentrations of air pollutants in the city. During the 2008 Beijing Olympics Games, the mean concentration of SO_2, $PM_{2.5}$ and NO_2 were reduced by 51.0%, 43.7% and 13% compared to the period before Olympic Games in Beijing and its surrounding area [7]. Wang et al. [8] demonstrated that the concentration of O_3, SO_2, CO and NOx decreased 23%, 61%, 25% and 21%, respectively, compared to previous years. Wang et al. [9] found that the average concentrations of $PM_{2.5}$, PM_{10}, SO_2 and NO_2 during the Asia-Pacific Economic Cooperation period decreased by 47%, 36%, 62% and 41% respectively, whereas concentrations of O_3 increased by 102%. Meanwhile, emission control measures which were implemented in Shanghai and Guangzhou also successfully improved air quality for the World Expo and the Asian Games [10,11].

In this study, we used air quality and air pollutants measurement data at nine sites in Nanjing City from July to September 2014 to quantify the efficiency of the emission control measures in YOG. The objectives of this study were (i) to show the temporal and spatial characteristics of air quality and air pollutants during the YOG, and (ii) to give some suggestions for controlling air pollution in large cities.

2. Data and Methods

In this paper, the AQI and hourly concentrations data for six air pollutants ($PM_{2.5}$, PM_{10}, SO_2, NO_2, CO and O_3) were provided by the China National Environmental Monitoring Center (CNEMC). The AQI approach is based on the National Ambient Air Quality Standards of China (NAAQS-2012) [12]. There are nine sites located in Nanjing, including Maigao Bridge (MG), Caochangmen (CC), Shanxi Road (SX), Zhonghuamen (ZH), Ruijin Road (RJ), Xuanwu Lake (XW), Pukou (PK), Olympic Stadium (OS) and Xianlin University Town (XL). To show the effectiveness of the implementation of control measures, we acquired hourly monitoring data from July to September 2014. We divided the data into three time periods, before the period of the emission reduction (1 to 31 July 2014), during the period of the emission reduction (1 to 31 August 2014) and after the period of the emission reduction (1 to 30 September 2014), to assess the impacts of emission reduction measures on air quality. In addition, we also recorded the data of the AQI in the surrounding cities, including Shanghai (SH), Hangzhou (HZ1), Huzhou (HZ2), Jiaxing (JX), Hefei (HF), Maanshan (MAS), Wuhu (WH), Xuancheng (XC), Chuzhou (CZ1), Bengbu (BB), Suzhou (SZ), Wuxi (WX), Changzhou (CZ2), Zhenjiang (ZJ), Yangzhou (YZ), Huaian (HA), Taizhou (TZ), Nantong (NT), Xuzhou (XZ), Yancheng (YC), Suqian (SQ) and Lianyungang (LYG), which are shown in Figure 1.

Meteorological data, including air temperature (T), relative humidity (RH), wind speed (WS) and atmospheric pressure (AP), were obtained from the China Meteorological Data Sharing Service

System Administration (http://data.cma.cn/site/index.html), and solar radiation (SR) was recorded by Watchdog weather station (Watchdog 2900ET, Spectrum Technologies, Inc., USA) on the campus of Nanjing University of Information Science & Technology (NUIST) (32°14′ N, 118°42′ E).

This study analyzed temporal variation and characteristics of air quality and air pollutants in different periods using the measured time-series. All graphs were created using origin 9.0 software (Origin Lab, Northampton, MA, USA).

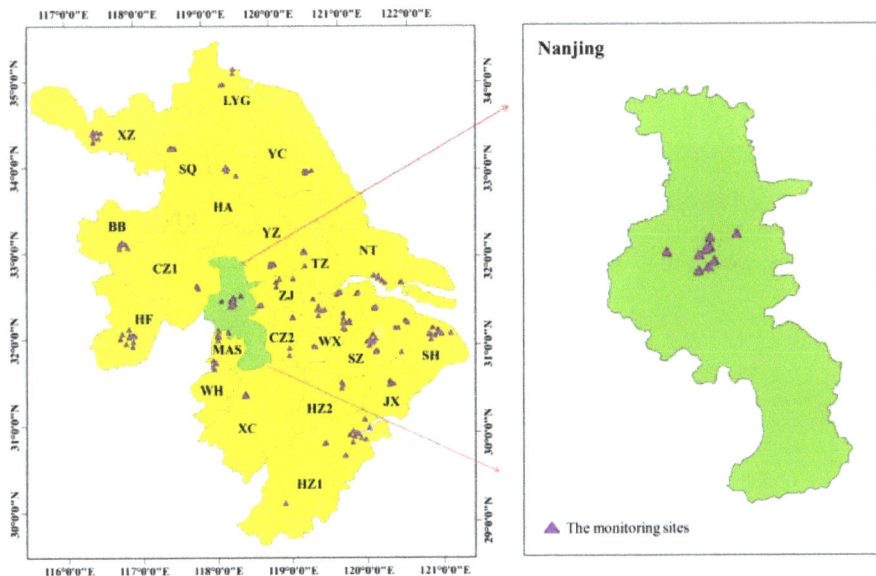

Figure 1. Distribution of air quality monitoring stations in Nanjing and surrounding cities.

3. Results and Discussion

3.1. Variation in Meteorological Conditions

Air temperature, relative humidity, wind speed, solar radiation and atmospheric pressure were important external factors affecting the concentration of pollutants. In order to evaluate the effectiveness of the implementation of control measures, we need to eliminate the interference of meteorological conditions, so we compare meteorological parameters at different periods. Table 1 shows the mean of meteorological parameters before the emission reduction (July, 2014), during the emission reduction (August, 2014) and during the same period in 2013 (August, 2013). The mean air temperature during the period of the emission reduction was 25.6 °C, which was slightly lower than the average level in July 2014. Relative humidity in August 2013 was far lower than the average level in August 2014, while relative humidity in July 2014 was close to the average level in August 2014. The average wind speed during the emission reduction period was 2.19 m/s, which was also very close to the period before the emission reduction. Solar radiation during the emission reduction was very much lower than that in 2013 for the same time, while solar radiation in July 2014 was higher than the average level in August 2014. Atmospheric pressure was not much different across the three periods. Overall, compared with the period during the emission reduction, solar radiation in July 2014 was higher, and other meteorological parameters were close to the average level during the emission reduction. By comparison, we found that the meteorological parameters in July 2014 were close to the average level during the emission reduction, but were very different from that in August 2013.

Thus, the meteorological conditions in July 2014 were more appropriate than those in August 2013 to evaluate the effect of air quality improvement during the period of the emission reduction.

Table 1. Comparison of meteorological conditions during the different periods.

	Air Temperature (°C)	Relative Humidity (%)	Wind Speed (m/s)	Solar Radiation (w/m^2)	Atmospheric Pressure (hPa)
August 2013	30.8	65.9	3.09	205.6	1000.9
July 2014	27.1	83.4	2.26	161.6	1001.5
August 2014	25.6	87.1	2.19	127.7	1003.5

3.2. Comparison of Air Quality and Air Pollutants in Different Periods

During the emission reduction, a number of control measures were adopted to guarantee good air quality in Nanjing. To understand the air quality and air pollutants variation characteristics during this period, the AQI and air pollutant concentrations were plotted against time using the average daily concentration of pollutants data in different periods (Figure 2). Table 2 shows the mean value of the AQI and the mean concentration of air pollutants during the different time periods in Nanjing.

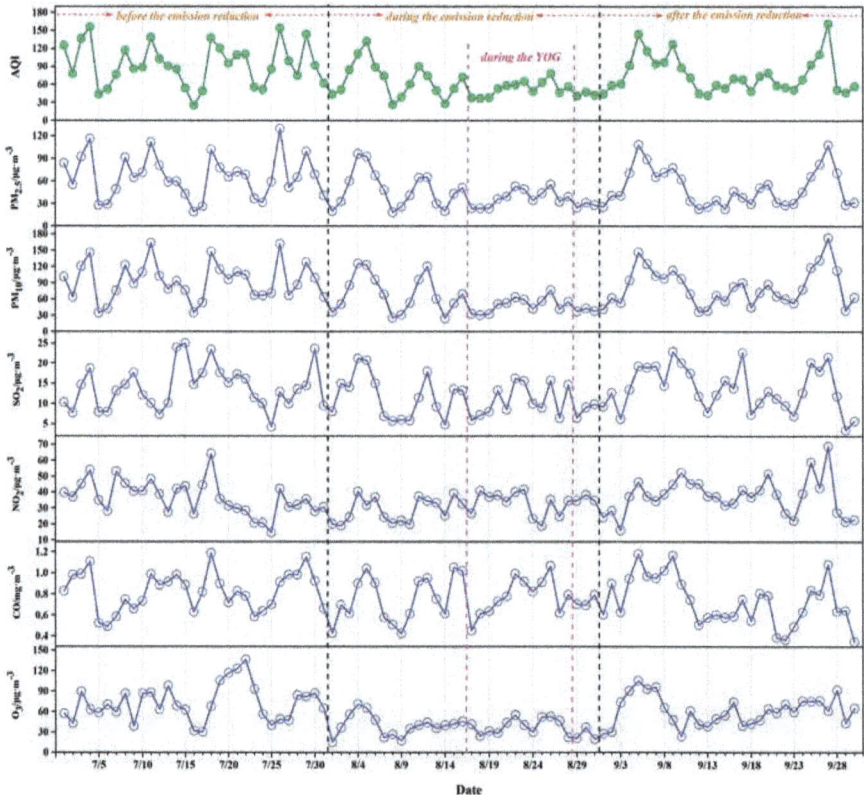

Figure 2. Daily variation of the AQI and air pollutant concentrations in different periods.

Table 2. Mean value of the AQI and mean air pollutant concentrations during the different periods.

Period		Air Quality Index	PM$_{2.5}$ ($\mu g \cdot m^{-3}$)	PM$_{10}$ ($\mu g \cdot m^{-3}$)	SO$_2$ ($\mu g \cdot m^{-3}$)	NO$_2$ ($\mu g \cdot m^{-3}$)	CO ($mg \cdot m^{-3}$)	O$_3$ ($\mu g \cdot m^{-3}$)
Before the emission reduction	1–31 July 2014	93.35	66.23	93.29	13.97	36.60	0.831	72.64
During the emission reduction	1–31 August 2014	59.65	42.44	59.01	11.12	31.09	0.760	38.39
During the Youth Olympic Games	17–28 August 2014	53.40	37.84	49.60	10.86	32.89	0.778	39.12
After the emission reduction	1–30 September 2014	76.20	51.27	83.70	13.66	37.67	0.732	61.78

The AQI is an indicator of air quality, which is used to describe the air quality level of a place and ranges from 0 to 500, where higher values represent worse air quality. The air quality was the best in August 2014 with the AQI of 59.65, an average decrease of 36.10% compared with July 2014, reflecting a significant improvement in the air quality during the emission reduction.

During the emission reduction period, the average PM$_{2.5}$ mass concentration was 42.44 $\mu g \cdot m^{-3}$. Compared with the period before the emission reduction, it exhibited a decrease of 35.92%. The PM$_{2.5}$ levels were much lower than the mean PM$_{2.5}$ concentration of 106 $\mu g \cdot m^{-3}$ in August 2012 in Nanjing [13].

From Figure 2, we can see that the trend and variability of PM$_{10}$ showed a similar trend to PM$_{2.5}$. PM$_{10}$ concentration declined from 93.29 $\mu g \cdot m^{-3}$ in July 2014 to 59.01 $\mu g \cdot m^{-3}$ during the emission reduction period, a 36.75% reduction. It is suggested that the pollution control measures of halting work at construction sites and shutting down heavy-industry factories during the emission reduction yielded a significant effect on the concentration of PM$_{10}$. In addition, we calculated the ratios between PM$_{2.5}$ and PM$_{10}$ and found that the higher ratios occurred in August 2014 (0.73) and the lower ratios occurred in August 2013 (0.50), indicating that PM$_{2.5}$ was a main source for PM$_{10}$ during the emission reduction period.

SO$_2$ concentration was reduced from 13.97 $\mu g \cdot m^{-3}$ in July 2014 to 11.12 $\mu g \cdot m^{-3}$ during the emission reduction period, a decrease of 20.40%. When the control measures were terminated, some factories and enterprises returned to normal production in Nanjing and the surrounding cities after the Youth Olympic Games, thus the SO$_2$ level began to rise to 13.66 $\mu g \cdot m^{-3}$. The variation of SO$_2$ concentration was affected mainly by the emission sources and weather conditions. Due to low emission sources, high temperature and strong convection in the summer, the concentration of SO$_2$ was much lower than in other seasons due to the influence of gas-to-particle conversion and wet scavenging.

NO$_2$ and CO concentrations were reduced by 15.1% and 8.54%, respectively, in comparison to the levels recorded for July 2014. However, they showed a tendency to increase during the YOG, thus the concentrations of NO$_2$ and CO were 32.89 $\mu g \cdot m^{-3}$ and 0.778 $mg \cdot m^{-3}$, respectively, compared with August 2014, which was closely related to meteorological conditions. NO$_2$ and CO are important precursors for O$_3$ formation, and they were affected by the photochemical reaction which enhanced with the increase of solar radiation. According to the comparison of meteorological parameters, we found that solar radiation during the emission reduction (127.7 $w \cdot m^{-2}$) was higher than that that during the YOG (120.31 $w \cdot m^{-2}$). This indicated that NO$_2$ and CO levels in August 2014 were more likely to convert to O$_3$ than those of 17–28 August 2014.

Ground-level O$_3$ is a greenhouse gas, a strong oxidant and a secondary pollutant mainly produced via photochemical reactions of nitrogen oxides, carbon monoxide and volatile organic compounds [14]. Its concentration is affected by solar radiation, air temperature and mixing/transport [15]. O$_3$ concentration in July 2014 was 72.64 $\mu g \cdot m^{-3}$, and was 38.39 $\mu g \cdot m^{-3}$ during the emission reduction period, showing a 47.15% decrease, which was significantly greater than the those other air pollutants. As a secondary pollutant, the decrease of O$_3$ was due to the reduction of precursors and the weather conditions during the period of the emission reduction.

3.3. Diurnal Evolution of Air Pollutant Concentrations

Figure 3 shows the diurnal variation of air pollutant concentrations in different periods. Although the air pollutant concentrations of each period are different, the basic pattern shows that the air pollutants diurnal variation curve in August 2014 was lower than those of the other two periods, except for CO.

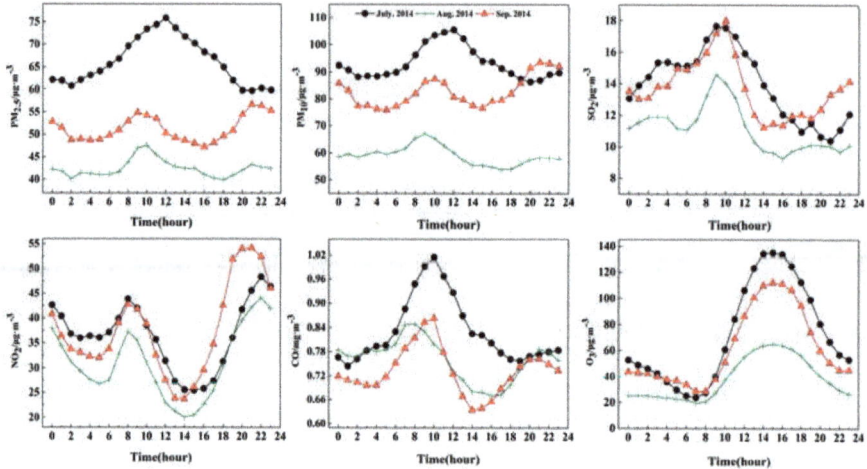

Figure 3. Diurnal variation of air pollutant concentrations in different periods.

$PM_{2.5}$ concentration was gradually reduced from midnight to 2:00–4:00, followed by a morning peak around 9:00–10:00, except for July 2014 (before the emission reduction). Concentrations then decreased until 16:00–18:00, after which they rose until 21:00–22:00. Therefore, two periods displayed a bimodal pattern with peaks between 9:00–10:00 and 21:00–22:00. The morning peak was attributed to enhanced anthropogenic activity during rush hour [16]. After sunrise, strong turbulence in the developing convective boundary layer led to lower $PM_{2.5}$ concentration. During the night, lower mixed layer height led to higher $PM_{2.5}$ concentration. The diurnal variation characteristic of PM_{10} was similar to that of $PM_{2.5}$ concentration, which reflected the positive effects of the emission reduction measures.

SO_2 concentration showed a unimodal pattern; the maximum SO_2 concentrations tended to appear around 9:00–10:00 during the day, which was similar to that of $PM_{2.5}$. The mixing layer was high and photochemical conversion was strong at this time, as SO_2 in the upper air was transferred to the ground which led to the peak [9]. The main source of NO_2 in Nanjing is automobile exhaust. Diurnal variation of NO_2 presented a double-peak curve. The first peak appeared at 08:00 after the morning peak traffic, then the concentration of NO_2 was reduced by photolysis due to the enhancement of solar radiation, and the minimum values of NO_2 concentration occurred at 13:00–14:00. With the arrival of the evening peak traffic, NO_2 concentration gradually increased from 13:00–14:00 to 21:00–22:00, and another significant peak appeared at 21:00–22:00, similar to that observed in Chengdu City [17].

The highest CO hourly concentration appeared at around 08:00–10:00 in the morning, followed by a sharply decreasing trend to the lowest value at around 14:00–15:00, except for July 2014 (before the emission reduction). A slight but continuous increase was observed from 14:00–15:00 to 21:00, and finally a second weaker peak was seen at night due to the decrease in boundary layer height, which resulted in the accumulation of pollutants. The diurnal variations of O_3 concentrations in three periods showed similar characteristics; the lowest ozone concentrations for three periods appeared at around 07:00–08:00 with values around 20–30 $\mu g \cdot m^{-3}$. With increasing temperature and solar radiation in the daytime, O_3 concentration increased from 07:00–08:00 until 15:00, when it reached its peak. The concentration then rapidly fell until midnight.

3.4. Spatial Variation of the AQI and Air Pollutant Concentrations

In order to analyze the spatial variation of air quality and air pollutants during the period of the emission reduction, the mean values and standard errors of the AQI and six air pollutant concentrations in nine different sites of Nanjing were compared with the period before the emission reduction, as shown in Figure 4. The AQI during the emission reduction was lower than that in July 2014 at all monitoring stations; the percentage of decrease of ranged from 31.54 to 44.90% on spatial variation, which indicated the effects on air quality of implementing the reduction measures.

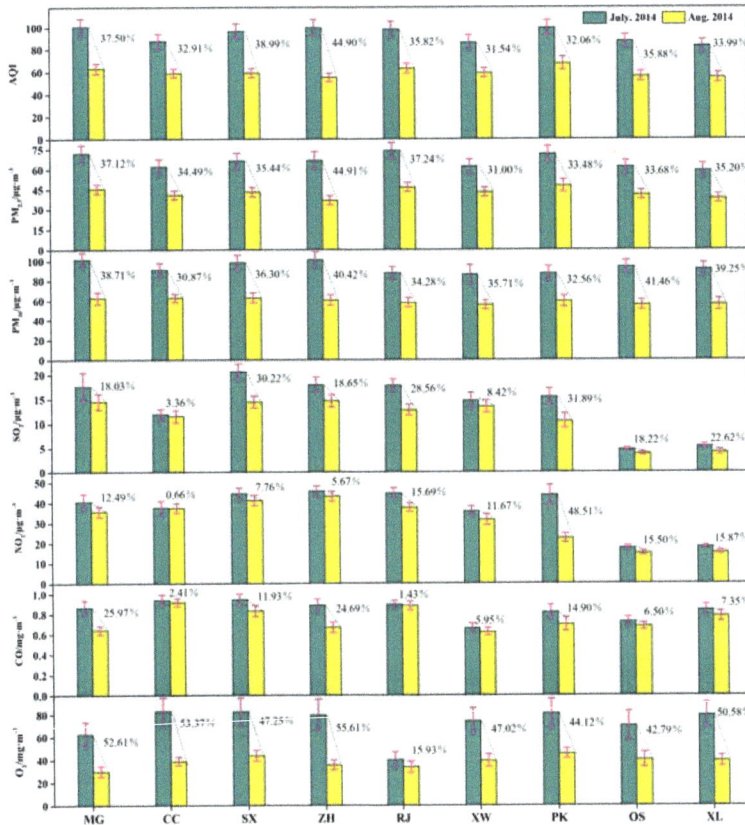

Figure 4. Spatial variation of the AQI and air pollutants at each site before and during the emission reduction.

Concentrations of $PM_{2.5}$, PM_{10} and O_3 during the emission reduction period decreased significantly at nine monitoring sites compared with those in July 2014. The percentage of decrease of $PM_{2.5}$ ranged from 31.00 to 44.91%, showing that the strict pollution control measures carried out in Nanjing could reduce the concentration of $PM_{2.5}$. The percentage of decrease of PM_{10} ranged from 30.87 to 41.46%, which was similar to $PM_{2.5}$ in the spatial variation. The percentage of decrease of O_3 ranged from 15.93 to 55.61%, which was the largest of the six air pollutants. The percentage of decrease of SO_2 ranged from 3.36 to 31.89%, which was lower than that observed in Beijing during the APEC period [9]. Most of the coal-burning factories were shut down during the period of the emission reduction in Nanjing. Thus, the decrease of SO_2 indicated the effects of the reduction of coal consumption [18]. The percentage of decrease of NO_2 ranged from 0.66 to 48.51%, which reflected the good influence of reducing motor vehicle exhaust and industrial production. The percentage of decrease of CO ranged from 1.43 to 25.97%, which was the smallest of the six air pollutants; moreover, the daily average of CO has not exceeded the Chinese national assessment standard in Nanjing in recent years.

The decrease of the AQI at the ZH site was the highest of all the sites. The decrease of $PM_{2.5}$ at the ZH site was the highest, while the decrease of $PM_{2.5}$ at the XW site was lower than the others. The decrease of PM_{10} at the OS site was higher than those at other monitoring sites, reflecting the effectiveness of control measures for PM_{10} in the main stadium of the Nanjing Youth Olympic Games. The PK site was located in the suburbs of Nanjing, and the decrease of SO_2 and NO_2 at the PK site was the highest of all the monitoring sites, indicating the effectiveness of controlling motor vehicles and industrial production. The decrease of O_3 exceeded 40% in the spatial variation, except for the RJ site, with a decrease of 55.61% at the ZH site, which was the largest of all the monitoring sites.

3.5. Variation of Air Quality in Nanjing and the Surrounding Cities

From 1 to 31 August 2014, Nanjing cooperated with 22 surrounding cities to establish an air pollution joint prevention group to guarantee the air quality in August in Nanjing. Figure 5 shows the AQI of Nanjing and 22 nearby cities in July and August 2014. In July 2014, the air quality was the worst in WX, compared with the other cities; the AQI for WX was the highest (105), followed by CZ2 and YZ. The air quality was the best in LYG, XZ and SH, where the AQI was the lowest (74). In August 2014, the maximum value of the AQI occurred in JX and the minimum value of the AQI appeared in Nanjing. We found that the difference between the AQI in July 2014 and August 2014 was very significant ($p < 0.05$). Figure 6 represents the percentage of decrease of the AQI during the emission reduction period compared with July 2014 in Nanjing and 22 nearby cities. The AQI of all cities showed a downward trend. The decrease in the AQI in Nanjing was 35.48%, which was the largest of the 23 cities, followed by YZ and HZ2, and was almost the same in ZJ, YC and WX. The decrease in the AQI in CZ2, NT and TZ were 19.15%, 18.60% and 18.18%, respectively. The decrease in the AQI in SH, HF, XZ, JX, HA and LYG were less than 10% and greater than 0%, and was the lowest in LYG with only a 2.70% reduction.

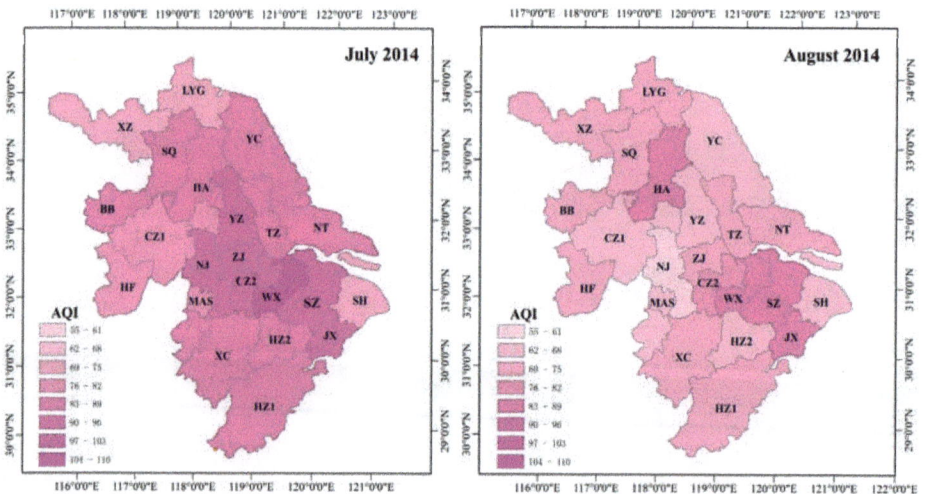

Figure 5. Variation of the AQI in Nanjing and its surrounding cities during the emission reduction period.

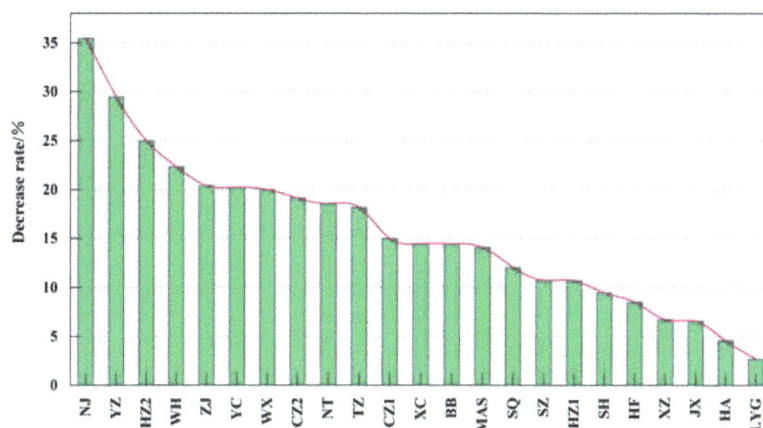

Figure 6. Decrease percentage of the AQI during the emission reduction period compared with July 2014 in Nanjing and its surrounding cities.

4. Discussion and Conclusions

Air pollutant concentrations are not only related to emissions from air pollution sources, but are also influenced by meteorological conditions. In order to eliminate the interference of meteorological conditions, we divided the meteorological data into three time periods: before, during the emission reduction and during the same period of the previous year (2013). According to the comparison of meteorological data in three periods, we found that most of the meteorological parameters in July 2014 were closer to the average level during the period of the emission reduction than that in August 2013. Therefore, we analyzed the concentrations of air pollutants before, during and after the emission reduction, which was more appropriate than the comparison with the year 2013 to evaluate the effectiveness of air pollution reduction actions.

According to our analysis and comparison, the concentrations of six air pollutants during the emission reduction period were significantly lower than those in the prophase mean before the emission reduction, especially O_3. Compared with the average levels in July 2014, the mean concentrations of $PM_{2.5}$, PM_{10}, SO_2, NO_2, CO and O_3 decreased by 35.92%, 36.75%, 20.40%, 15.05%, 8.54% and 47.15%, respectively. All these reductions in air pollutants indicated that the emission control measures were successful during the YOG, and little related to the local meteorological conditions. After the YOG, air pollutants increased significantly due to the cancellation of the emission control measures, except for CO. In addition, we also analyzed the diurnal variations of air pollutant concentrations in different periods. Overall, the diurnal variation curve of air pollutants during the period of the emission reduction was lower than that in July 2014, except for CO. In the spatial variation of pollutants, the AQI during the emission reduction was lower than that in July 2014 at all monitoring stations, the decrease of which ranged from 31.54 to 44.90% in spatial variation. Meanwhile, the decrease of $PM_{2.5}$ ranged from 31.00% to 44.91% at different sites, the decrease of PM_{10} ranged from 30.87% to 41.46%, the decrease of SO_2 ranged from 3.36% to 31.89%, the decrease of NO_2 ranged from 0.66% to 48.51%, the decrease of CO ranged from 1.43% to 25.97%, and the decrease of O_3 ranged from 15.93% to 55.61%. Our results suggested that the AQI of Nanjing and 22 nearby cities showed a downward trend when compared with July 2014, which reflected the effects on air quality of implementing the regional emission reduction measures.

In recent years, China has successfully hosted many international events. In order to improve air quality during the events, the local government has implemented some short-term periods with temporary emission control measures to reduce the emissions of pollutants. During the 2008 Olympic Games, the Beijing municipal government took many air pollution control measures, which included relocating some heavily polluting enterprises, encouraging natural gas instead of coal-fired boilers and domestic stoves, limiting the use of cars and so on [19]. During the 2010 Asian Games, the Guangzhou government made great efforts to improve the air quality, which mainly included controlling emissions from industries and transportation restrictions; vehicles could be driven only on alternate days depending on license plate numbers, and all construction activities were put on hold during the Games [20]. During the APEC period, the Chinese government took more stringent emission reduction measures in Beijing and the surrounding region, including banning heavily polluting vehicles, closing heavily polluting factories, slowing down construction activities and so on [21]. Some studies found that the mass concentration of air pollutants during the events decreased significantly, except for O_3 [9,22]. The results of our studies showed that the emission control measures were the most effective on O_3 out of the six air pollutants, indicating that the major O_3 precursors (nitrogen oxides and volatile organic compounds) were well controlled, and that meteorological conditions also played a positive role during the YOG.

The Nanjing government carried out temporary strict environmental regulations to ensure good air quality during the YOG in 2014. Other surrounding cities cooperated with Nanjing to guarantee good air quality during the Games. Approximately 2630 construction sites were halted. Heavy-industry factories, including the petrochemical industry, iron and steel industry, building materials industry and so on were required to reduce manufacturing by 20%. High-emission or yellow-labeled vehicles were not allowed to drive on the road. Open space barbecue restaurants were closed. Over 900 electric buses and 500 electric taxis have been put into operation [23]. In addition, 22 surrounding cities in the Yangtze River Delta region were asked to cooperate with Nanjing to close industries with high pollution emissions during the emission reduction period [24].

According to the analysis of this study, several suggestions have been made to improve the air quality of Nanjing. On the one hand, as regional emission reduction measures were the main factor in improving the air quality in Nanjing, regional cooperation and joint defense and control are considered the future direction in air pollution control. On the other hand, greater efforts should be exerted to control dust, coal-burning, vehicle emissions and industrial production around Nanjing, which would play a significant role in improving the air quality of Nanjing.

In order to protect human health and improve air quality, we may consider taking the following several measures and recommendations for air pollution reduction over long-term period:

1. To reduce emissions. Use more non-pollution energy, such as solar energy, wind energy and hydropower. Reform energy structure and use low-polluting energy, such as natural gas, biogas and alcohol. In addition, before the pollutants enter the atmosphere, use technology such as dust and smoke abatement to eliminate the partial pollutant in the waste gas, as it can reduce the quantity of pollutants entering the atmosphere.
2. To curb emissions and make full use of the atmosphere's self-purification ability. Due to different meteorological conditions, the efficiency of the removal of pollutants in the atmosphere is different. The pollutants concentration may be different in different regions, even when the same amount of pollutants has been released into the atmosphere. Therefore, respond to different regions at different times for the effective control of emissions.
3. Factors such as the location of factory sites, chimney designs, city planning and the location of industrial districts must be given proper attention so as to not cause redundant iterations of pollution, which can in turn result in serious local pollution events.
4. Afforestation enables more plants to absorb pollutants, thus reducing the degree of air pollution.

Acknowledgments: This work was financially supported by the National Natural Science Fund (41475108).

Author Contributions: Hui Zhao and Youfei Zheng conceived and designed the experiments as well as writing the paper; Hui Zhao analyzed the data; Ting Li helped perform the statistical analysis.

Conflicts of Interest: The authors declare no conflict of interest.

References

1. Chan, C.K.; Yao, X.H. Air pollution in mega cities in China. *Atmos. Environ.* **2008**, *42*, 1–42. [CrossRef]
2. Gao, J.J.; Tian, H.Z.; Cheng, K.; Lu, L.; Zheng, M.; Wang, S.X.; Hao, J.M.; Wang, K.; Hua, S.B.; Zhu, C.Y.; Wang, Y. The variation of chemical characteristics of $PM_{2.5}$ and PM_{10} and formation causes during two haze pollution events in urban Beijing, China. *Atmos. Environ.* **2015**, *107*, 1–8. [CrossRef]
3. Leiva, M.A.; Santibanez, D.A.; Ibarra, S.; Matus, P.; Seguel, R. A five-year study of particulate matter ($PM_{2.5}$) and cerebrovascular diseases. *Environ. Pollut.* **2013**, *181*, 1–6. [CrossRef] [PubMed]
4. Pascal, M.; Falq, G.; Wagner, V.; Chatignoux, E.; Corso, M.; Blanchard, M.; Host, S.; Pascal, L.; Larrieu, S. Short-term impacts of particulate matter (PM_{10}, $PM_{10-2.5}$, $PM_{2.5}$) on mortality in nine French cities. *Atmos. Environ.* **2014**, *95*, 175–184. [CrossRef]
5. Shankardass, K.; Jerrett, M.; Dell, S.D.; Foty, R.; Stieb, D. Spatial analysis of exposure to traffic-related air pollution at birth and childhood atopic asthma in Toronto, Ontario. *Health Place* **2015**, *34*, 287–295. [CrossRef] [PubMed]
6. Yang, Y.R.; Liu, X.G.; Qu, Y.; Wang, J.L.; An, J.L.; Zhang, Y.H.; Zhang, F. Formation mechanism of continuous extreme haze episodes in the megacity Beijing, China, in January 2013. *Atmos. Res.* **2015**, *155*, 192–203. [CrossRef]
7. Xin, J.Y.; Wang, Y.S.; Tang, G.Q.; Wang, L.L.; Sun, Y.; Wang, Y.H.; Hu, B.; Song, T.; Ji, D.S.; Wang, W.F.; et al. Variability and reduction of atmospheric pollutants in Beijing and its surrounding area during the Beijing 2008 Olympic Games. *Chin. Sci. Bull.* **2010**, *55*, 1937–1944. [CrossRef]
8. Wang, W.T.; Primbs, T.; Tao, S.; Simonich, S.M. Atmospheric Particulate Matter Pollution during the 2008 Beijing Olympics. *Environ. Sci. Technol.* **2009**, *15*, 5314–5320. [CrossRef]
9. Wang, Z.S.; Li, Y.T.; Chen, T.; Li, L.J.; Liu, B.X.; Zhang, D.W.; Sun, F.; Wei, Q.; Jiang, L.; Pan, L.B. Changes in atmospheric composition during the 2014 APEC conference in Beijing. *J. Geophys. Res. Atmos.* **2015**, *120*, 695–707. [CrossRef]
10. Hao, N.; Valks, P.; Loyola, D.; Cheng, Y.F.; Zimmer, W. Space-based measurements of air quality during the World Expo 2010 in Shanghai. *Environ. Res. Lett.* **2011**, *6*, 67–81. [CrossRef]
11. Wu, F.C.; Xie, P.H.; Li, A.; Chan, K.L.; Hartl, A.; Wang, Y.; Si, F.Q.; Zeng, Y.; Qin, M.; Xu, J.; et al. Observations of SO2 and NO2 by mobile DOAS in the Guangzhou eastern area during the Asian Games 2010. *Atmos. Meas. Tech.* **2013**, *6*, 2277–2292. [CrossRef]
12. Hu, J.L.; Ying, Q.; Wang, Y.G.; Zhang, H.L. Characterizing multi-pollutant air pollution in China: Comparison of three air quality indices. *Environ. Int.* **2015**, *84*, 17–25. [CrossRef] [PubMed]
13. Shen, G.F.; Yuan, S.Y.; Xie, Y.N.; Xia, S.J.; Li, L.; Yao, Y.K.; Qiao, Y.Z.; Zhang, J.; Zhao, Q.Y.; Ding, A.J.; et al. Ambient levels and temporal variations of PM2.5 and PM10 at a residential site in the mega-city, Nanjing, in the western Yangtze River Delta, China. *J. Environ. Sci. Health. A* **2014**, *49*, 171–178.
14. Stella, P.; Personne, E.; Loubet, B.; Lamaud, E.; Ceschia, E.; Beziat, P.; Bonnefond, J.M.; Irvine, M.; Keravee, P.; Mascher, N.; et al. Predicting and partitioning ozone fluxes to maize crops from sowing to harvest: The Surfatm-O3 model. *Biogeosciences* **2011**, *8*, 2869–2886. [CrossRef]
15. Leung, L.R.; Gustafson, W.I. Potential regional climate change and implications to U.S. air quality. *Geophys. Res. Lett.* **2005**, *32*, 367–384. [CrossRef]
16. Zhao, X.J.; Zhang, X.L.; Xu, X.F.; Xu, J.; Meng, W.; Pu, W.W. Seasonal and diurnal variations of ambient PM2.5 concentration in urban and rural environments in Beijing. *Atmos. Environ.* **2009**, *43*, 2893–2900. [CrossRef]
17. Xie, Y.Z.; Pan, Y.P.; Ni, C.J.; Chen, Z.H.; Wei, X. Temporal and spatial variations of atmospheric pollutants in urban Chengdu during summer [in Chinese]. *Acta. Sci. Circumst.* **2015**, *35*, 975–983.
18. Apergis, N.; Loomis, D.; Payne, J.E. Are fluctuations in coal consumption transitory or permanent? Evidence from a panel of US states. *Atmos. Environ.* **2010**, *87*, 2424–2426. [CrossRef]
19. Wang, S.X.; Zhao, M.; Xing, J.; Wu, Y.; Zhou, Y.; Lei, Y.; He, K.B.; Fu, L.X.; Hao, J.M. Quantifying the air pollutants emission reduction during the 2008 Olympic Games in Beijing. *Environ. Sci. Technol.* **2010**, *44*, 2490–2496. [CrossRef] [PubMed]
20. Liu, H.; Wang, X.M.; Zhang, J.P.; He, K.B.; Wu, Y.; Xu, J.Y. Emission controls and changes in air quality in Guangzhou during the Asian Games. *Atmos. Environ.* **2013**, *76*, 81–93. [CrossRef]
21. Li, R.P.; Mao, H.J.; Wu, L.; He, J.J.; Ren, P.P.; Li, X.Y. The evaluation of emission control to PM concentration during Beijing APEC in 2014. *Atmos. Pollut. Res.* **2016**, *7*, 363–369. [CrossRef]

22. Wang, Z.S.; Li, Y.T.; Zhang, D.W.; Chen, T.; Sun, F.; Li, L.J.; Li, J.X.; Sun, N.D.; Chen, C.; Wang, B.Y. Analysis on air quality in Beijing during the 2014 APEC conference [in Chinese]. *J. Environ. Sci.* **2016**, *36*, 675–683.

23. Ding, J.; Vander, A.R.J.; Mijling, B.; Levelt, P.F.; Hao, N. NOx emission estimates during the 2014 Youth Olympic Games in Nanjing. *Atmos. Chem. Phys.* **2015**, *15*, 6337–6372. [CrossRef]

24. Li, S.W.; Li, H.B.; Luo, J.; Li, H.M.; Qian, X.; Liu, M.M.; Bi, J.; Cui, X.Y.; Ma, L.Q. Influence of pollution control on lead inhalation bioaccessibility in $PM_{2.5}$: A case study of 2014 Youth Olympic Games in Nanjing. *Environ. Int.* **2016**, *94*, 69–75. [CrossRef] [PubMed]

atmosphere

MDPI

Article

Characterization of Particulate Matter (PM$_{2.5}$ and PM$_{10}$) Relating to a Coal Power Plant in the Boroughs of Springdale and Cheswick, PA

Casey D. Bray *, William Battye, Pornpan Uttamang, Priya Pillai and Viney P. Aneja

Department of Marine, Earth, and Atmospheric Sciences, North Carolina State University, Raleigh, NC 27695, USA; whbattye@ncsu.edu (W.B.); puttama@ncsu.edu (P.U.); prpillai@ncsu.edu (P.P.); vpaneja@ncsu.edu (V.P.A.)
* Corresponding: cdbray@ncsu.edu; Tel.: +1-919-515-3690

Received: 7 August 2017; Accepted: 18 September 2017; Published: 23 September 2017

Abstract: Ambient concentrations of both fine particulate matter (PM$_{2.5}$) and particulate matter with an aerodynamic diameter less than 10 micron (PM$_{10}$) were measured from 10 June 2015 to 13 July 2015 at three locations surrounding the Cheswick Power Plant, which is located between the boroughs of Springdale and Cheswick, Pennsylvania. The average concentrations of PM$_{10}$ observed during the periods were 20.5 ± 10.2 μg m^{-3} (Station 1), 16.1 ± 4.9 μg m^{-3} (Station 2) and 16.5 ± 7.1 μg m^{-3} (Station 3). The average concentrations of PM$_{2.5}$ observed at the stations were 9.1 ± 5.1 μg m^{-3} (Station 1), 0.2 ± 0.4 μg m^{-3} (Station 2) and 11.6 ± 4.8 μg m^{-3} (Station 3). In addition, concentrations of PM$_{2.5}$ measured by four Pennsylvania Department of Environmental Protection air quality monitors (all within a radius of 40 miles) were also analyzed. The observed average concentrations at these sites were 12.7 ± 6.9 μg m^{-3} (Beaver Falls), 11.2 ± 4.7 μg m^{-3} (Florence), 12.2 ± 5.3 μg m^{-3} (Greensburg) and 12.2 ± 5.5 μg m^{-3} (Washington). Elemental analysis for samples (blank – corrected) revealed the presence of metals that are present in coal (i.e., antimony, arsenic, beryllium, cadmium, chromium, cobalt, lead, manganese, mercury, nickel and selenium).

Keywords: PM$_{2.5}$; PM$_{10}$; coal-fired power plant; particulate matter emissions

1. Introduction

With over 600 active coal-fired power plants in the US, 39% of the country's total electricity generation is attributed to coal (Available online: http://www.eia.gov). While these plants may bring jobs and prosperity to their surrounding regions, they also emit dangerous pollutants into the atmosphere. According to the US Environmental Protection Agency (EPA) National Emissions Inventory, coal-fired power plants emit sulfur dioxide, oxides of nitrogen and particulate matter into the atmosphere. They also emit 84 of the 187 Hazardous Air Pollutants (HAP) regulated by the US EPA [1]. In addition to the stack emissions from the coal-fired power plant, coal handling may also emit pollutants into the atmosphere and thus degrade the air quality in the vicinity near the power plant [2]. Several studies have been conducted on the relationship between particulate matter and emissions from coal-fired power plants [2–14]. Particulate matter measured near a coal-fired power plant is known to contain a number of harmful chemicals that are present in coal and are also known to have carcinogenic properties, such as antimony, arsenic, beryllium, cadmium, chromium, cobalt, lead, manganese, mercury, nickel, selenium and polycyclic organic matter (POM) [2–4]. All of these chemicals are known to be hazardous and are thus regulated by the US EPA as HAPs under the Clean Air Act, as amended in 1990 [15]. The emission of these pollutants into the atmosphere is known to be dangerous to both human health and welfare. Exposure to high concentrations of HAPs can lead to a number of adverse health effects such as damage to the eyes, skin, lungs, kidneys and the nervous system, and can even cause cancer, pulmonary disease and cardiovascular disease [14,16]. Particulate matter ($PM_{2.5}$), which is particulate matter with an aerodynamic diameter of 2.5 microns or less, is also a dangerous atmospheric pollutant due to its small size, which can travel deep into people's lungs and lead to a number of severe health effects. Elevated concentrations of $PM_{2.5}$ are known to be associated with cardiovascular issues (heart disease, heart attacks, etc.) as well as respiratory issues, reproductive issues and even cancer [16,17]. In addition to harming human health, coal-fired power plants can also lead to a number of environmental impacts as well, such as acidification of the environment, bioaccumulation of toxic metals, the contamination of water sources, reduced visibility due to haze as well as degradation of buildings and monuments.

The Cheswick Power Plant is located in the southwestern part of Pennsylvania along the Allegheny River right between the boroughs of Springdale (40.5414° N, 79.7821° W) and Cheswick (40.5416° N, 79.8002° W) (Figure 1). While the Cheswick Power Plant has brought economic benefits to the small boroughs of Springdale and Cheswick, PA, the plant also brought with it a multitude of harmful effects on both human health and welfare. According to the Allegheny County Health Department's Point Source Emissions Inventory Report, the Cheswick Power Station is the largest point source emitter of both criteria pollutants and hazardous air pollutants (HAP) in Allegheny County, PA [18]. In 2006, the plant was listed as the 17th Dirtiest Power Plant for sulfur dioxide [1]. In 2010, the plant once again made headlines, ranking 41st in the Top Power Plant Hydrochloric Acid Emitters and 91st in the Top US Power Plant Lead Emitters [19]. Since then, significant efforts have been made to reduce emissions [18]. Overall, emissions of carbon monoxide and sulfur dioxide did decrease in 2011, while emissions of nitrogen oxides, particulate matter, NO_x, and volatile organic compounds (VOC) increased [18]. However, the plant remains Allegheny County's largest point source emitter of both criteria pollutants and hazardous air pollutants. The objective of this study was to measure the air quality at three different locations within the two boroughs surrounding the power plant in order to determine concentrations of both $PM_{2.5}$ and PM_{10} due to emissions from the power plant. Particulate matter measurements measured by the PA Department of Environmental Protection (DEP) were also analyzed for the study period and compared against the measurements taken for this study.

2. Experiments

2.1. NCSU Monitoring Stations

Three experimental sites were set up within Springdale and Cheswick, PA, each within a mile of the Cheswick power plant (Figure 1). There were two sampling periods in this study. Two of the three sites took samples from 10 June 2015 to 27 June 2015 and then from 30 June 2015 to 13 July 2015. Station 1 was located at 244 Center St., Springdale, PA (40.5402° N, 79.7884° W), which is less than half of a mile from the power plant. Samples were taken at this location during both sampling periods. Station 2 was located at 200 Hill Avenue, Cheswick, PA (40.542° N, 79.80° W), which is approximately one mile from the power plant. The sampling period for this study was from 10 June 2015 to 27 June 2015. The third station was located at 1212 Fairmont St, Cheswick, PA (40.5465° N, 79.8031° W). The sampling period for this site was from 30 June 2015 to 13 July 2013. All three of these sites are in residential areas, very close to the power plant. The monitoring sites were chosen considering the prevailing wind direction (west southwesterly) such that during both sampling periods, one site was upwind of the power plant (those sites in Cheswick) and one site was downwind of the power plant (the site in Springdale). However, it must be noted that the wind during the sampling period was highly variable.

Each experimental site was equipped to measure $PM_{2.5}$ and PM_{10} using Reference Ambient Air Sampler (RAAS) high volume air samplers by Anderson Instruments as well as meteorological data measured by Met One Model Automet portable weather stations equipped with an onboard data logger. The sampling period for each instrument was set to 24 h, and the air volumetric flow on both types of instruments was set at the standard flow rate of 16.67 Liters/minute. The sampling filters used to collect both $PM_{2.5}$ and PM_{10} were 47 mm diameter Teflon filters. For $PM_{2.5}$ sampling, there was an additional smaller filter used (~39 mm diameter) in the WINS impactor, which was wetted with impactor oil, in order to restrict particles larger than 2.5 micron from reaching the 47 mm Teflon filter. All the filters were pre-tared and numbered.

Figure 1. Location of monitoring stations in relation to the Cheswick Power Plant and the two boroughs of Springdale and Cheswick.

High Volume (Hi Vol) 47 mm MTL Teflon filters were used to analyze mass concentration of PM_{10} while Hi Vol 39 mm Teflon filters were used to analyze mass concentration of $PM_{2.5}$. Filter weighing measurements took place inside a temperature/relative humidity-controlled ISO Class 6 (<1000 $PM_{0.5}$/cf) clean room employing a draft-shielded microbalance with anti-static wand. Temperature and relative humidity were controlled to 21 °C and 35%, respectively. The filter mass was determined through five weightings of the filter, with each weight bracketed by a reading of the internal zero of the balance. A buoyancy correction was applied to the mass and the average of the zeros bracketing the mass was then subtracted from the result. This was repeated four additional times, with the average of these five results being reported as the final mass. The gravitational analysis was performed by the Research Triangle Park's (RTP) office of Applied Research Associates, Inc. (ARA).

Ten samples were chosen from this study to undergo an inorganic analysis, i.e., five for $PM_{2.5}$ and five for PM_{10}. In addition, two sample blanks were analyzed for calibration. Samples were analyzed from all three monitoring locations. The samples were selected to represent a mix of particulate matter concentrations collected during times when the plant was running. The inorganic analysis was then performed by X-Ray Fluorescence (XRF) and then Ion Chromatography (IC) analysis.

2.2. PA $PM_{2.5}$ Monitoring Sites

In addition to the measurement sites used in the study, data from four monitoring stations run by the state of PA (PA Department of Environmental Protection) was also used to compare against measurements taken in the field campaign. The four monitoring sites used in this study (Figure 2) were Beaver Falls (40.7478° N, 80.3157° W), Florence (40.4454° N, 80.4212° W), Greensburg (40.3043° N, 79.5060° W) and Washington (40.1706° N, 80.2617° W). These four sites contain continuous Met One BAM 1020 $PM_{2.5}$ monitors [20]. It is important to note that while the NCSU monitoring sites were located fairly close to the Cheswick Power Plant (i.e., less than 5 miles), the PA DEP monitoring sites were located 20–40 miles from the power plant.

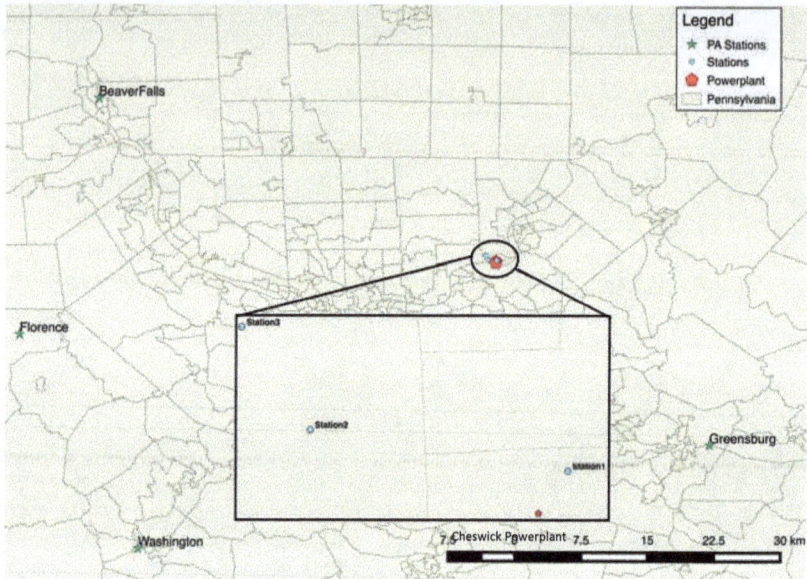

Figure 2. Location of the PA DEP monitoring sites in relation to the NCSU monitoring stations. The PA DEP monitoring sites are represented by the green stars while the NCSU monitoring sites are represented by the blue dots.

3. Results and Discussion

The observed average 24-hour concentrations of both $PM_{2.5}$ (0.02–22.76 μg m^{-3}) and PM_{10} (0.9−44.15 μg m^{-3}) from the NCSU monitoring sites were found to be lower than the US EPA National Ambient Air Quality Standard of 35 μg m^{-3} and 150 μg m^{-3}, respectively (Figure 3).

The average concentrations of PM_{10} observed during the periods were 20.5 $\pm$ 10.2 μg m^{-3} (Station 1), 16.1 $\pm$ 4.9 μg m^{-3} (Station 2) and 16.5 $\pm$ 7.1 μg m^{-3} (Station 3). The average concentrations of $PM_{2.5}$ observed at the stations were 9.1 $\pm$ 5.1 μg m^{-3} (Station 1), 0.2 $\pm$ 0.4 μg m^{-3} (Station 2) and 11.6 $\pm$ 4.8 μg m^{-3} (Station 3). Station 1 observed the maximum concentrations of both $PM_{2.5}$ (22.76 μg m^{-3}) and PM_{10} (44.15 μg m^{-3} μg m^{-3}). While Station 3 also observed higher concentrations of particulate matter (both $PM_{2.5}$ and PM_{10}), the results of a t test (at α = 0.05) indicated that the concentrations at Station 1 were indeed higher than those observed at Stations 2 and 3. There are several plausible causes for this: the location of Station 1 versus Stations 2 and 3 in relation to the meteorological conditions as well as additional potential sources of particulate matter into the atmosphere (i.e., traffic emissions, roadway dust, transport). Concentrations of particulate matter tended to be higher when conditions were warm and sunny. Conversely, the lowest concentrations of particulate matter were primarily observed during rainy conditions. This was expected due to the removal of the particulates from the atmosphere through wet deposition. However, there were a few days where elevated concentrations were observed when the plant was running during rainy conditions. This may have resulted from the stability of the atmosphere in these conditions, which allows for the atmospheric accumulation of particulate matter. Nevertheless, comparisons of particulate matter concentrations with meteorological conditions (Figures 4 and 5) failed to show any strong correlations.

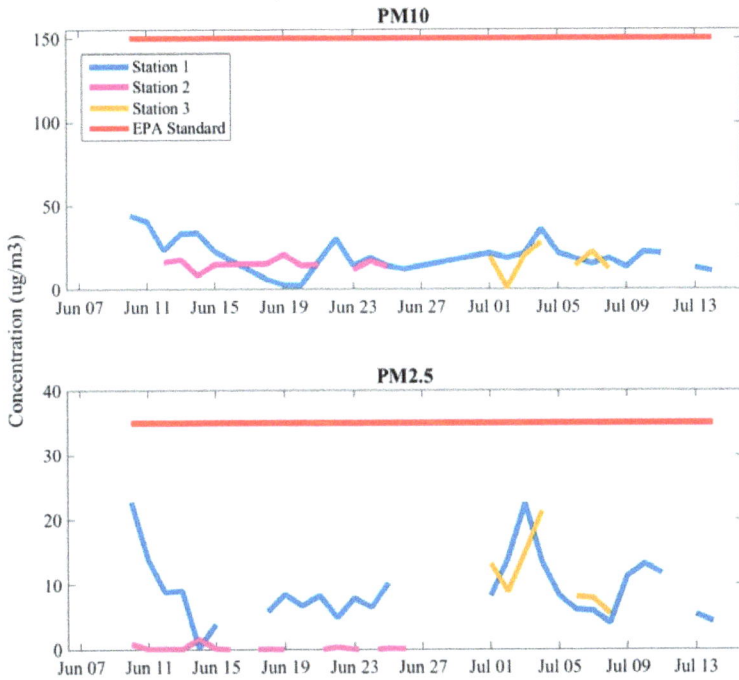

Figure 3. The daily 24-hour concentration of PM_{10} and $PM_{2.5}$ at Station 1 (blue), Station 2 (magenta) and Station 3 (gold) plotted against the US EPA National Ambient Air Quality Standards value for 24-hour average PM_{10} concentration (red).

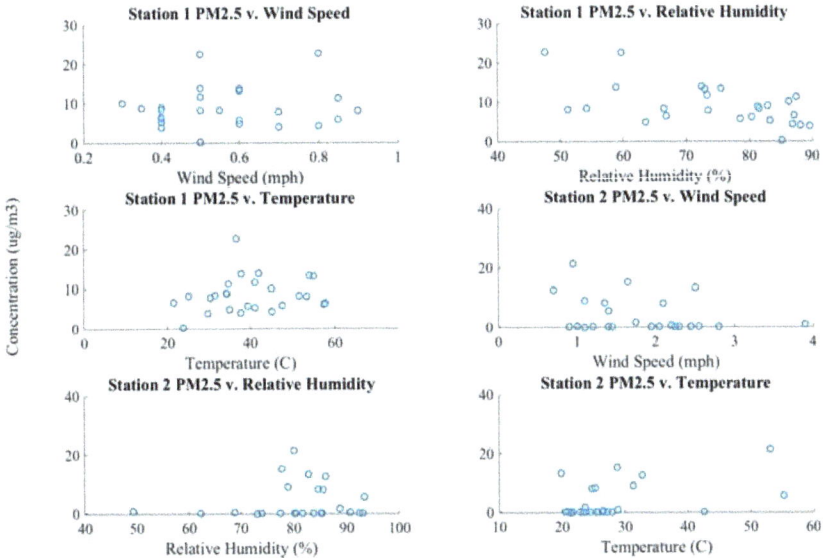

Figure 4. Comparing concentrations of $PM_{2.5}$ with meteorological variables for both Station 1 and Stations 2 and 3, denoted by Station 2 in this figure.

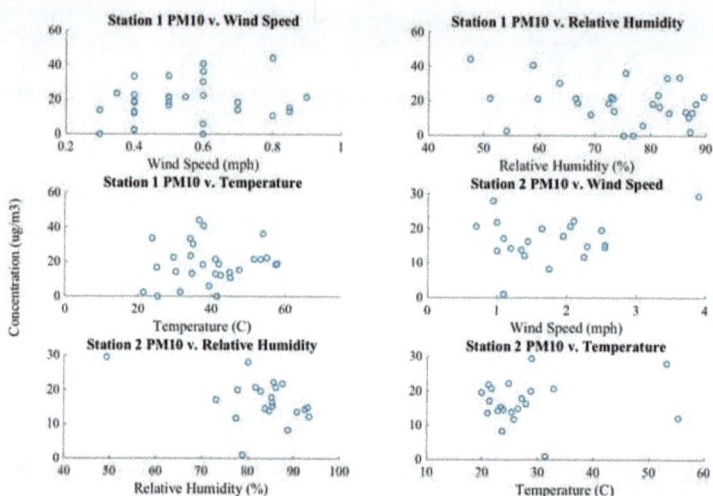

Figure 5. Comparing concentrations of PM$_{10}$ with meteorological variables for both Station 1 and Stations 2 and 3, denoted by Station 2 in this figure.

When comparing the median wind speed versus the concentration of PM$_{10}$ at Stations 1 and 2, positive correlations of 0.2 and 0.27, respectively, were observed. When comparing the median relative humidity with the concentration of PM$_{10}$, negative correlations of 0.3 and 0.44 were observed, respectively, while minor positive correlations of 0.08 and 0.12 were observed for temperature and concentration at Stations 1 and 2, respectively. When comparing meteorological conditions with concentrations of PM$_{2.5}$ at Stations 1 and 2, a negative correlation of 0.32 between wind speed and concentration of PM$_{2.5}$ was observed at Station 2 while Station 1 observed a positive correlation of 0.12. The strongest meteorological correlation was a negative correlation of -0.56 between relative humidity and the concentration of PM$_{2.5}$ at Stations 1, which is nominally downwind from the plant. In contrast, the correlation between relative humidity and PM$_{2.5}$ was weakly positive, at 0.09, for Station 2. Positive correlations of 0.54 and 0.46 were observed when comparing temperature versus concentrations of PM$_{2.5}$ at Stations 1 and 2, respectively. However, when considering these correlations, it is important to acknowledge the short sample period and thus small sample.

The median wind direction at upwind versus downwind stations was compared with the average 24-hour concentrations of particulate matter, where the categorizations of upwind and downwind were denoted based on the daily median wind direction. The results of a t test provide $p = 0.5$ and $p = 0.4$, for concentrations of PM$_{10}$ and PM$_{2.5}$, respectively, which suggests that the relationship is not statistically significant.

Despite the fact that the median wind direction was not significantly correlated with the concentrations of particulate matter, the local hourly wind direction likely did impact concentrations of particulate matter. Furthermore, it is likely that some of the relatively high concentrations of both particulate matter observed at Station 1 were found to have come from the direction of the coal site on the power plant's property (Figures 6 and 7). In addition to observing elevated concentrations coming directly from the plant, elevated concentrations of both PM$_{2.5}$ and PM$_{10}$ were also observed coming from other directions. This can likely be attributed to a number of sources, such as emissions from vehicles, road dust. In addition, the elevated concentrations can also be attributed to transport due to local and regional meteorological phenomena. Furthermore, fairly elevated concentrations of PM$_{10}$ are observed at faster wind speeds. While wind speeds would generally reduce ambient concentrations of particulate matter in the atmosphere, the observed elevated concentrations could simply be due to an emission source located extremely close to the monitoring site (i.e., a stalled truck on the side of the road).

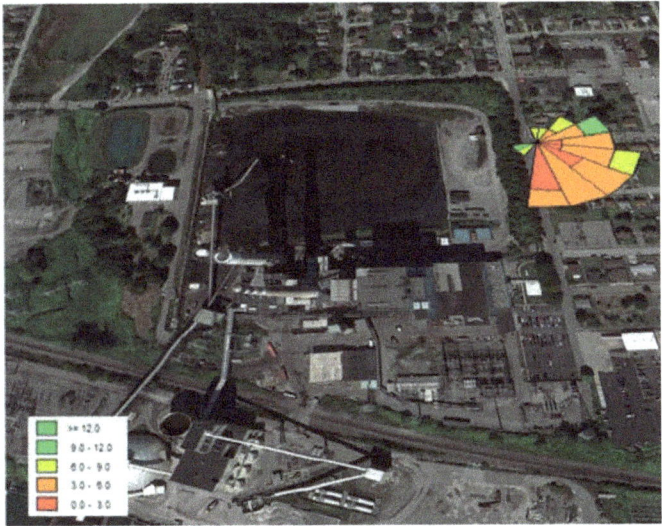

Figure 6. Wind rose created for $PM_{2.5}$ concentrations measured at Station 1. Green colors represent the highest concentrations of particulate matter while the orange and red colors represent lower concentrations of particulate matter, as shown in the legend. Units of particulate matter concentration are $\mu g\ m^{-3}$. The upper side of the wind rose represents North. The length of the wind rose vectors represents average wind speed, where the longer vectors represent a higher average wind speed.

Figure 7. Wind rose created for PM_{10} concentrations measured at Station 1. Green colors represent the highest concentrations of particulate matter while the orange and red colors represent lower concentrations of particulate matter, as shown in the legend. Units of particulate matter concentration are $\mu g\ m^{-3}$. The upper side of the wind rose represents North. The length of the wind rose vectors represents average wind speed, where the longer vectors represent a higher average wind speed.

In contrast to this, neither Stations 2 nor 3 (not pictured) showed extremely high concentrations coming from the power plant. However, this is not entirely surprising for two reasons: the first reason is that the winds were primarily such that those stations were upwind of the power plant and that both of these stations were twice as far as Station 1 for the plant, thus allowing for some dispersion of pollutants emitted from the plant and the coal site to occur.

In addition to this, concentrations of both $PM_{2.5}$ and PM_{10} were compared at calm wind conditions (median wind speeds less than 1 kt) and non-calm wind conditions (Table 1). When comparing average concentrations of $PM_{2.5}$ during calm winds (9.29 ± 5.59 µg m^{-3}) with average concentrations during not calm winds (3.53 ± 5.18 µg m^{-3}), it is clear that the average concentrations during calm winds are higher than the average concentrations during not calm winds. Similarly, average concentrations of PM_{10} observed during calm winds (19.83 ± 10.29 µg m^{-3}) was higher than the average concentrations observed when the winds were not calm (16.25 ± 4.87 µg m^{-3}). Two *t* tests were also conducted in order to determine whether or not the comparison between concentrations of $PM_{2.5}$ and PM_{10} during calm and not calm winds was statistically significant. The results of the t tests provide $p = 0.052$ and $p = 0.001$, for concentrations of PM_{10} and $PM_{2.5}$, respectively. This suggests that while the comparison between concentrations of $PM_{2.5}$ during calm and not calm winds is statistically significant, the comparison for concentrations of PM_{10} is not quite statistically significant.

Concentrations of particulate matter were also compared with the plant's gross load during the period (power plant gross load obtained via: EPA. Available online: http://www.ampd.epa.gov/ampd). Figure 8 compares the Cheswick Power Plant's daily gross load with the daily 24-hour average concentration of PM_{10} and $PM_{2.5}$. When comparing the daily gross load with the particulate matter concentrations, the correlation coefficients (ranging between -0.23 and 0.15) suggest that the concentrations of particulate matter at each station are not statistically correlated with the plant gross load.

Table 1. Comparison of concentrations of $PM_{2.5}$ and PM_{10} (µg m^{-3}) during both calm and not calm wind conditions. Calm wind conditions are considered to be conditions where the median wind speed is less than 1 kt.

	Calm Winds (µg m^{-3})		Not Calm Winds (µg m^{-3})	
	$PM_{2.5}$	PM_{10}	$PM_{2.5}$	PM_{10}
Maximum	22.76	44.15	15.05	29.31
Minimum	0.16	0.91	0.02	8.30
Mean	9.29	19.83	3.53	16.25
Median	8.26	20.55	0.35	15.03
Standard Deviation	5.59	10.29	5.18	4.87
Number of Samples	31	33	15	16

Five PM_{10} samples and five $PM_{2.5}$ samples were subjected to XRF and IC analysis. The XRF analysis represents the PM samples as elements, while the IC analysis represents the PM samples as ions. When comparing the results of the analyses, it is evident that they are consistent. The results of the XRF analysis (Figure 9) showed that the primary constituents of PM_{10} are sulfur (0.66–1.77 µg m^{-3}), silicon (0.14–2.47 µg m^{-3}), aluminum (0.03–1.06 µg m^{-3}) and iron (0.08–0.95 µg m^{-3}). Similarly, the primary constituents of $PM_{2.5}$ were found to be sulfur (0.41–1.17 µg m^{-3}), silicon (0.02–0.23 µg m^{-3}) and iron (0.03–0.06 µg m^{-3}). Particulate matter is typically composed of a complex mixture of chemicals that are strongly dependent on source characteristics. Inorganic analysis of the particulate matter revealed the presence of antimony, arsenic, beryllium, cadmium, chromium, cobalt, lead, manganese, mercury, nickel and selenium (Figure 9). All of these metals present in the PM samples are known to be present in coal [21]. While these elements are all found within coal and coal ash, they are also present in crustal material. Based on the small number of samples analyzed, it was not possible to discern differences between the upwind and downwind samples. The highest concentration of sulfur observed in a sample during the period was observed at Station 1. In this case, it is possible that this can be attributed to the coal pit, which is located less than half of a mile away from the station. However, it is also important to note that there are other coal-fired power plants located within 50 miles of the Cheswick Power plant, which could also contribute to the elevated sulfur concentrations.

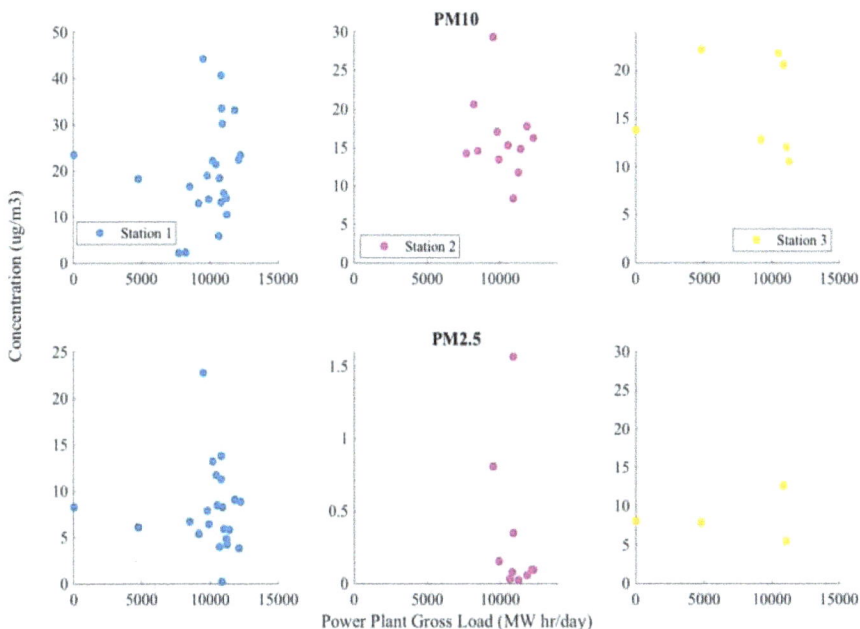

Figure 8. Comparing concentrations of PM_{10} and $PM_{2.5}$ at each station with the power plant gross load. The blue dots represent the concentrations at Station 1, the magenta dots represent the concentrations at Station 2 and the yellow/gold dots represent the concentrations at Station 3.

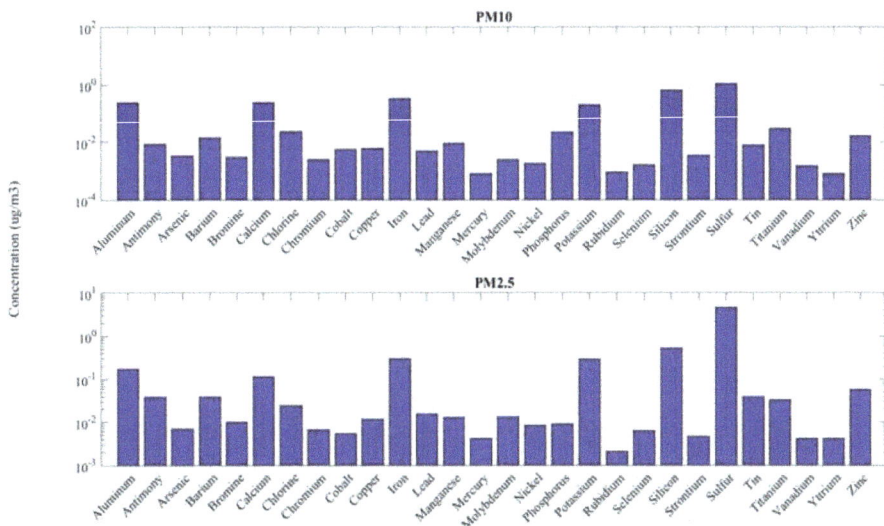

Figure 9. The results of the XRF analysis plotted on a log scale.

The results of the IC analysis (Figure 10) showed that the dominant species of particulate matter in this region are primarily sulfate, nitrate and ammonium. Within the $PM_{2.5}$ samples analyzed, concentrations of sulfate ranged from 2.31 µg m^{-3} to 3.08 µg m^{-3}, nitrate concentrations ranged from 0.03 µg m^{-3} to 0.58 µg m^{-3}, and concentrations of ammonium ranged from 0.82 µg m^{-3} to 1.15 µg m^{-3}. Within the analyzed PM_{10} samples, concentrations of sulfate ranged from 1.05 µg m^{-3} to 4.73 µg m^{-3}, concentrations of nitrate ranged from 0.01 µg m^{-3} to 0.22 µg m^{-3} and concentrations of ammonium ranged from 0.38 µg m^{-3} to 1.69 µg m^{-3}. As described above, it was not possible to discern differences between the upwind and downwind samples based on the small sample size. It is also important to note that sulfate, nitrate and ammonium are also common constituents of particulate matter and thus cannot be attributed entirely to the power plant.

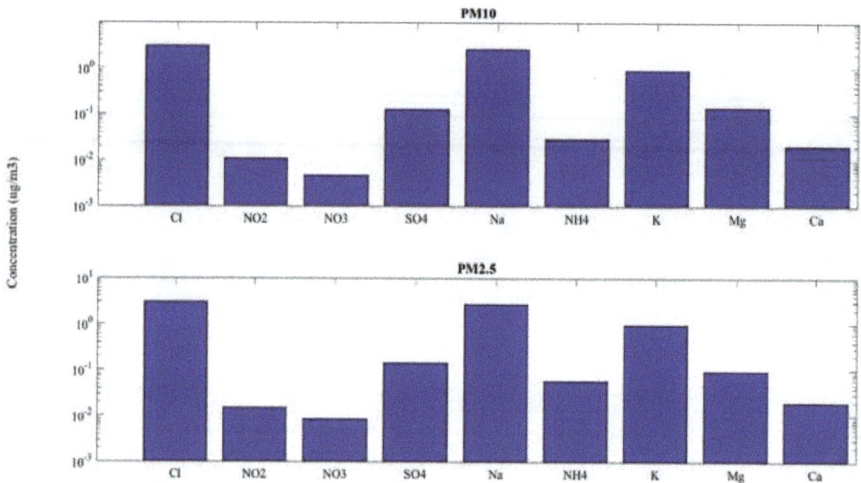

Figure 10. The results of the IC analysis, plotted on a log scale.

The concentrations of $PM_{2.5}$ measured by four PA Department of Environmental Protection air quality monitors were also analyzed in this study. The observed average 24-hour concentrations at these sites were 12.7 ± 6.9 µg m^{-3} (Beaver Falls), 11.2 ± 4.7 µg m^{-3} (Florence), 12.2 ± 5.3 µg m^{-3} (Greensburg) and 12.2 ± 5.5 µg m^{-3} (Washington). Similar to what was observed at the NCSU monitoring sites, the PA DEP monitors also observed 24-hour average concentrations of fine particulate (2–28.2 µg m^{-3}) matter below the US EPA 24-hour NAAQS (35 µg m^{-3}), with the exception of the Beaver Falls monitoring stations, which observed a concentration of 42.6 µg m^{-3} on 4 July 2015 (Figure 11). However, this elevated concentration can likely be explained by the presence of fireworks and other combustion processes due to the holiday.

When comparing the average PA DEP monitoring station $PM_{2.5}$ concentrations [20] with the average NCSU monitor $PM_{2.5}$ concentrations (Figure 12), it is evident that the monitors used in this study observed lower concentrations than the concentrations observed at the PA DEP monitoring sites. However, the general trend in concentrations for both monitoring networks is the same. There are several possible explanations for this discrepancy. One potential cause for the differences in observations could be due to differences in the instrumentation used to measure the particulate matter. In addition, another potential cause for the observed differences could be due to location. Since the monitors are farther away, it is possible that the pollutants were transported aloft before mixing down to the surface level. Furthermore, it is also possible that there are other potential sources of particulate matter located near the sites.

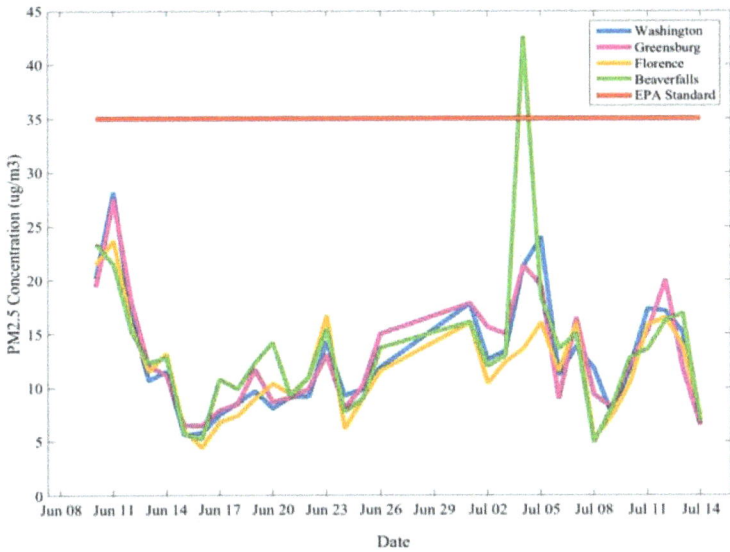

Figure 11. PM$_{2.5}$ measurements taken by the PA DEP monitoring sites during the study period. The red line represents the US EPA 24-hour NAAQS, the blue line represents the Washington monitoring site, the magenta represents the Greensburg monitoring site, the gold line represents the Florence monitoring site and the green line represents the Beaver Falls monitoring site.

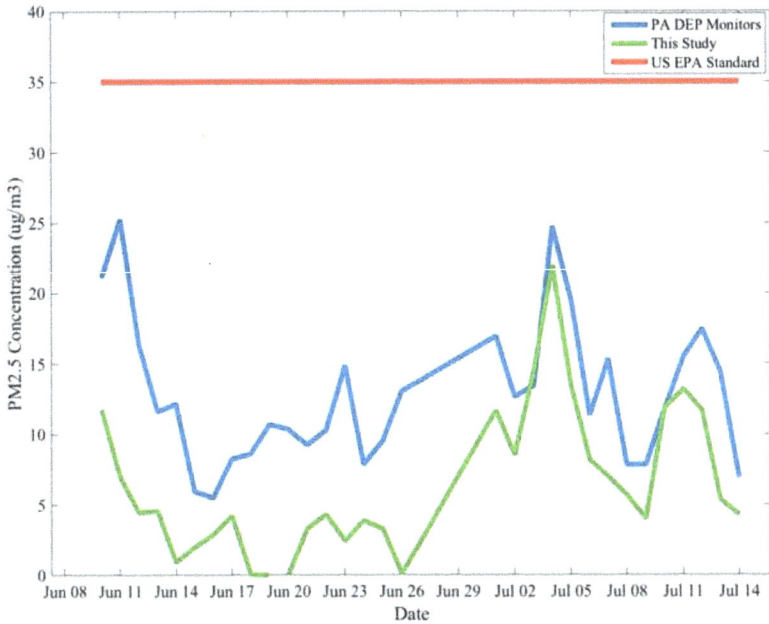

Figure 12. The average concentration measured by the NCSU monitors (this study, green line) compared with the average concentration measured by the PA DEP monitors (blue line line) and the US EPA 24-hour NAAQS (red line) for PM$_{2.5}$.

4. Conclusions

The ambient 24-hour average concentrations of both $PM_{2.5}$ (measured in this study as well as by the PA DEP) and PM_{10} did not reach levels higher than what is permitted by the US EPA 24-hour NAAQS. The average concentrations of PM_{10} observed during the periods were 20.5 ± 10.2 µg m^{-3} for Station 1, 16.1 ± 4.9 µg m^{-3} for Station 2 and 16.5 ± 7.1 µg m^{-3} for Station 3. The average concentrations of $PM_{2.5}$ observed at the stations were 9.1 ± 5.1 µg m^{-3} for Station 1, 0.2 ± 0.4 µg m^{-3} for Station 2 and 11.6 ± 4.8 µg m^{-3} for Station 3. The highest average concentration was observed at Station 1 for PM_{10} and at Station 3 for $PM_{2.5}$. However, Station 1 observed the highest daily average concentration for both $PM_{2.5}$ and PM_{10}. Not only was Station 1 the closest to the power plant, but the wind rose analysis showed that some of the elevated concentrations potentially came directly from the power plant's coal pit, in addition to other PM sources. However, other elevated concentrations were observed coming from other directions, suggesting that there are several sources for particulate matter located in the region.

The IC analysis showed that the dominant species of particulate matter were primarily sulfate, nitrate and ammonium. The results of the XRF analysis showed that the primary constituents of PM_{10} and $PM_{2.5}$ were sulfur, silicon, aluminum and iron. While these are all constituents that are observed from coal combustion emissions, these constituents are also prominent in most particulate matter speciation and therefore cannot be contributed directly to emissions from the Cheswick Power Plant.

The low particulate matter emissions and results of the speciation analyses could be attributed to a number of factors. Because the study period was so short, it is likely that the local scale meteorological conditions led to bias in the results. In addition, it is also possible that there were errors in the instrumentation that led to such low concentrations, particularly for $PM_{2.5}$ concentrations observed at Station 2. Furthermore, the low concentrations of particulate matter observed at the sites could also be attributed to the emission control technology that is added to the Cheswick power plant, which includes wet lime flue gas desulfurization, low NO_x burner technology with separated over fire air selective catalytic reduction and an electrostatic precipitator (Available online: http://www.ampd.epa. gov/ampd), which would explain the reduced concentrations of pollutants observed in this study.

Based on the results of this study, is not possible to determine a concrete conclusion on the role of the Cheswick Power Plant in concentrations of particulate matter in the Cheswick and Springdale boroughs. These inconclusive results can be attributed to a number of factors, including unfavorable meteorological conditions, potential issues with the measurement equipment as well as an extremely short sampling period. It must be emphasized that this study was conducted over a limited time period. Therefore, it is recommended that further work be done on this matter, with longer sampling periods occurring in each season in order to capture a seasonal profile of concentrations of particulate matter in this region.

Acknowledgments: The authors acknowledge the support from the Sierra Club. We want to thank Marty Blake, Karen Petito, and Natalie Bick for their help with the monitoring stations. We also want to thank Karsten Baumann for help with the sample mass analysis, John Walker for help with the inorganic analysis of PM samples, and Dennis Mikel for his help with the X-Ray Fluorescence analysis of the PM samples. We would also like to thank Sean Nolan and the PA Department of Environmental Protection, Bureau of Air Quality, for their assistance in obtaining the data for the PA DEP monitoring sites. In addition, we would also like to thank Zachary Fabish for his assistance with this project.

Author Contributions: Viney Aneja was the PI for this project. Priya Pillai led the measurement process. Pornpan Uttamang helped take measurements for the project and helped in research for the manuscript. William Battye helped in the data analysis and in the revision and improvement of the manuscript. Casey Bray assisted in the measurement process, analyzed the data and wrote the manuscript.

Conflicts of Interest: The authors report no conflict of interest.

References

1. Dirty Kilowatts. America's most Polluting Power Plants 2006. Available online: http://www.dirtykilowatts. org/dirty_kilowatts.pdf (accessed on 15 September 2015).

2. Aneja, V.P.; Isherwood, A.; Morgan, P. Characterization of Particulate Matter (PM_{10}) Related to Surface Coal Mining Operations in Appalachia. *Atmos. Environ.* **2012**, *54*, 496–501. [CrossRef]

3. Nielsen, M.T.; Livbjerg, H.; Fogh, C.L.; Jensen, J.N.; Simonsen, P.; Lund, C.; Poulsen, K.; Sander, B. Formation and Emission of Fine Particles from Two Coal-Fired Power Plants. *Combust. Sci. Technol.* **2002**, *174*, 79–113. [CrossRef]

4. Tian, H.; Wang, Y.; Xue, Z.; Qu, Y.; Chai, F.; Hao, J. Atmospheric emissions estimation of Hg, As, and Se from coal-fired power plants in China. *Sci. Total Environ.* **2007**, *409*, 3078–3081. [CrossRef] [PubMed]

5. Contini, D.; Cesari, D.; Conte, M.; Donateo, A. Application of PMF and CMB Receptor Models for the Evaluation of the Contribution of a Large Coal-Fired Power Plant to PM_{10} Concentrations. *Sci. Total Environ.* **2016**, *560*, 131–140. [CrossRef] [PubMed]

6. Chow, J.C.; Watson, J.G. Review of $PM_{2.5}$ and PM_{10} Apportionment for Fossil Fuel Combustion and Other Sources by the Chemical Mass Balance Receptor Model. *Energy Fuels* **2002**, *16*, 222–260. [CrossRef]

7. Cohen, D.D.; Crawford, J.; Stelcer, E.; Atanacio, A.J. Application of Positive Matrix Factorization, Multi-Linear Engine and Back Trajectory Techniques to the Quantification of Coal–Fired Power Station Pollution in Metropolitan Sydney. *Atmos. Environ.* **2012**, *61*, 204–211. [CrossRef]

8. Gladney, E.S.; Small, J.A.; Gordon, G.E.; Zoller, W.H. Composition and Size Distribution of in-Stack Particulate Material at a Coal-Fired Power Plant. *Atmos. Environ.* **1976**, *10*, 1071–1077. [CrossRef]

9. Lee, S.W. Source Profiles of Particulate Matter Emissions from a Pilot-Scale Boiler Burning North American Coal Blends. *J. Air Waste Manag. Assoc.* **2001**, *51*, 1568–1578. [CrossRef]

10. Liu, X.; Xu, M.; Yao, H.; Yu, D.; Zhang, Z.; Lü, D. Characteristics and Composition of Particulate Matter from Coal-Fired Power Plants. *Sci. China Ser. E* **2009**, *52*, 1521–1526. [CrossRef]

11. Wu, H.; Pedersen, A.J.; Glarborg, P.; Frandsen, F.; Dam-Johansen, K.; Sander, B. Formation of Fine Particles in Co-Combustion of Coal and Solid Recovered Fuel in a Pulverized Coal-Fired Power Station. *Proc. Combust. Inst.* **2011**, *33*, 2845–2852. [CrossRef]

12. Young, G.S.; Fox, M.A.; Trush, M.; Kanarek, N.; Glass, T.A.; Curriero, F.C. Differential Exposure to Hazardous Air Pollution in the United States: A Multilevel Analysis of Urbanization and Neighborhood Socioeconomic Deprivation. *Int. J. Environ. Res. Public Health* **2012**, *9*, 2204–2225. [CrossRef] [PubMed]

13. Zhou, W.; Cohan, D.; Pinder, R.; Neuman, J.; Holloway, J.; Peischl, J.; Ryerson, T.; Nowak, J.; Flocke, F.; Zheng, W. Observation and Modeling of the Evolution of Texas Power Plant Plumes. *Atmos. Chem. Phys.* **2012**, *12*, 455–468. [CrossRef]

14. Anderson, H.R.; Atkinson, R.W.; Bremner, S.A.; Marston, L. Particulate air pollution and hospital admissions for cardiorespiratory diseases: Are the elderly at greater risk? *Eur. Pespir. J.* **2003**, *21*, 39–46. [CrossRef]

15. Congress.gov. Clean Air Act Amendments of 1990. Available online: https://www.congress.gov/bill/101st-congress/senate-bill/1630/amendments (accessed on 24 July 2017).

16. MacIntosh, D.; Spengler, J. Emissions of Hazardous Air Pollutants from Coal Fired Power Plants. 2011. Available online: http://www.lung.org/assets/documents/healthy-air/coal-fired-plant-hazards.pdf (accessed on 15 September 2015).

17. Pope, C.A., III; Dockery, D.W. Health Effects of Fine Particulate Air Pollution: Lines that Connect. *J. Air Waste Manag. Assoc.* **2006**, *56*, 709–742. [CrossRef]

18. Kelly, M.; Fischman, G. Point Source Emission Inventory Report. 2011. Available online: http://www.achd.net/air/pubs/pdf/2011_emissions_inventory_report.pdf (accessed on 15 September 2015).

19. Levin, I. America's Top Power Plant Toxic Polluters 2011. Available online: http://www.environmentalintegrity.org/documents/report-topuspowerplanttoxicairpolluters.pdf (accessed on 15 September 2015).

20. Nolan, S.; PA DEP. PADEP $PM_{2.5}$ Air Quality Monitoring Data. Personal Communication, 2017.

21. Finkelman, R.B. Modes of occurrence of environmentally-sensitive trace elements in coal. In *Environmental Aspects of Trace Elements in Coal; Anonymous*; Springer: Dordrecht, The Netherlands, 1995; pp. 24–50.

![atmosphere logo] *atmosphere*

Article

Comparing CMAQ Forecasts with a Neural Network Forecast Model for PM$_{2.5}$ in New York

Samuel D. Lightstone *, Fred Moshary and Barry Gross

Optical Remote Sensing Lab, City College of New York, New York, NY 10031, USA;
moshary@ccny.cuny.edu (F.M.); gross@ccny.cuny.edu (B.G.)
* Correspondence: slights01@citymail.cuny.edu

Received: 30 June 2017; Accepted: 26 August 2017; Published: 29 August 2017

Abstract: Human health is strongly affected by the concentration of fine particulate matter (PM$_{2.5}$). The need to forecast unhealthy conditions has driven the development of Chemical Transport Models such as Community Multi-Scale Air Quality (CMAQ). These models attempt to simulate the complex dynamics of chemical transport by combined meteorology, emission inventories (EI's), and gas/particle chemistry and dynamics. Ultimately, the goal is to establish useful forecasts that could provide vulnerable members of the population with warnings. In the simplest utilization, any forecast should focus on next day pollution levels, and should be provided by the end of the business day (5 p.m. local). This paper explores the potential of different approaches in providing these forecasts. First, we assess the potential of CMAQ forecasts at the single grid cell level (12 km), and show that significant variability not encountered in the field measurements occurs. This observation motivates the exploration of other data driven approaches, in particular, a neural network (NN) approach. This approach makes use of meteorology and PM$_{2.5}$ observations as model predictors. We find that this approach generally results in a more accurate prediction of future pollution levels at the 12 km spatial resolution scale of CMAQ. Furthermore, we find that the NN is able to adjust to the sharp transitions encountered in pollution transported events, such as smoke plumes from forest fires, more accurately than CMAQ.

Keywords: air quality model; Air Quality System (AQS); Community Multi-Scale Air Quality (CMAQ) model; fine particulate matter (PM$_{2.5}$); Aerosol Optical Depth (AOD)

1. Introduction

Fine particulate matter air pollution (PM$_{2.5}$) is an important issue of public health, particularly for the elderly and young children. The study by Pope et al. suggests that exposure to high levels of PM$_{2.5}$ is an important risk factor for cardiopulmonary and lung cancer mortality [1,2]. Furthermore, increased risk of asthma, heart attack and heart failure have been linked to exposure to high PM$_{2.5}$ concentrations [3].

PM$_{2.5}$ levels are dynamic and can fluctuate dramatically over different time scales. In addition to local emission sources, pollution events can be the result of aerosol plume transport and intrusion into the lower troposphere. When there is a potential high pollution event, the local air quality agencies must alert the public, and advise the population on proper safety measures, as well as direct the reduction of emission producing activities. Therefore, accurately measuring and predicting fine particulate levels is crucial for public safety.

The U.S. Environmental Protection Agency (EPA) established the National Ambient Air Quality Standards (NAAQS), which regulate levels of pollutants such as fine particulate matter. The New York State Department of Environment Conservation (NYSDEC) operates ground stations for monitoring PM$_{2.5}$ and speciation throughout NY State [4]. However, surface sampling is expensive and existing networks are limited and sparse. This results in data gaps that can affect the ability to forecast $PM_{2.5}$ over a 24-h period. The EPA developed the Models-3 Community Multi-scale Air Quality system (CMAQ), to provide 24–48 h air quality forecasts. CMAQ provides an investigative tool to explore proper emission control strategies. CMAQ has been the standard for modeling air pollution for nearly two decades because of its ability to independently model different pollutants while describing the atmosphere using "first-principles" [5].

In their studies, McKeen et al. and Yu et al. evaluate the accuracy of CMAQ forecasts [6,7]. To do so, they use the CMAQ 1200 UTC (Version 4.4) forecast model. They observe the midnight-to-midnight local time forecast and compare the hourly and daily average forecasts to the ground monitoring stations. McKeen et al. [6] observed minimal diurnal variations of $PM_{2.5}$ at urban and suburban monitor locations, with a consistent decrease of PM values between 0100 and 0600 local time. However, the CMAQ model showed significant diurnal variations, leading McKeen et al. to conclude that aerosol loss during the late night and early morning hours has little effect on $PM_{2.5}$ concentrations, while the CMAQ model does not account for this. Therefore, in addition to testing the hourly CMAQ forecast for a 24-h period, we focus on the daytime window for two reasons: (1) to assess the accuracy of CMAQ when aerosols do not play a reduced roll in forecasting; (2) the forecast should predict the air quality during the time of maximum human exposure.

While these studies make a distinction between rural and urban locations, they take the average results for all rural and urban locations respectively; thereby, their assessment of the CMAQ model was as at a regional scale, rather than a localized one. In addition to regional emissions, these studies also considered extreme pollution events such as the wildfires in western Canada and Alaska, which occurred during the observation period for the studies by Yu et al. and McKeen et al. The results of this assessment concluded that due to insufficient representation of transport pollution associated with the burning of biomass, CMAQ significantly under predicted the $PM_{2.5}$ values for these events.

In the study by Huang et al. [8], the bias corrected CMAQ forecast was assessed for both the 0600 and 1200 UTC release times. The study revealed a general improvement of forecasting skill for the CMAQ model. However, it was observed that the bias correction was limited in predicting extreme events, such as wildfires, and new predictors must be included in the bias correction to predict these events. In this study, CMAQ was assessed as a regional forecasting tool, taking 551 sites, and evaluating the average results in six sub-regions.

In our present assessment of the current operational CMAQ forecast model (Version 4.6), we differ from the regional studies above in the following ways: Firstly, in addition to the 1200 UTC forecast, we evaluated the 0600 UTC forecast for the same period to determine if release time affects the CMAQ forecast. Second, we focused on specific locations, both rural and urban, to assess the potential of CMAQ as a localized forecasting tool. In addition, we revisited the forecast potential of CMAQ for high pollution events, to determine if these events are generally caused by transport, or by local emissions. Finally, we tailor the forecast comparisons to focus on the potential of providing next day forecasts using data prior to 5 p.m. of the previous day, since this is an operational requirement for the state environmental agencies.

In focusing on both rural and urban areas in New York State, previous studies have shown anomalies in $PM_{2.5}$ from CMAQ forecasts. For example, in [9], using CMAQ (Version 4.5) with various planetary boundary layer (PBL) parameterizations, $PM_{2.5}$ forecasts during the summer pre-dawn and post-sunset periods were often highly overestimated in New York City (NYC). Further analysis of these cases demonstrated that the most significant error was the retrieval of the PBL height, which was often compressed by the CMAQ model, and did not properly take into account the Urban Heat Island mechanisms that expand the PBL layer [10]. This study showed the importance of PBL height dynamics and meteorological factors that motivated the choice of meteorological forecast inputs used during the NN development.

The objective of this paper is to determine the best method to forecast $PM_{2.5}$ by direct comparison with CMAQ output products. In particular, using the CMAQ forecast model, as a baseline, we explore the performance of a NN based data driven approach with suitable meteorological and prior $PM_{2.5}$ input factors.

Paper Structure

Our present paper is organized in the following manner: In Section 2, we analyze CMAQ as the baseline forecaster. We briefly describe the CMAQ model and the forecast schedules that are publically available, as well as the relevant ground stations we use for comparison. We then describe and perform a number of statistical tests using both the direct, as well as the bias compensated, CMAQ outputs. In this section, we show the large dispersion in using the direct results without bias correction.

In Section 3, we present our NN data driven strategy. This includes a description of all the relevant input factors used, including a combination of present and predicted meteorology, as well as diurnal trends of prior $PM_{2.5}$ levels. We present our first statistical results for the comparisons between CMAQ and the NN for a variety of experiments in order to highlight the conditions in which the NN results are generally an improvement. Then we explore the forecast performance for high pollution multiday transport events, which result in the highest surface $PM_{2.5}$ levels during the observed time period. In this comparison, analyzed by combining a sequence of next-day forecasts together, we find that the neural network seems to follow the trends in $PM_{2.5}$ more accurately than the CMAQ model.

In Section 4, we summarize our results and describe potential improvements.

2. CMAQ Local and Regional Assessment

2.1. Datasets

2.1.1. Models

The CMAQ V4.6 (CB05 gas-phase chemistry) with 12 km horizontal resolution was used for this paper. The CMAQ product for meteorology predictions used is the North American Model Non-hydrostatic Multi-scale Model (NAM-NMMB). This version was made available starting February 2016. The CMAQ data used for this paper is from 1 February 2016 until 31 October 2016. The station names and locations are listed in Table A1. The data can be accessed from [11], and the model description can be found in [12,13].

The CMAQ model used has a few different configurations: release times of 0600 UTC and 1200 UTC, and each release time has a standard forecast as well as a bias corrected forecast. The analog ensemble method is used for bias corrections. The idea is to look at similar weather patterns for the forecast period, and statistically correct the numerical $PM_{2.5}$ forecast based on historical errors. The analog ensemble method is described in detail in Huang et al. [8]. For each release time, CMAQ provides a 48-h forecast. The release time of 0600 UTC and 1200 UTC (2 a.m. and 8 a.m. EDT) does not give the public enough time to react to the forecast on the same day as the release. Consequently, for the 0600 UTC release time, the forecast hours 22–45 were used, and for the release time of 1200 UTC the forecast hours of 16–39 were used. This allowed us to construct a complete 24-h diurnal period for the forecast time window, which facilitated comparison with the field station data.

2.1.2. Ground-Based Observations

$PM_{2.5}$ ground data is collected from the EPA's AirNow, which collects NYSDEC monitoring station measurements in real time. The station data used for the forecast experiments in this article are from the New York State stations listed in Table A2, from 1 January 2011 until 31 December 2016. To assess the accuracy of CMAQ model forecasts, matching the model to the ground monitoring station is necessary. To do this, we use the ground NYSDEC stations that lay within the CMAQ grid cell only. Ground stations that are not found in a CMAQ grid cell were not used for comparison; therefore, no spatial interpolation was done on the model results while mapping the model or meteorological data to the AirNow ground stations. This matching method is widely used for comparing the CMAQ model to ground monitoring stations [6,7,14]. The locational data-points are depicted in Figure A1, the CMAQ grid cell information can be found in Table A1, and the NYSDEC station information can be found in Table A2.

2.2. Methods

Assessing Accuracy of CMAQ Forecasting Models

The forecasting skill of the different models were evaluated by computing the R^2 and the root mean square (RMSE) values from a regression analysis comparing the model to the AirNow observations. High R^2 values and low RMSE values indicated a good match between the prediction and the observations. Finally, to directly assess potential biases in the regression assessment, residual plots are provided to show significant concentration bias.

2.3. Results

2.3.1. Effects of Bias and Release Time

Figure 1 shows the regression plots for the hourly CMAQ model output compared to the ground station data for the City College of New York (CCNY Station) to illustrate the general behavior of the CMAQ model, and how the forecast is affected by different forecast release times, and by the bias corrections applied. The results of the R^2 analysis for all ground stations can be found in the supplementary materials.

All forecasts from the CMAQ model over CCNY have a positive correlation to the ground data. The effect on the forecast for different release times, if any, is minimal.

As seen in Figure 1a,c, the standard model generally overestimates the ground. While the bias correction improves the over-prediction, the results are more dispersed. This can be verified from the fact that the bias correction decreases the root mean square error (RMSE), but it also decreases the R^2 value for both release times.

In Figure 1 we assess the overall skill for a 24-h CMAQ forecast. In Figure 2, we determine if the CMAQ model could be improved by simply moving the forecast release time to a later point in the day, thereby including the most up-to-date inputs in the model. To do this, we make a direct comparison between CMAQ forecasts with different release times. In Figure 2, the R^2 value is computed for each hour of the day. The release time of 0600 UTC, with forecast hours of 22–45, is compared to the 1200 UTC release time, with forecast hours 16–39, to determine if the lower number of forecast hours yields more accurate predictions. It is clear from Figure 2 that the later release time does not lead to a significant improvement in the accuracy of the forecast, and this is true for both urban and non-urban test sites.

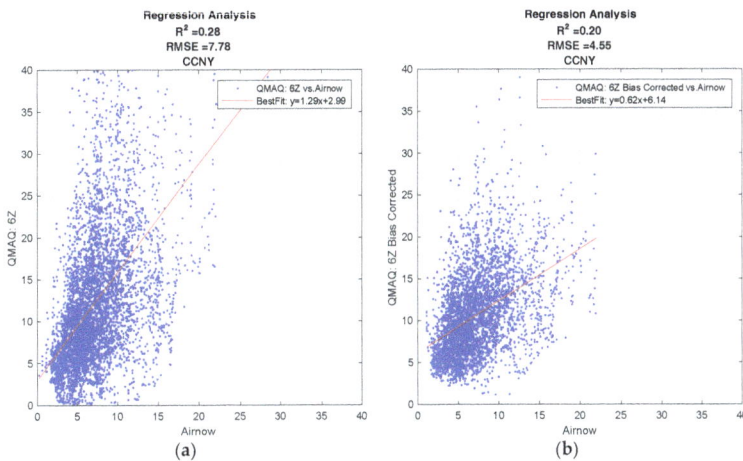

Figure 1. *Cont.*

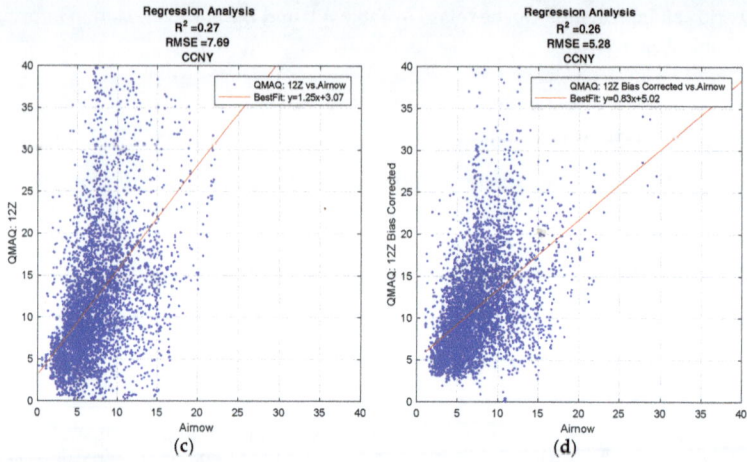

Figure 1. Community Multi-Scale Air Quality (CMAQ) regression analysis. (**a**) Standard, 06Z release time; (**b**) Bias Corrected, 06Z release time; (**c**) Standard, 12Z release time; (**d**) Bias Corrected, 12Z release time.

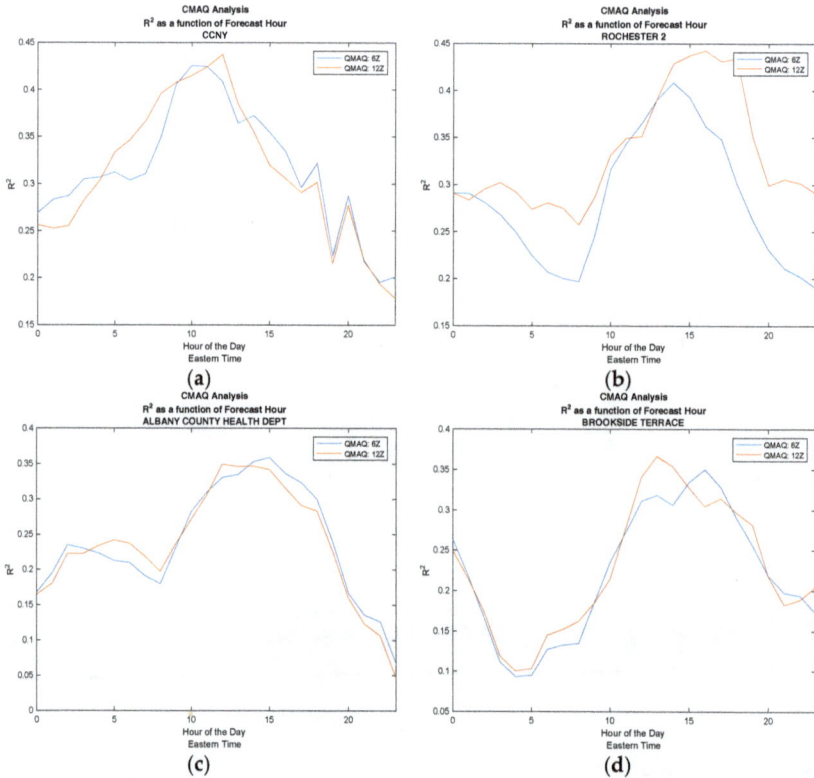

Figure 2. Comparing the effect of different release times for CMAQ by plotting the R^2 value as a function of time of day. (**a**) City College of New York (CCNY); (**b**) Rochester; (**c**) Albany; (**d**) Brookside Terrace.

It can be seen from this analysis that the CMAQ model performs best for midday hours, which is reasonable, since this is the period when convective mixing is most dominant. As discussed in reference [9], PBL modeling is very complex during the predawn/post-sunset period and errors in the PBL height clearly are a significant concern for further model development.

2.3.2. Differences between Urban and Non-Urban Locations

To get a better understanding of the spatial performance of the model, a multi-year time-series of daily averaged $PM_{2.5}$ observations from ground monitoring is used to compare the relationship between $PM_{2.5}$ values in New York City to the rest of New York State. Figure 3a is the regression analysis for this time period, and shows how the $PM_{2.5}$ values for NYC are strongly correlated to non-NYC areas, R^2 ~0.6. This indicates that while $PM_{2.5}$ values in NYC are generally higher than the rest of the state, the $PM_{2.5}$ level in NYC are still correlated to the levels in the rest of the state.

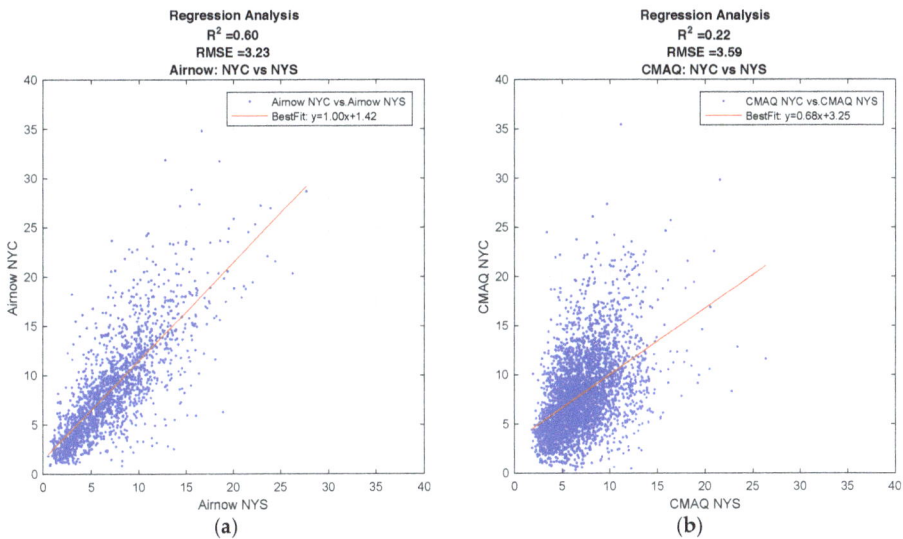

Figure 3. Regression analysis comparing $PM_{2.5}$ levels between NYC and the rest of NYS (non-NYC sites). (**a**) Multi-year day-averaged $PM_{2.5}$ analysis from NYSDEC ground observations; (**b**) CMAQ model comparison between NYC and NYS.

The same analysis comparing NYC to the rest of NYS was done with CMAQ forecast values as seen in Figure 3b. In this case, the correlation between NYC and NYS is not so strong, R^2 ~0.2. From this analysis alone, we can only speculate the reason for a low correlation between CMAQ forecasts for NYC and the rest of NYS is due to strong spatial differences in the National Emission Inventory (NEI) entries. However, the strong correlation in ground observations between NYC and NYS shows that while urban source emission may be a significant cause for somewhat higher levels of $PM_{2.5}$, there is still a strong correlation between NYC and NYS, and an accurate forecasting model must take this into account.

The limitations of CMAQ forecasting on a local pixel level indicate that other approaches should be explored. In particular, we explore the potential of data-driven models for localized forecasting in the next section.

3. Data Driven (Neural Network) Development

3.1. Datasets

3.1.1. Ground-Based Observations

PM$_{2.5}$ data collected from NYSDEC ground-monitoring stations is used for inputs in the neural network. These are the same ground stations listed above, in Section 2.1.2.

3.1.2. Models

The meteorological data was collected from the National Centers for Environmental Prediction (NCEP) North American Regional Reanalysis (NARR). NARR has high-resolution reanalysis of the North American region, 0.3 degrees (32 km) at the lowest latitude, including assimilated precipitation. The NARR makes available 8-times-daily and monthly means respectively. The data collected for this paper is the 8-times-daily means for the duration 1 January 2011 until 31 December 2016. Figure A1 shows the proximity of the meteorological data and the CMAQ model outputs to the ground stations.

The NN network was created and tested using historical data. In this paper, meteorology "forecast" data refers to NARR data that was observed the day of the PM$_{2.5}$ forecast. "Observed" or "measured" meteorology refers to NARR data that was observed before the forecast release time.

3.2. Methods

3.2.1. Development of the Neural Network

As stated above, the accurate prediction of PM$_{2.5}$ values is crucial for air quality agencies, so that they could alert the public of the severity and duration of a high pollution event. Therefore, it is imperative that the forecast predictions are released to the public the day before the event. For this paper, we chose 5 p.m. as a target for the forecast release time. Therefore, we ensure that all the methods tested, utilize factors that are available to the state agency prior to 2100 UTC (5 p.m. EDT).

Input Selection Scenarios

The NN input includes the following NARR meteorological data: surface air temperature, surface pressure, planetary boundary layer height (PBLH), relative humidity, and horizontal wind (10 m). To account for the seasonal variations, the month is also used as an input in the neural network. The PM input variables for the NN are the PM$_{2.5}$ measurements averaged over a three-hour frequency to match the meteorological dataset. The NN output is the next day PM$_{2.5}$ values.

In order to optimize the performance of the neural network, preliminary tests were done to determine the optimum utilization of the meteorological input variables. These test were done to determine if the "forecast" or the "observed" meteorology, or a combination of the two, should be used as input variables.

The forecast time window is midnight-to-midnight EDT for the forecast day, while the time window with the observed data is midnight to 5 p.m. EDT the day the forecast is released.

For the PBLH, the forecast value is always used as the input. One NN design employed only the forecast meteorological values as inputs. The second design utilized a combination of the forecast and the observed data, by subtracting the eight observation datasets from the eight forecast datasets. This first NN architecture uses the meteorological values as predictors, while the second design uses metrological trends as predictors. We note that this comparison does not affect the number of inputs used, allowing for a direct comparison of information content.

In scenario 1, where only the MET forecasts are used, we use the following inputs, where *i* represents the indices for time windows for the observation day, and *j* represents the indices for time windows for the forecast day (from the NARR forecasts), the NN inputs design is:

$PM_{2.5}(i)$	$i = 1 : 5$	$time\ window\ (i) = (i-1) \times 3 : i \times 3$	(Field measurements)
$MET_{forecast}(j)$	$j = 1 : 8$	$time\ window\ (j) = (j-1) \times 3 : j \times 3$	(NARR Forecasts)
$PBLH(j)$	$j = 1 : 8$	$time\ window\ (j) = (j-1) \times 3 : j \times 3$	(NARR Forecasts)

In scenario 2, where the differential between the observation day and forecast day of the MET variables are used, the architecture for the NN inputs is:

$PM_{2.5}(i)$	$i = 1 : 5$	$time\ window\ (i) = (i-1) \times 3 : i \times 3$	(Field measurements)
$MET_{forecast}(j) -$	$j = 1 : 8$	$time\ window\ (j) = (j-1) \times 3 : j \times 3$	(NARR Forecasts)
$MET_{observed}(i)$	$i = 1 : 8$	$time\ window\ (i) = (i-1) \times 3 : i \times 3$	(NARR Observations)
$PBLH(j)$	$j = 1 : 8$	$time\ window\ (j) = (j-1) \times 3 : j \times 3$	(NARR Forecasts)

To show the robustness of the NN, the data used for training the neural networks came from 2011–2015 alone, while the network was tested with data from 2016. In both scenarios, the targets for the NN were taken to be the complete set of $PM_{2.5}$ over all time windows of the forecast day:

Targets: $PM_{2.5}(j)$	$j = 1 : 8$	$time\ window\ (j) = (j-1) \times 3 : j \times 3$	(Field measurements)

Neural Network Training Approach

In developing a NN $PM_{2.5}$ forecast for all of New York State (NYS), we needed to take into account the very different emission sources, and to a lesser extent the meteorological conditions, between New York City (NYC) and the other sites in NYS. We found that the best solution is to design two different neural networks. The first is trained only over NYC sites, while the second is trained for the rest of NYS. It is important to note that we do not try to build a unique NN for every station, since this is not a useful approach for local agencies. PM and Meteorological data from 2011–2015, were used for training.

For NYC, since the stations are very close to each other, the NN was trained with spatial mean values of the ground PM monitors and NARR meteorological datasets. For NYS, all the PM and meteorological data from each site outside of NYC were used. Some site-specific information was implicitly included by using the surface pressure as inputs, which provides some indicator of surface elevation.

The neural network was developed using the MATLAB Neural Network Toolbox [15]. The Levenberg-Marquardt network was deployed using 10 hidden nodes. The break down for the NN input data is: 70% training, 15% validation, and 15% testing. Because the sample set of training, validation, and testing is divided randomly over the entire dataset, accuracy of the NN was determined by testing each network over 2016 data only, a time window that was not included in training. Once the NN function was created, the 2016 meteorological and PM data was passed through the network, and the outputs were stored with the date-time and station location as indices.

Neural Network Scenario Results

Figure 4a shows the performance of the NN using the forecast metrological data as inputs, while Figure 4b shows the performance of the NN using the difference between the forecast and the current days measurements. The NN utilizing the difference configuration is clearly better, with a higher R^2 value, 0.44 compared to 0.36, and a lower root-mean-square value, 3.09 compared to 4.59. In addition, there are substantially less anomalous high $PM_{2.5}$ forecasts. Since this improvement was seen in all test cases, we only used scenario 2, (differential meteorology) NN configuration. From these results, we see that meteorological trends are better indicators of $PM_{2.5}$ than meteorology alone. This appears to us to be a reasonable result since the meteorology trend better isolates particular mesoscale conditions, which is known to be a significant factor in boundary layer dynamics.

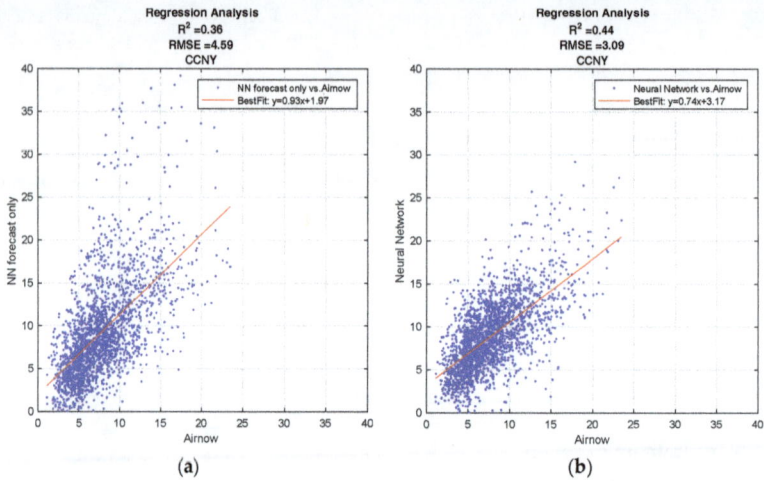

Figure 4. Results from the regression analysis to maximize Neural Network performance for the different scenarios. (**a**) NN designed with the forecast meteorological data (Scenario 1); (**b**) NN designed by taking the difference between the forecast and the current days measurements (Scenario 2).

3.3. Results

3.3.1. Neural Network and CMAQ Comparison

The R^2 value for CMAQ and the NN, both compared to AirNow observations, is computed for each forecast model and for each location. As a representative example of the overall performance, the R^2 value for NYC, represented by CCNY, is compared to NYS, represented by Brookside Terrace, a non-NYC, non-urban station, and these results are displayed in Figure 5. The individual results for each location can be found in the supplementary materials.

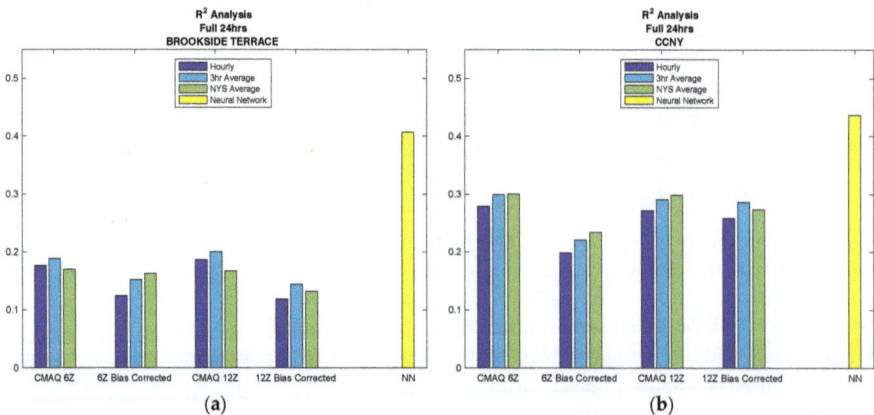

Figure 5. Regression analysis is computed for the comparison between AirNow observations and the various prediction models, for the complete CMAQ database time period (February 2016, through October 2016). The R^2 value for each model is plotted in the figure above to compare CMAQ to the NN. The CMAQ model includes the different release times as well as bias compensated vs. uncompensated runs. In addition, different time and spatial averaging of CMAQ is considered at each location. (**a**) Brookside Terrace, representative of non-NYC; (**b**) CCNY, representative of NYC.

From Figure 5 above, it can be seen that the most accurate forecast model is the neural network for both NYS and NYC over any of the CMAQ forecasts studied. Regarding CMAQ, we note better performance for NYC than for non-urban areas. This is in contrast to the neural network, where there is very little variation in the results for locations that are urban versus non-urban, indicating that locational inputs in the model, such as the surface pressure, improves forecasting skill.

In addition, for all cases, it can be seen that taking the time average improves the CMAQ results. Furthermore, the spatial averaging over NYS (with 1-h time sampling) shows more improvement in most NYC cases and some non-NYC cases as well. These results indicate the possibility that the best use for CMAQ forecasting is on a regional level. This is supported from the 12 km grid cell resolution for CMAQ, a cell size typical for regional analysis.

We note again that the different release times for CMAQ has almost no effect on the forecast accuracy. In Figure 6, we compared the diurnal performance of the NN to the CMAQ model. The most apparent result is the dramatic improvement of the NN during the night and morning hours, where the CMAQ model has the most difficulty. This is clearly due to the machine learning approach where the time differences, the inputs, and forecast periods have a dramatic effect on output performance.

This also explains the general downward trend, where performance tails off in the late afternoon and becomes closer to the CMAQ performance. This can be expected, because larger time delays should lead to more dispersion between the outputs and input PM levels.

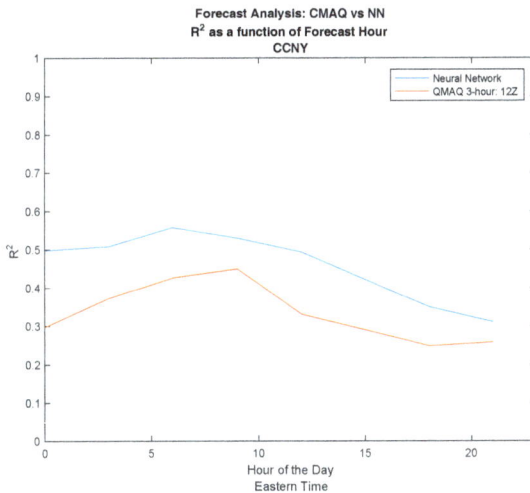

Figure 6. Comparing the effect of different release times for the NN in comparison to CMAQ by plotting the R^2 value as a function of time of day.

Figure 7 below shows the residual results for CMAQ in comparison to the neural network. For CMAQ, as noted above, there seems to exist a non-random bias pattern, where CMAQ generally over predicts for low and high PM values, and under predicts for medium values. This pattern seems to indicate that the CMAQ model may not capture all of the underlying variability factors. On the other hand, for the neural network, the behavior of the residuals is clearly stochastic in nature.

We find that an optimized NN approach generally results in a more accurate prediction of future pollution levels, as compared to CMAQ, for a single grid cell (resolution 12 km).

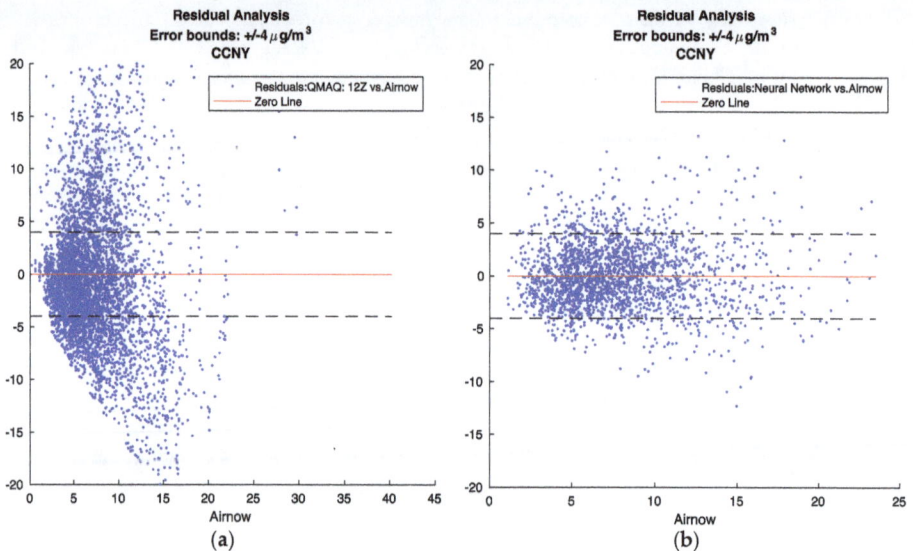

Figure 7. Residual analysis. The standard deviation of $PM_{2.5}$ from the AirNow ground monitoring sites was calculated to be 4 µg/m^3, therefore, +/−4 µg/m^3 was used for the error bounds (**a**) CMAQ; (**b**) Neural Network.

3.3.2. Heavy Pollution Transport Events

Because the neural network is data-driven, the network performs better when the most up-to-date inputs are used. This explains the degradation of performance with time, as seen in Figure 6. In the current design of the neural network, we only used five $PM_{2.5}$ inputs, instead of maximum possible in a 24-h period, eight. In the training of the NN, there were very few extreme event cases, $PM_{2.5} > 25$ µg/m^3. The lack of suitable training statistics for these events causes the NN approach to have difficultly in adjusting to the sharp contrast with the onset of the event.

Therefore, a second neural network was trained with the same design as the neural network illustrated above; however, this neural network produces a 24-h forecast at 5 p.m. for the time period, 5 p.m.–5 p.m. (instead of a next day 24-h midnight-to-midnight forecast). This neural network uses all eight PM measurements, because there is no lag time between the release time and the first forecast hour. This neural network, referred to as NN Continuous, was not used in the statistical analysis for the different forecast models (because the 24-h forecast period is different than the forecast analysis above), but is being explored in the extreme event cases. The reason for developing this continuous neural network is to determine if the continuous nature of the network produces better results in extreme pollution events.

To explore the behavior of the different models under high pollution transport conditions, the forecasts coinciding with the wildfires of Fort McMurray in Alberta, Canada were analyzed. The wildfire started on 1 May 2016, and was declared under control on 5 July 2016. Although the wildfire lasted for over two months, evidence of increased $PM_{2.5}$ surface levels in NYC resulting from the wildfire were detected on 9 May, and on 25 May. On these dates, instances of aloft plume intrusions and the mixing down into the planetary-boundary layer were observed by a ceilometer and a Raman-Mie Lidar [16]. In Figure 8, we plot the CMAQ and NN model forecasts, focusing on the transport intrusions into NYC on 25 May.

The first thing to notice in Figure 8a, is the oscillations in the CMAQ model, and to notice how these oscillations smooth out in Figure 8b,c, where the three-hour time average and the New York State spatial average are tested respectively. It is logical that for heavy transport cases, domain averaging helps decrease oscillations; however, we still see significant underestimation of the event.

This is the first case where we analyze the behavior of the continuous neural network. Looking at Figure 8c,d, it is clear that the continuous neural network is able to respond to the trend of the high pollution event faster, and more accurately, then the standard neural network.

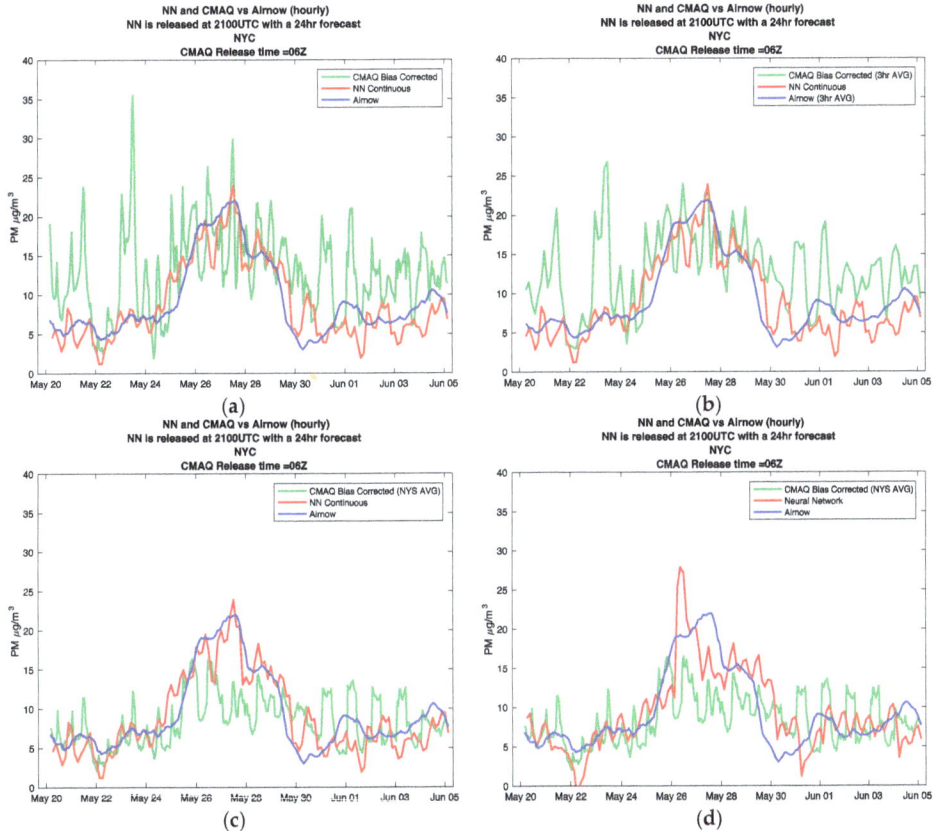

Figure 8. NYC surface $PM_{2.5}$ levels affected from the wildfires of Alberta Canada 2016. The plots focus on the aloft plumes mixing down into the PBL on May 25. The plots show different models vs. AirNow observation (**a**) CMAQ Biased hourly, NN continuous; (**b**) CMAQ Biased 3-h average, NN continuous; (**c**) CMAQ Biased state average, NN continuous; (**d**) CMAQ Biased state average, standard Neural Network.

4. Conclusions

In this paper, we first made a baseline assessment of the V4.6 CMAQ forecasts, and found significant dispersion as well as a tendency for the model to overestimate the ground truth field measurements. Even in the bias corrected case, the residuals error in the model was found to have significant bias patterns, indicating that there are predictors not included in the model that could significantly improve the results.

These results motivated the development of data driven approaches such as a NN. In developing a data driven NN next day forecast model, we found a general improvement of performance when using prior $PM_{2.5}$ inputs together with the difference between present and next day meteorological parameter forecasts. This "differential NN" approach performed significantly better than if we used only the future forecast variables, indicating that meteorological pattern trends are important indicators.

Using this NN architecture, we then made extensive regression based comparisons between CMAQ next day forecast models and regionally trained NN next day forecasts for the NYS and NYC regions. In general, we found that the NN results are a significant improvement over the CMAQ forecasts in all cases. These comparisons were made to be consistent with state agencies where forecasts should be available by 5 p.m. In addition, we also made a diurnal comparison, which illustrated that; the NN approach had superior forecasting skills during the early part of the day but degraded smoothly as the forecast time increased. By mid-day, the differences between the two approaches was much closer.

To improve the CMAQ forecasts, we found limited improvement when spatial averaging is extended beyond the single pixel 12 km resolution to all of New York State. Even in this case, the NN results were generally more accurate.

Finally, we focused on forecast performance for transported high pollution events such as Canadian wildfires. In these cases, we found that the CMAQ forecasts had large temporal fluctuations, which could hide most of the event. In this case, significant improvement was obtained when using state averaged bias corrected outputs; however, in general, the smoothed results underestimate the local $PM_{2.5}$ measurements.

In this application, we found the neural network approach provides a reasonably smooth forecast, although the transition from a clean state to a polluted state is very poor. Nevertheless, the standard NN performed better than CMAQ in this scenario. Further improved results for the NN were obtained in the transition period when the forecast time of the NN was reduced (NN continuous), making the transition from training to testing continuous.

Future Work

While the continuous NN does adjust quickly to the sharp contrast in transport events, this design limits the scope of the forecast period. Clearly, local data alone is not ideal for this application. Non-local data that can identify high pollution events and assesses their potential mixing with our region is needed. As a preliminary analysis, we explored the use of a combination of HYSPLIT Air Parcel Trajectories with GOES satellite Aerosol Optical Depth (AOD) retrievals to improve the NN. In particular, we analyzed the use of these tools to quantify the relative AOD levels for all air parcels that reach our target area. We found that by properly counting the trajectories weighted by the AOD, a good correlation was seen between the relative AOD and the $PM_{2.5}$ levels. Therefore, we believe that using the relative AOD metric as an additional input factor can make improvements in the NN approach. When GOES-R AOD retrievals, with high data latency and multispectral inversion capabilities [17,18], become available, we plan to incorporate these AOD metrics as predictors in the NN.

Supplementary Materials: The following are available online at www.mdpi.com/2073-4433/8/9/161/s1, Figure S1: Regression Analysis.

Acknowledgments: The authors gratefully acknowledge the NOAA Air Resources Laboratory (ARL) for the provision of the HYSPLIT transport and dispersion model used in this publication. We further gratefully acknowledge Jeff McQueen for providing the V4.6 CMAQ forecasts. Finally, S. Lightstone would like to acknowledge partial support of this project through a NYSERDA EMEP Fellowship award.

Author Contributions: Samuel Lightstone was responsible for the NN experiments and the CMAQ assessment; Lightstone also wrote the paper. Barry Gross was responsible for design of the different experiments as well as conceiving the use of the GOES AOD and HYSPLIT trajectories to estimate transported AOD as a predictor of $PM_{2.5}$. Fred Moshary provided significant critical assessment and suggestions of all results.

Conflicts of Interest: The authors declare no conflict of interest.

Appendix A. Datasets

Table A1. CMAQ Grid Cell Information.

Name	Abbreviation	Latitude	Longitude	Land Type
Amherst	AMHT	42.99	−78.77	Suburban
CCNY	CCNY	40.82	−73.95	Urban
Holtsville	HOLT	40.83	−73.06	Suburban
IS 52	IS52	40.82	−73.90	Suburban
Loudonville	LOUD	42.68	−73.76	Urban
Queens College 2	QC2	40.74	−73.82	Suburban
Rochester Pri 2	RCH2	43.15	−77.55	Urban
Rockland County	RCKL	41.18	−74.03	Rural
S. Wagner HS	WGHS	40.60	−74.13	Urban
White Plains	WHPL	41.05	−73.76	Suburban

Table A2. NYSDEC Station Information.

NYSDEC ID	Station Name	Latitude	Longitude	Land Type
360010005	Albany County Health Dept	42.6423	−73.7546	Urban
360050112	IS 74	40.8155	−73.8855	Suburban
360291014	Brookside Terrace	42.9211	−78.7653	Suburban
360551007	Rochester 2	43.1462	−77.5482	Urban
360610135	CCNY	40.8198	−73.9483	Urban
360810120	Maspeth Library	40.7270	−73.8931	Suburban
360850055	Freshkills West	40.5802	−74.1983	Suburban
360870005	Rockland County	41.1821	−74.0282	Rural
361030009	Holtsville	40.8280	−73.0575	Suburban
361192004	White Plains	41.0519	−73.7637	Suburban

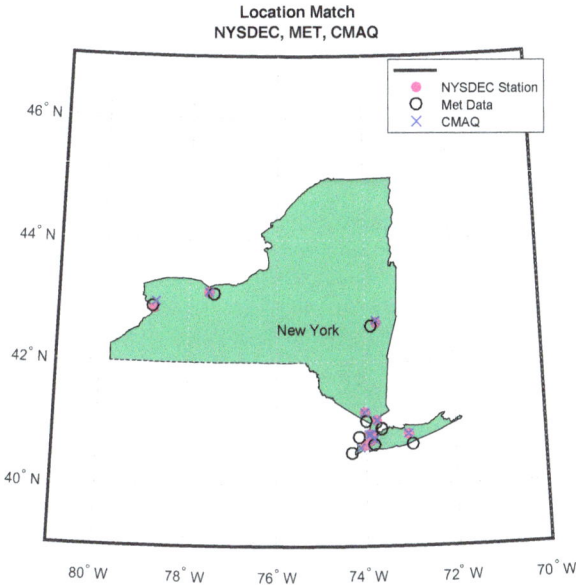

Figure A1. This map shows the proximity of the ground NYSDEC stations to the NARR meteorological data, and the CMAQ forecast data.

References

1. Pope, C.A., III; Dockery, D.W. Health effects of fine particulate air pollution: Lines that connect. *J. Air Waste Manag. Assoc.* **2006**, *56*, 709–742. [CrossRef] [PubMed]
2. Pope, C.A., III; Burnett, R.T.; Thun, M.J.; Calle, E.E.; Krewski, D.; Ito, K.; Thurston, G.D. Lung cancer, cardiopulmonary mortality, and long-term exposure to fine particulate air pollution. *JAMA* **2002**, *287*, 1132–1141. [CrossRef] [PubMed]
3. Weber, S.A.; Insaf, T.Z.; Hall, E.S.; Talbot, T.O.; Huff, A.K. Assessing the impact of fine particulate matter (PM$_{2.5}$) on respiratory-cardiovascular chronic diseases in the New York City Metropolitan area using Hierarchical Bayesian Model estimates. *Environ. Res.* **2016**, *151*, 399–409. [CrossRef] [PubMed]
4. Rattigan, O.V.; Felton, H.D.; Bae, M.; Schwab, J.J.; Demerjian, K.L. Multi-year hourly PM$_{2.5}$ carbon measurements in New York: Diurnal, day of week and seasonal patterns. *Atmos. Environ.* **2010**, *44*, 2043–2053. [CrossRef]
5. Byun, D.; Schere, K.L. Review of the governing equations, computational algorithms, and other components of the Models-3 Community Multiscale Air Quality (CMAQ) modeling system. *Appl. Mech. Rev.* **2006**, *59*, 51–77. [CrossRef]
6. McKeen, S.; Chung, S.H.; Wilczak, J.; Grell, G.; Djalalova, I.; Peckham, S.; Gong, W.; Bouchet, V.; Moffet, R.; Tang, Y.; et al. Evaluation of several PM$_{2.5}$ forecast models using data collected during the ICARTT/NEAQS 2004 field study. *J. Geophys. Res. Atmos.* **2007**, *112*. [CrossRef]
7. Yu, S.; Mathur, R.; Schere, K.; Kang, D.; Pleim, J.; Young, J.; Tong, D.; Pouliot, G.; McKeen, S.A.; Rao, S.T. Evaluation of real-time PM$_{2.5}$ forecasts and process analysis for PM$_{2.5}$ formation over the eastern United States using the Eta-CMAQ forecast model during the 2004 ICARTT study. *J. Geophys. Res. Atmos.* **2008**, *113*. [CrossRef]
8. Huang, J.; McQueen, J.; Wilczak, J.; Djalalova, I.; Stajner, I.; Shafran, P.; Allured, D.; Lee, P.; Pan, L.; Tong, D.; et al. Improving NOAA NAQFC PM$_{2.5}$ Predictions with a Bias Correction Approach. *Weather Forecast.* **2017**, *32*, 407–421. [CrossRef]
9. Doraiswamy, P.; Hogrefe, C.; Hao, W.; Civerolo, K.; Ku, J.Y.; Sistla, G. A retrospective comparison of model-based forecasted PM$_{2.5}$ concentrations with measurements. *J. Air Waste Manag. Assoc.* **2010**, *60*, 1293–1308. [CrossRef] [PubMed]
10. Gan, C.-M.; Wu, Y.; Madhavan, B.L.; Gross, B.; Moshary, F. Application of active optical sensors to probe the vertical structure of the urban boundary layer and assess anomalies in air quality model PM$_{2.5}$ forecasts. *Atmos. Environ.* **2011**, *45*, 6613–6621. [CrossRef]
11. Files in /mmb/aq/sv/grib. Available online: http://www.emc.ncep.noaa.gov/mmb/aq/sv/grib/ (accessed on 1 December 2016).
12. NCEP Operational Air Quality Forecast Change Log. Available online: http://www.emc.ncep.noaa.gov/mmb/aq/AQChangelog.html (accessed on 1 May 2017).
13. Lee, P.; McQueen, J.; Stajner, I.; Huang, J.; Pan, L.; Tong, D.; Kim, H.; Tang, Y.; Kondragunta, S.; Ruminski, M.; et al. NAQFC developmental forecast guidance for fine particulate matter (PM$_{2.5}$). *Weather Forecast.* **2017**, *32*, 343–360. [CrossRef]
14. McKeen, S.; Grell, G.; Peckham, S.; Wilczak, J.; Djalalova, I.; Hsie, E.; Frost, G.; Peischl, J.; Schwarz, J.; Spackman, R.; et al. An evaluation of real-time air quality forecasts and their urban emissions over eastern Texas during the summer of 2006 Second Texas Air Quality Study field study. *J. Geophys. Res. Atmos.* **2009**, *114*. [CrossRef]
15. Demuth, H.; Beale, M. *MATLAB and Neural Network Toolbox Release 2016a*; The MathWorks, Inc.: Natick, MA, USA, 1998.
16. Wu, Y.; Pena, W.; Diaz, A.; Gross, B.; Moshary, F. Wildfire Smoke Transport and Impact on Air Quality Observed by a Multi-Wavelength Elastic-Raman Lidar and Ceilometer in New York City. In Proceedings of the 28th International Laser Radar Conference, Bucharest, Romania, 25–30 June 2017.

Atmosphere **2017**, *8*, 161

17. Anderson, W.; Krimchansky, A.; Birmingham, M.; Lombardi, M. The Geostationary Operational Satellite R Series SpaceWire Based Data System. In Proceedings of the 2016 International SpaceWire Conference (SpaceWire), Yokohama, Japan, 25–27 October 2016.
18. Lebair, W.; Rollins, C.; Kline, J.; Todirita, M.; Kronenwetter, J. Post launch calibration and testing of the Advanced Baseline Imager on the GOES-R satellite. In Proceedings of the SPIE Asia-Pacific Remote Sensing, New Delhi, India, 3–7 April 2016.

atmosphere

MDPI

Article

Multi-Year (2013–2016) PM$_{2.5}$ Wildfire Pollution Exposure over North America as Determined from Operational Air Quality Forecasts

Rodrigo Munoz-Alpizar [1,*], Radenko Pavlovic [1], Michael D. Moran [2], Jack Chen [2],
Sylvie Gravel [2], Sarah B. Henderson [3], Sylvain Ménard [1], Jacinthe Racine [1], Annie Duhamel [1],
Samuel Gilbert [1], Paul-André Beaulieu [1], Hugo Landry [1], Didier Davignon [1], Sophie Cousineau [1]
and Véronique Bouchet [1]

[1] Canadian Meteorological Centre Operations, Environment and Climate Change Canada,
    2121 Rte Transcanadienne, Montreal, QC H9P 1J3, Canada; radenko.pavlovic@canada.ca (R.P.);
    sylvain.menard@canada.ca (S.M.); jacinthe.racine@canada.ca (J.R.); annie.duhamel@canada.ca (A.D.);
    samuel.gilbert@canada.ca (S.G.); paul-andre.beaulieu@canada.ca (P.-A.B.); hugo.landry@canada.ca (H.L.);
    didier.davignon@canada.ca (D.D.); sophie.cousineau@cafnada.ca (S.C.); veronique.bouchet@canada.ca (V.B.)
[2] Air Quality Research Division, Environment and Climate Change Canada, Toronto, ON M3H 5T4, Canada;
    mike.moran@canada.ca (M.D.M.); jack.chen@canada.ca (J.C.); sylvie.gravel@canada.ca (S.G.)
[3] Environmental Health Services, British Columbia Centre for Disease Control, Vancouver, BC V5Z 4R4,
    Canada; sarah.henderson@bccdc.ca
*   Correspondence: rodrigo.munoz-alpizar@canada.ca; Tel.: +1-514-421-7266

Received: 30 June 2017; Accepted: 14 September 2017; Published: 19 September 2017

Abstract: FireWork is an on-line, one-way coupled meteorology–chemistry model based on near-real-time wildfire emissions. It was developed by Environment and Climate Change Canada to deliver operational real-time forecasts of biomass-burning pollutants, in particular fine particulate matter (PM$_{2.5}$), over North America. Such forecasts provide guidance for early air quality alerts that could reduce air pollution exposure and protect human health. A multi-year (2013–2016) analysis of FireWork forecasts over a five-month period (May to September) was conducted. This work used an archive of FireWork outputs to quantify wildfire contributions to total PM$_{2.5}$ surface concentrations across North America. Different concentration thresholds (0.2 to 28 $\mu g/m^3$) and averaging periods (24 h to five months) were considered. Analysis suggested that, on average over the fire season, 76% of Canadians and 69% of Americans were affected by seasonal wildfire-related PM$_{2.5}$ concentrations above 0.2 $\mu g/m^3$. These effects were particularly pronounced in July and August. Futhermore, the analysis showed that fire emissions contributed more than 1 $\mu g/m^3$ of daily average PM$_{2.5}$ concentrations on more than 30% of days in the western USA and northwestern Canada during the fire season.

Keywords: air quality modeling; wildfire smoke; fine particulate matter; wildfire pollution exposure

---

1. Introduction

Wildfires are large, uncontrolled vegetation fires that result from natural processes or anthropogenic activities. In North America they are a major natural hazard, with high interannual variability in both the number of fires and the total burned area. Every year, wildfires consume millions of hectares of forest in North America, resulting in several community evacuations due to the direct threat of fire or the indirect threat of heavy smoke [1]. According to the 2016 report of the Canadian Interagency Forest Fire Centre (CIFFC), during the last decade an average of 7000 wildfires occurred each year in Canada and burned an average of 2.6 million hectares per year [2]. Annual costs of wildfire suppression in Canada have ranged from about $0.5 billion to $1 billion in the last decade [3]. As an extreme example, one wildfire in western Canada from 1 May to 4 July 2016 burned an area of 590,000 ha, roughly the size of the Canadian province of Prince Edward Island, and forced the evacuation of nearly 90,000 people from the city of Fort McMurray in northeastern Alberta. The damages caused by this fire were estimated to be on the order of $9.5 billion [4].

In addition to economic impacts, wildfires can adversely affect both air quality (AQ) and human health. The AQ impacts depend on the amount and chemical composition of the emissions from these fires, the smoke plume dynamics, and the meteorological conditions that drive the transport and diffusion of wildfire smoke. Biomass burning from wildfires can release significant amounts of pollutants into the atmosphere, including particulate matter (PM), ammonia (NH_3), and ozone (O_3) precursors such as nitrogen oxides (NOx), volatile organic compounds (VOCs), and carbon monoxide (CO) [5]. Although many of these species are harmful to human health, the population health impacts of wildfire smoke have been attributed mainly to short-term concentrations of PM less than 2.5 µm in aerodynamic diameter ($PM_{2.5}$). Two recent systematic reviews found that short-term smoke exposure is strongly associated with small increases in daily mortality from all causes, and with acute respiratory outcomes ranging in severity from increased reporting of symptoms through to increased risk of hospital admissions. Associations were weaker for acute cardiovascular morbidity and birth outcomes, but suggestive of effects in both cases [6,7]. In addition, new evidence about wildfire smoke is emerging rapidly given the severity of fires across North America over the past decade. Furthermore, wildfire smoke is playing an increasingly important role in long-term air pollution as fires get larger and other sources such as motor vehicles and industry come under increasingly strict regulation [8]. Long-term exposure to $PM_{2.5}$ is associated with the development of a wide and growing range of chronic diseases [9].

During wildfire events, $PM_{2.5}$ concentration at the ground level may be significantly increased, such that it exceeds the levels established by regulatory agencies to protect the environment and human health. In Canada, for example, the established Canadian Ambient Air Quality Standards (CAAQS) for $PM_{2.5}$ concentration are an annual mean of 10 µg/m^3 and a daily mean of 28 µg/m^3 [10]. The CAAQS metric for annual concentration of $PM_{2.5}$ is based on the three-year average of the annual average concentrations, and on the three-year average of the annual 98th percentile of the daily 24-h average concentrations. Furthermore, the AQ impacts of wildfire emissions are not limited to the local or regional scales. Under some meteorological conditions, wildfire smoke plumes can disperse widely and travel thousands of kilometers, affecting people living far away from the fire location [11]. Observational evidence indicates that the long-range transport of wildfire smoke can episodically increase PM and O_3 ground-level concentrations at regional and continental scales. For example, smoke from Canadian wildfires was associated with high concentrations of PM in areas great distances from the fire source, such as Baltimore and Washington, D.C. in the eastern USA and as far away as Europe [11–14]. Moreover, Canadian wildfires have also been linked to increased O_3 concentrations in Houston, TX and the northeastern USA, as well as Europe [15–19]. On the other hand, long-range transport of Siberian wildfire smoke has also contributed to exceedances in O_3 and $PM_{2.5}$ on the west coast of Canada [20,21].

The assessment of human exposure to smoke from wildfires is challenging because such smoke episodes are typically sporadic and short-lived, with highly variable concentrations in both space and time [22,23]. Furthermore, the spatial variability of population exposure to wildfire smoke cannot be correctly represented based solely on regulatory monitoring data because these data provide limited spatial coverage. For example, impacts are often observed in populated non-urban areas where regulatory monitoring networks are sparse or not available. On the other hand, remote sensing data from satellites can be used over very large areas, covering locations where the monitoring networks are missing. However, these measurements provide information about the total atmospheric column of air pollutants rather than the ground-level concentrations. They are also generally not available at night and can be masked by the occurrence of clouds. Additionally, satellite overpasses may occur only once a day or every few days, resulting in large amounts of missing information. Therefore, deterministic AQ forecast models have become a useful tool to fill the temporal and spatial gaps in available measurements and provide guidance about AQ over the coming hours and days [24–31].

AQ forecasting systems consisting of 3D numerical weather prediction (NWP) models with on-line or off-line chemical transport models (CTMs) have become a valuable tool in the past 15 years. They can provide guidance in the production of AQ forecasts and assist public health authorities in understanding pollutant exposures and developing public actions to protect populations against those exposures. The accuracy of pollutant exposure estimates using modeled AQ data depends on the ability of the forecast systems to reproduce observed concentrations of air pollutants. The differences among the current AQ forecast systems that consider anthropogenic emissions of pollutants have been reviewed recently [24,25], but to date only a few AQ forecast systems have been developed that combine information from wildfires and meteorology to retrospectively or prospectively estimate the emissions, transport, and diffusion of wildfire smoke [26,27].

In order to provide guidance to regional AQ forecasters, first responders, and public health decision-makers about the dispersion of smoke from large wildfires, Environment and Climate Change Canada (ECCC) has developed FireWork [27], an on-line, one-way coupled meteorology–chemistry model based on near-real-time wildfire emissions. FireWork was built on the existing ECCC operational AQ forecast system, and was first deployed in 2013 during the Canadian wildfire season to deliver real-time forecasts of wildfire smoke plumes over North America. Previous studies have shown the ability of FireWork to forecast $PM_{2.5}$ in terms of statistical scores and spatial distributions, as well as public health impacts [27–29]. Observed trends and seasonal variability of $PM_{2.5}$ are well captured by the model. However, FireWork's ability to simulate the emission and dispersion of wildfire smoke is currently limited by factors such as the accuracy of wildfire emission factors, the treatment of fire behavior, and the suitability of plume-rise algorithms.

Here we conduct a multi-year (2013–2016) analysis of FireWork forecasted $PM_{2.5}$ concentrations from biomass burning over North America to provide estimates of the population exposure to $PM_{2.5}$ from wildfires for several concentration thresholds. The number of days that exceed these thresholds as well as the magnitude of the area in exceedance was estimated. The goal of this work is to help public health professionals, policymakers, and the general public better understand the human health impacts of wildfire-related $PM_{2.5}$ pollution.

2. Methodology

2.1. North American Wildfire AQ Forecasting System

The FireWork system was first run in an experimental mode beginning in 2013 at ECCC's Canadian Centre for Meteorological and Environmental Prediction (CCMEP). The system became operational in April 2016. The FireWork system is identical to the ECCC operational Regional Air Quality Deterministic Prediction System (RAQDPS) [30–32], except for the inclusion of satellite-derived, near-real-time biomass burning emissions from natural, prescribed, and agricultural fires [27]. The on-line RAQDPS modeling system relies on the GEM-MACH (Global Environmental Multi-scale-Modelling Air quality and Chemistry) model, an on-line, one-way coupled CTM (i.e., meteorology affects chemistry, but chemistry does not affect meteorology), embedded within the Global Environmental Multi-scale (GEM) model. Both the RAQDPS and FireWork systems input the same hourly anthropogenic gridded emissions fields based on processing the 2010 Canadian national Air Pollutant Emission Inventory (APEI), the 2011 U.S. National Emissions Inventory (NEI), and the 1999 Mexican emissions inventory, as well as biogenic and sea-salt emissions from natural sources [31]. Each of the three national anthropogenic inventories accounts for emissions of at least seven criteria air pollutants: SO_2, NO_x, VOC, CO, NH_3, $PM_{2.5}$, and PM_{10}.

The calculation of the near-real-time biomass burning emissions required by FireWork starts with the Canadian Forest Service's operational Canadian Wildland Fire Information System (CWFIS), which provides fire activity and fire danger conditions across Canada and the continental United

States during the active wildfire season [33]. The primary data used by the CWFIS to capture fire activity come from satellite-based detection systems: NASA's Moderate Resolution Imaging Spectroradiometer (MODIS) instrument, NOAA's Advanced Very High Resolution Radiometer (NOAA/AVHRR), and Visible Infrared Imaging Radiometer Suite (VIIRS) imagery through NASA and the U.S. Forest Service Remote Sensing Applications Center [34]. During the fire season, fire activity is updated six times daily in the CWFIS, corresponding to the frequency of available satellite-based retrievals. Relevant fire information estimated by CWFIS for each fire hotspot includes fuel type, surface fuel consumption, crown fuel consumption, total fuel consumption, and forest floor fuel consumption. Estimates of daily biomass-burning emissions for individual hotspots are then obtained using fuel consumption values from the CWFIS and emission factors from the Fire Emission Production Simulator (FEPS), a component of the BlueSky Modeling Framework [26]. More detailed information about the FireWork modelling system framework and its data flow is provided in other recent publications [27,28].

In the current operational setup, the seasonal FireWork system runs twice per day at 00 UTC and 12 UTC during the North American fire season from 1 April to 31 October. FireWork simulation results provide numerical AQ forecast guidance over North America with a 48-h lead time. In 2013, 2014, and 2015, when FireWork was run as an experimental model version at CCMEP, the period from April to October was only partially covered (see Table 1). In April 2016, however, when the FireWork System became operational [35], FireWork forecasts were extended to cover the full wildfire season.

The seasonal peak for wildfire events in Canada occurs in the months of June, July, and August [2], and initially our analysis focused on this three-month period. However, due to the extreme wildfires that occurred in northern Alberta in May 2016, we decided to extend our analysis to a five-month period from May to September. In order to backfill this five-month period for the years 2013–2015 (Table 1), FireWork was rerun using the operational forecasting approach [27] with the same FireWork version that had been used each year. This required three older versions of FireWork to be run because new, updated versions of the RAQDPS had been introduced before each fire season [31,35].

Table 1. 2013–2016 experimental and operational FireWork start/end forecast periods together with additional periods for which FireWork was rerun retrospectively.

Year	Experimental/Operational FireWork		Added Periods
	Start	End	
2013	June 1	August 31	May 1–31; Sep. 1–31
2014	June 9	October 1	May 1–June 8
2015	May 21	October 31	May 1–20
2016	April 1	October 31	-

Seasonal fire emissions for May 1 through September 30 were estimated for North America using the FireWork emissions system [27]. The total fire emissions of key trace gases and particulate species for each year from 2013–2016 show a significant variation in the seasonal totals (Table S1; see Supplementary Materials). Maximum emissions were observed for the 2014 season, and there was substantial spatial variability in the regional estimates (data not shown). The mean seasonal FireWork primary $PM_{2.5}$ emissions for North America for 2013–2016 of 1.4 Tg/season are lower than but comparable to previous estimates of North American annual $PM_{2.5}$ emissions from wildfires (1.9 and 2.2 Tg/y) [36]. For context, total U.S. anthropogenic $PM_{2.5}$ emissions in 2010 were estimated to be 4.1 Tg/y [37], so fire emissions are an important source of $PM_{2.5}$.

The FireWork domain covers most of Canada and the USA (including Alaska), as well as northern Mexico, with a 10 km × 10 km grid (Figure 1). A new operational version of FireWork with a new domain and new 10 km × 10 km grid (Figure 1) was introduced during the 2016 wildfire season [35]. Results presented in Section 3 are calculated on the original grid used by FireWork prior to 7 September 2016. FireWork results for the period after the grid change (approximately three weeks)

Atmosphere **2017**, *8*, 179

were interpolated to the original grid to allow for consistent analysis. The area covered by the original FireWork domain was 33,085,648 km^2.

Figure 1. FireWork domain boundaries before (green) and after (red) 7 September 2016. The 10 km × 10 km grid is not shown.

2.2. Wildfire Emissions' Contribution to PM$_{2.5}$ Pollution

We used FireWork forecasts to analyze the contribution of fire-originated fine particulate matter (fire-PM$_{2.5}$) to PM$_{2.5}$ pollution over North America. In order to estimate the direct contribution of fire-PM$_{2.5}$ to the total PM$_{2.5}$ concentration forecasted by FireWork, the RAQDPS forecast PM$_{2.5}$ concentration field valid at the same hour was subtracted from the FireWork field. This simple strategy removes the contribution of the anthropogenic sources and other natural sources considered by both the RAQDPS and FireWork, and makes it possible to isolate wildfire smoke plume locations and follow their evolution over time [27]. The analysis of forecasted wildfire smoke presented in this paper is based on the set of hourly PM$_{2.5}$ concentration fields generated by this subtraction. Note that fire-PM$_{2.5}$ includes contributions from both primary PM$_{2.5}$ emissions and secondary aerosol formation from primary gas-phase emissions.

An essential part of characterizing the impacts of exposure to wildfire pollution is to understand both long-term (monthly to yearly) and short-term (hourly to daily) exposures. We used multi-year FireWork simulations (2013–2016) to characterize both long- and short-term wildfire pollution exposure over North America by calculating averages based on multi-year, seasonal, monthly, daily, and hourly concentrations and assessing areas affected by different concentration thresholds. Furthermore, we compared these averages with the PM$_{2.5}$ CAAQS of 10 µg/m^3 (annual standard) and 28 µg/m^3 (daily standard) [10], and with lower thresholds of 0.2, 1, and 5 µg/m^3. The 0.2 µg/m^3 threshold was based on the U.S. Environmental Protection Agency (EPA) Significant Impact Level (SIL) guidance document, which defines 0.2 µg/m^3 as the threshold below which any annual PM$_{2.5}$ change is considered negligible [38]. Also, from our own experience with FireWork, 0.2 µg/m^3 is the lowest value not susceptible to numerical noise that can be considered when analyzing the contributions of fire-PM$_{2.5}$ to total forecasted PM$_{2.5}$ concentrations. The 1 and 5 µg/m^3 thresholds were considered to transition between the minimal 0.2 µg/m^3 threshold and the 10 µg/m^3 threshold.

2.3. Population Exposure Estimation

Statistics on population exposure to wildfire smoke for Canada and the USA can be calculated by combining FireWork fields of direct contributions of fire-$PM_{2.5}$ to total surface $PM_{2.5}$ concentration with population data. We used population data from the 2016 Canadian census [39,40] and from the 2010 U.S. census [41]. For the 2016 Canadian census, we used population reported at the dissemination area (DA) level, where each DA typically has a population of 200 to 1000 people. For the 2010 U.S. census, we used population reported at the block-group level (Figure 2). Although U.S. population projections are available for 2016, they are only available at the coarser census-tract level rather than the more finely resolved block-group level (see Figure S1 in the Supplementary Materials).

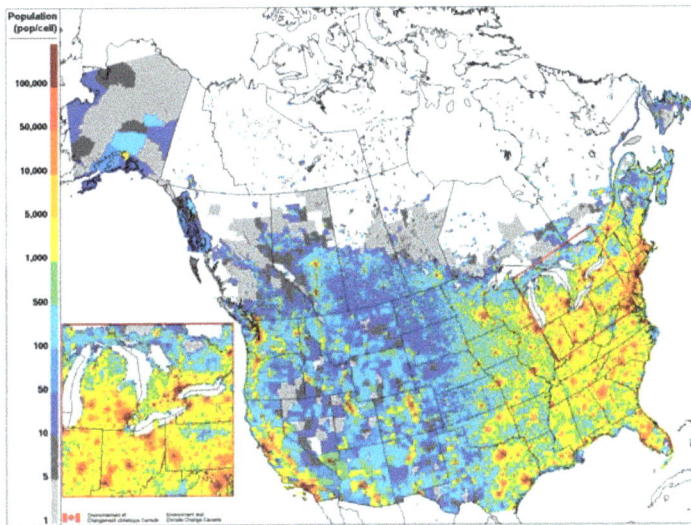

Figure 2. Population count per FireWork grid cell (10 km × 10 km) based on the 2016 Canadian census and the 2010 U.S. census. The red box over the four southern Great Lakes marks the location of the inset.

A number of steps were required to estimate the population affected at different wildfire $PM_{2.5}$ concentration thresholds. The first step was to determine the population for each 10 km × 10 km grid cell on the FireWork domain. To do so, 2016 Canadian population data reported by DA [39] were incorporated into a shapefile containing DA polygons [40]. The same step was performed at the sub-county level for 2010 U.S. population data [41]. The two population shapefiles were then interpolated separately to the 10 km by 10 km FireWork grid using a normalized conservative approach that preserved population within polygons. This interpolation approach divided the population value of each polygon between the grid cells wholly or partly contained within the polygon based on fractional area and assuming uniform population density within a DA or sub-county (Figure 2). The total populations contained in the older FireWork domain (Figure 1) for these two censuses were 35,148,512 in Canada and 305,744,285 in the USA.

The next step was to identify the aggregate population for the set of FireWork grid cells above a $PM_{2.5}$ concentration threshold. The Canadian and U.S. populations were processed separately. For grid cells along the Canada-USA border, it was necessary to determine in which country each cell was mainly located. This was done using a mask indicating the country associated with each grid cell. The population values for the set of FireWork grid cells associated with each $PM_{2.5}$ threshold were then summed together. The final step was to assess population exposure to different $PM_{2.5}$ concentrations.

To do this, five $PM_{2.5}$ thresholds (0.2, 1, 5, 10, and 28 $\mu g/m^3$) were considered and applied to monthly and seasonal (May to September) contribution of fire-$PM_{2.5}$ to total surface $PM_{2.5}$ concentrations.

2.4. Exposure Frequency Estimation

The temporal frequency of $PM_{2.5}$ pollution exposure is another critical factor in determining health impacts from fire-$PM_{2.5}$. One way to characterize this factor is to examine the number of occurrences of hourly or daily fire-$PM_{2.5}$ concentrations above specific thresholds for individual grid cells. This is done by counting the number of hours or days in a wildfire season with wildfire-related $PM_{2.5}$ concentrations above four different levels: 1, 5, 10 and 28 $\mu g/m^3$. In this case the lowest threshold of 0.2 $\mu g/m^3$ was not considered because it is less meaningful over shorter averaging periods.

3. Results

3.1. Area Affected by Wildfire Smoke

In 2013–2016, almost all areas of Canada and the USA included in the FireWork domain were affected by wildfire smoke based on a seasonal fire-$PM_{2.5}$ > 0.2 $\mu g/m^3$ exceedance at least once per grid cell (Figures 3–6). Western North America was more affected by wildfire smoke than eastern North America in all four years. From a continental perspective, 2014 had the most intense wildfire season, based on the seasonal values of the area affected by wildfire pollution for $PM_{2.5}$ concentration thresholds from 1 $\mu g/m^3$ to 28 $\mu g/m^3$ (Table 2). This year was followed by 2015 and 2013, while 2016 was the year least affected by wildfire smoke. Based on the average seasonal concentrations, the percentage areas of the FireWork domain (including land and water areas) above the 0.2 $\mu g/m^3$ threshold were 52%, 49%, 44%, and 22% for the years 2013–2016, respectively (Table 2). Above the 1 $\mu g/m^3$ threshold the corresponding percentage areas were 14%, 17%, 13%, and 6%, and above the 5 $\mu g/m^3$ threshold the values were 0.8%, 1.9%, 1.1%, and 0.4%.

Table 2. FireWork domain area affected (km^2 and percentages) by wildfire pollution above five $PM_{2.5}$ concentration thresholds based on average monthly and seasonal fire-$PM_{2.5}$ contributions to total average monthly and seasonal surface $PM_{2.5}$ concentrations. For reference, the total area of North America is 24.71 million km^2 and for the FireWork domain is 33.09 million km^2. The numbers in parentheses correspond to the percentage of the area affected.

$\mu g/m^3$	>0.2		>1		>5		>10		>28	
					2013					
May	754,020	(2.3%)	30,662	(0.1%)	1377	(0.0%)	196	(0.0%)	0	(0.0%)
June	12,478,900	(37.7%)	2,532,748	(7.7%)	184,058	(0.6%)	48,991	(0.1%)	7378	(0.0%)
July	17,235,170	(52.1%)	7,279,618	(22.0%)	861,986	(2.6%)	194,966	(0.6%)	13,326	(0.0%)
August	20,155,062	(60.9%)	9,873,087	(29.8%)	1,783,453	(5.4%)	604,476	(1.8%)	71,844	(0.2%)
September	9,351,929	(28.3%)	2,072,752	(6.3%)	84,693	(0.3%)	39,022	(0.1%)	10,611	(0.0%)
					2014					
May	1,746,428	(5.3%)	61,310	(0.2%)	6507	(0.0%)	2137	(0.0%)	200	(0.0%)
June	3,411,109	(10.3%)	872,761	(2.6%)	57,926	(0.2%)	13,461	(0.0%)	1686	(0.0%)
July	16,196,537	(49.0%)	7,260,984	(21.9%)	1,562,298	(4.7%)	541,468	(1.6%)	146,616	(0.4%)
August	19,536,859	(59.0%)	8,436,851	(25.5%)	2,560,671	(7.7%)	934,005	(2.8%)	157,462	(0.5%)
September	13,305,055	(40.2%)	4,531,984	(13.7%)	444,316	(1.3%)	137,840	(0.4%)	24,680	(0.1%)
					2015					
May	2,837,325	(8.6%)	252,707	(0.8%)	12,768	(0.0%)	2950	(0.0%)	197	(0.0%)
June	9,375,291	(28.3%)	2,604,739	(7.9%)	291,593	(0.9%)	42,184	(0.1%)	2552	(0.0%)
July	14,245,102	(43.1%)	5,173,825	(15.6%)	1,033,576	(3.1%)	306,687	(0.9%)	30,776	(0.1%)
August	15,581,769	(47.1%)	5,918,077	(17.9%)	1,236,416	(3.7%)	636,671	(1.9%)	143,120	(0.4%)
September	11,999,087	(36.3%)	1,629,073	(4.9%)	102,071	(0.3%)	45,570	(0.1%)	14,173	(0.0%)
					2016					
May	3,829,030	(11.6%)	972,851	(2.9%)	167,935	(0.5%)	58,168	(0.2%)	11,955	(0.0%)
June	2,532,347	(7.7%)	334,211	(1.0%)	35,392	(0.1%)	13,575	(0.0%)	3599	(0.0%)
July	7,866,765	(23.8%)	2,709,907	(8.2%)	223,084	(0.7%)	57,209	(0.2%)	8215	(0.0%)
August	7,807,827	(23.6%)	2,839,017	(8.6%)	477,835	(1.4%)	146,340	(0.4%)	31,658	(0.1%)
September	4,914,590	(14.9%)	1,715,135	(5.2%)	204,326	(0.6%)	71,075	(0.2%)	15,620	(0.0%)

Table 2. *Cont.*

µg/m³	>0.2		>1		>5		>10		>28	
				Seasonal Exceedances (Area)						
2013	17,151,407	(51.8%)	4,534,747	(13.7%)	269,933	(0.8%)	62,226	(0.2%)	9267	(0.0%)
2014	16,335,869	(49.4%)	5,616,429	(17.0%)	625,489	(1.9%)	216,897	(0.7%)	28,417	(0.1%)
2015	14,507,744	(43.8%)	4,203,396	(12.7%)	373,565	(1.1%)	90,825	(0.3%)	16,248	(0.0%)
2016	7,229,006	(21.8%)	1,931,583	(5.8%)	121,897	(0.4%)	40,841	(0.1%)	9388	(0.0%)

In terms of total area burned, in 2013 Canada had its seventh most intense wildfire season of the past 34 years [2], whereas the USA was 37% below its 2006–2016 average [42]. The wildfire season in 2013 effectively started in June, with maximum intensity reached in July and August. During these two months, the area for which the monthly average of fire-$PM_{2.5}$ exceeded the 0.2 µg/m³ threshold covered most of North America (Figure 3). In July, an area of 194,966 km² (the size of South Dakota) had average monthly fire-$PM_{2.5}$ above 10 µg/m³ and 13,326 km² were above 28 µg/m³ (Table 2). In August, using the same thresholds, these areas were 604,476 km² (larger than California; close to the size of Manitoba and Texas) and 71,844 km², respectively. August 2013 was the most active month of the year, with intense activity in northwestern and western Canada and the northwestern USA.

In 2014, Canada had its fifth most intense wildfire season of the past 34 years in terms of total area burned, while in the USA the value was 48% below its 2006–2016 average [2,42]. The extreme wildfire event in the Northwest Territories near the city of Yellowknife started in June and peaked in July (Figure 4). In July, intense wildfires began burning in British Columbia, Alberta, Washington, Oregon, California, and Idaho, bringing the total area with average monthly fire-$PM_{2.5}$ > 10 µg/m³ to 541,468 km², and > 28 µg/m³ to 146,616 km². These fires persisted into August, making August 2014 the most extreme month of the 2013–2016 period in terms of area affected by wildfire smoke. In this month, average monthly fire-$PM_{2.5}$ values above the 5, 10, and 28 µg/m³ thresholds covered 2,560,671 km² (larger than Alaska or Nunavut), 934,005 km² (the size of British Columbia), and 157,462 km² (the size of Georgia), respectively (Table 2).

The 2015 season was the sixth most intense of the past 34 years for Canada in terms of area burned [2]. In the USA, it was the peak year of the 2006–2016 period, with more than 10 million acres (40,500 km²) burned, 45% above the period average [42]. The 2015 fire season was marked by two intense wildfire periods: the first from 15 June to 15 July and the second from 1 to 15 August [27] (Figure 5). In the first period, most of the wildfires occurred in northwestern Canada (Alberta and Saskatchewan), whereas in the second period most of the wildfires occurred in the western USA (Washington, Oregon, Idaho, and California). In July and August areas of 306,687 km² and 636,671 km² had average monthly fire-$PM_{2.5}$ > 10 µg/m³, respectively, and areas of 30,776 km² and 143,120 km² were >28 µg/m³ (Table 2).

The 2016 fire season included unprecedented impacts in Canada on both people and the national economy. The entire city of Fort McMurray, Alberta, with a population of nearly 90,000, was evacuated in May when it was overrun by a large, fast-moving wildfire (Figure 6). Estimated insured fire damages to Fort McMurray were 9.6 billion dollars, the costliest insured natural disaster in Canadian history [4]. Despite this disaster, 2016 was the least intense wildfire season among the four years analyzed in terms of area burned [2] and area affected by wildfire smoke across Canada (Table 2). In the USA, the 2016 area burned was 21% below its 2006–2016 average [42]. On the other hand, the early start to the wildfire season in Canada made the month of May 2016 the most affected by wildfire pollution among the four Mays assessed. For May 2016 the area with average monthly fire-$PM_{2.5}$ > 10 µg/m³ and > 28 µg/m³ was 58,168 km² and 11,955 km² respectively (Table 2).

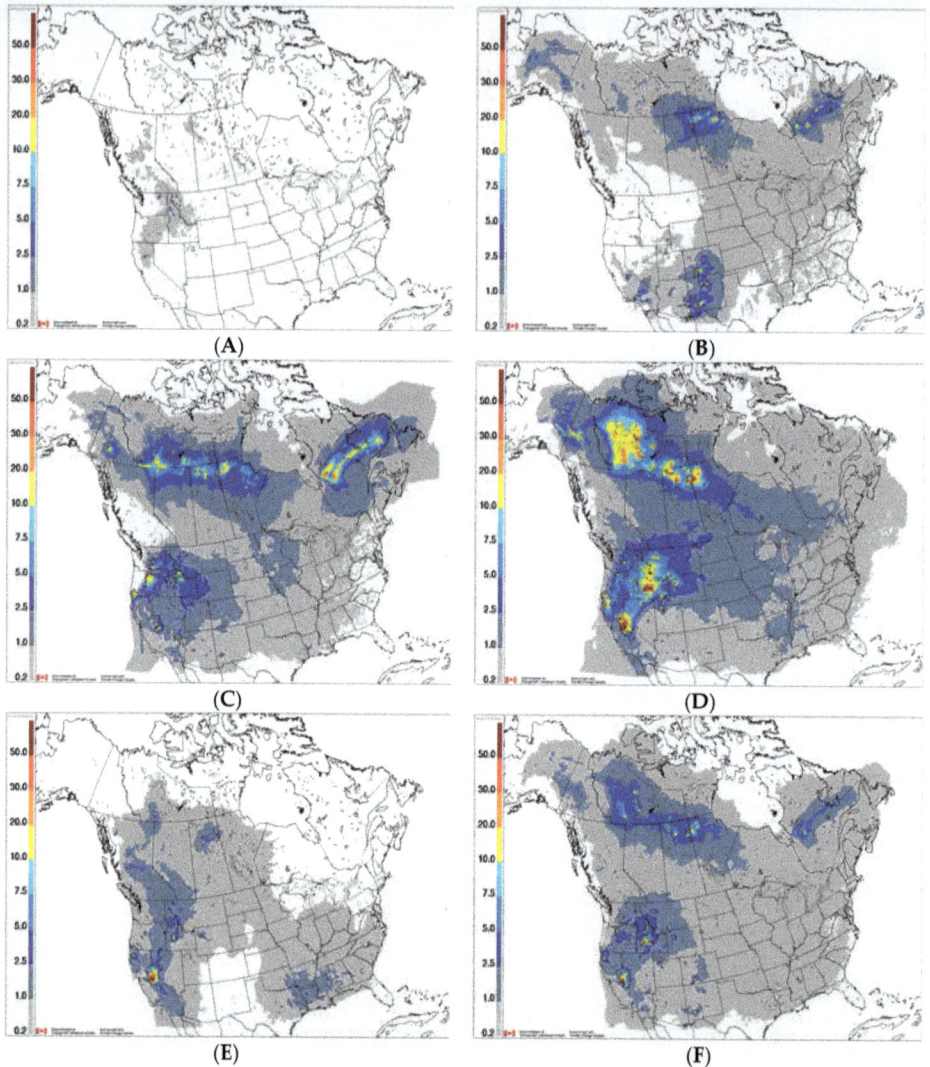

Figure 3. 2013 average monthly fire-PM$_{2.5}$ contribution to total forecasted surface PM$_{2.5}$ concentrations ($\mu g/m^3$) for (**A**) May, (**B**) June, (**C**) July, (**D**) August, (**E**) September, and (**F**) seasonal (May–September average). Note that the color scale is non-linear and white areas indicate values < 0.2 $\mu g/m^3$.

It is also of interest to examine the importance of fire-PM$_{2.5}$ relative to other sources of PM$_{2.5}$. Figure S2 shows the seasonal fire-PM$_{2.5}$ contribution to total PM$_{2.5}$ as a percentage for each of the four years. For 2013 to 2015, the seasonal contribution of fire-PM$_{2.5}$ to total PM$_{2.5}$ was 50% or more over much of northwestern North America and parts of the U.S. mountain west. In 2014 the seasonal wildfire contribution was 90% or greater for a large part of the Northwest Territories and parts of the interior of British Columbia. These results are not surprising considering that these areas have relatively few inhabitants and low anthropogenic emissions.

It is difficult to compare the monthly model forecasted fire-PM$_{2.5}$ directly with PM$_{2.5}$ measurements, as measurements are influenced by all PM$_{2.5}$ sources, not just biomass burning emissions. As an indirect comparison, however, we note that archived near-real-time measurement from the AirNow data feed (www.airnow.gov) include at least one U.S. or Canadian AQ station located close to wildfires having monthly $PM_{2.5}$ concentrations above 30, 50, and even 150 $\mu g/m^3$ for each of the four years analyzed here. The most extreme month was August 2015, when 13 $PM_{2.5}$ measurement stations reported mean monthly $PM_{2.5}$ concentrations of 30 $\mu g/m^3$ or above, including six stations in Idaho, two stations each in Oregon, Washington State, and British Columbia, and one station in Montana (see Table S2). FireWork also forecasted mean monthly fire-$PM_{2.5}$ concentrations above 30 $\mu g/m^3$ for two regions of California in August 2015; although no stations with available measurements were located in these regions, one nearby station in Shasta county in northern California had a mean monthly $PM_{2.5}$ concentration of 24.5 $\mu g/m^3$ and 100% data completeness.

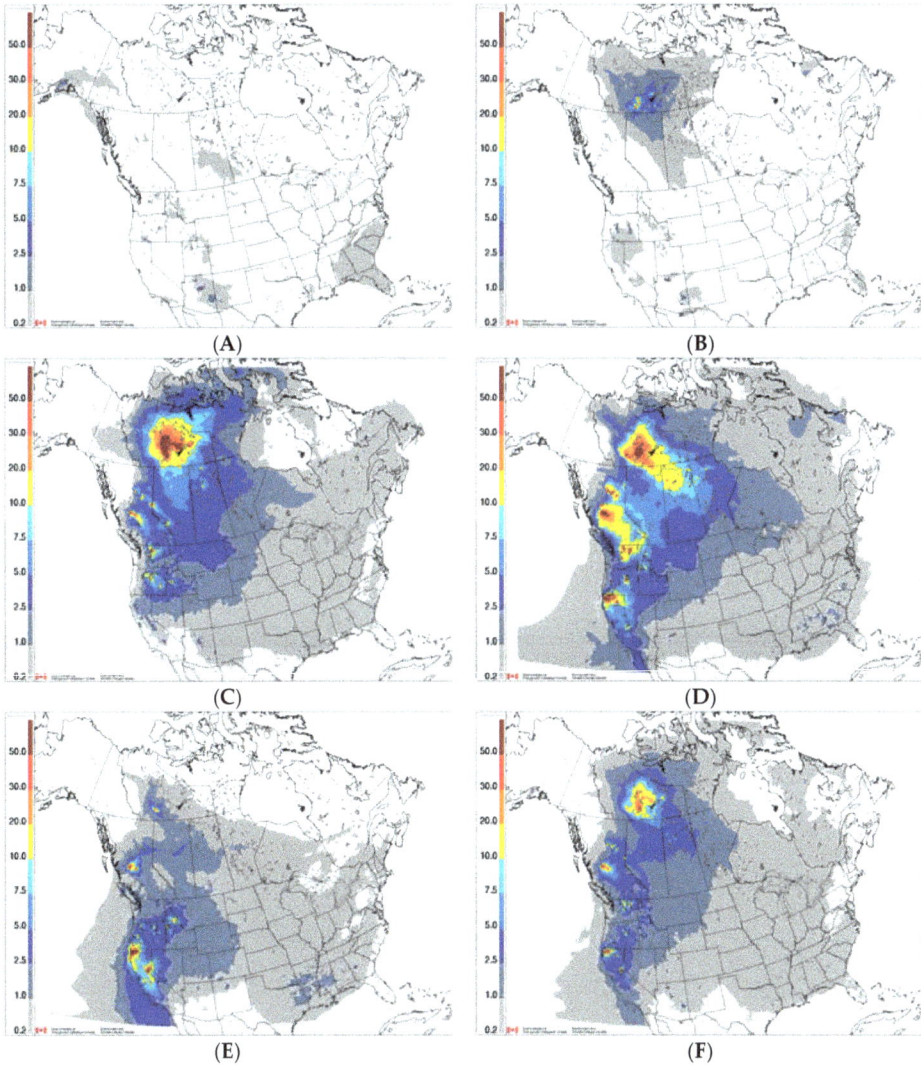

Figure 4. Same as Figure 3 but for 2014.

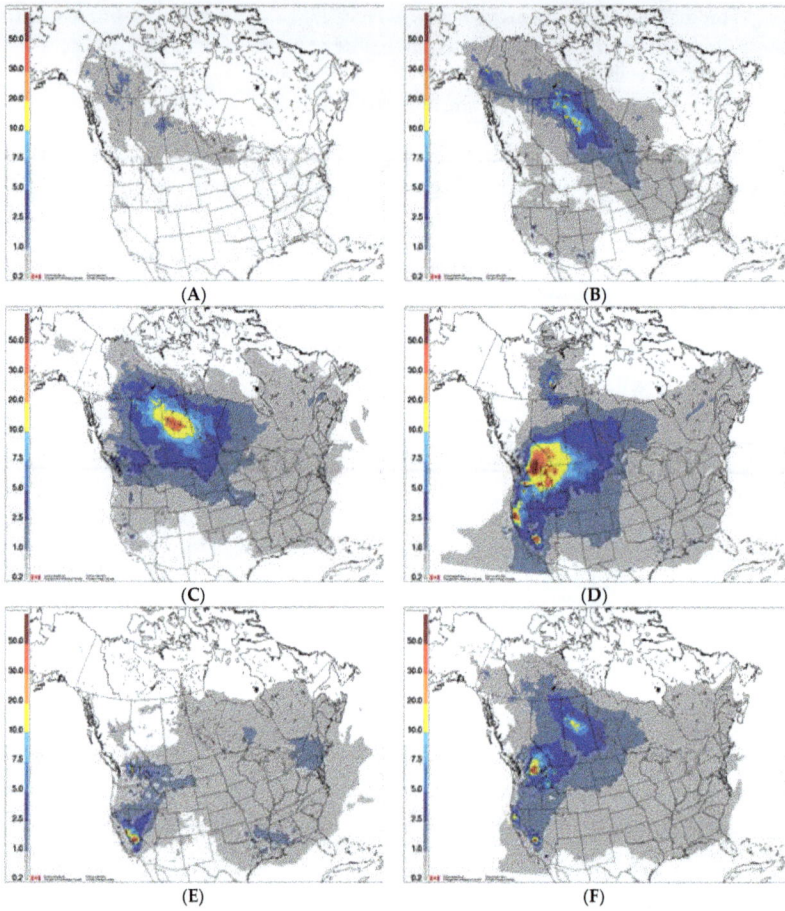

Figure 5. Same as Figure 3 but for 2015.

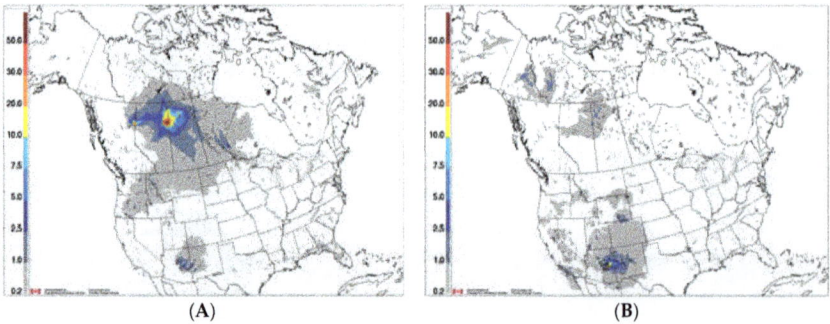

Figure 6. *Cont.*

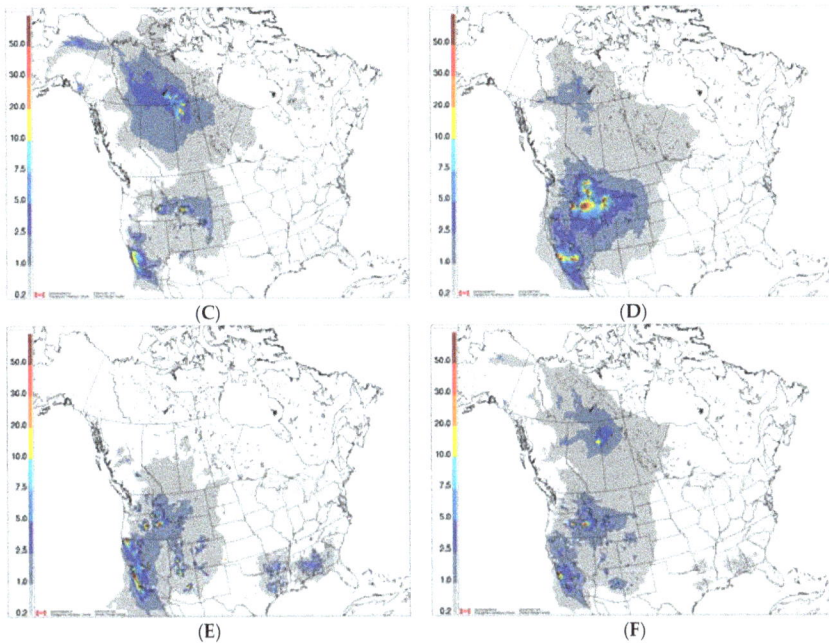

Figure 6. Same as Figure 3 but for 2016.

3.2. Population Exposure to Wildfire Pollution

In terms of population exposure to wildfire-related $PM_{2.5}$ pollution, for 17 of the 20 months considered, more than 1 million Canadians (3% of the population) were estimated to have been affected by average monthly fire-$PM_{2.5}$ > 0.2 $\mu g/m^3$ (Figure 7). During the same period, more than 14 million Canadians (39% of the population) were affected by average monthly fire-$PM_{2.5}$ > 0.2 $\mu g/m^3$ in 10 of the 20 months, and over 32 million Canadians (90% of the population) were affected in 7 months. The months affecting the most people were July and August (Figure 8 and Table S3), with August 2015 being the worst for average monthly fire-$PM_{2.5}$ > 10 $\mu g/m^3$. During August 2015 the proportion of the Canadian population affected by fire-$PM_{2.5}$ above thresholds of 0.2, 1, 5, 10, and 28 $\mu g/m^3$ were 97%, 21%, 11%, 8%, and 1.2%, respectively. The three periods in which the most Canadians were exposed to >28 $\mu g/m^3$ were: (a) August 2015 (417,171 people or 1.2% of the population) due to extreme wildfires in British Columbia and the northwestern USA; (b) May 2016 (69,909 people or 0.2% of the population) due to extreme wildfires in northern Alberta near Fort McMurray; and (c) August 2014 (27,160 people or 0.1% of the population) due to extreme wildfires in northwestern Canada. In all other months, less than 0.1% of the Canadian population was affected at this very high threshold (Table S3).

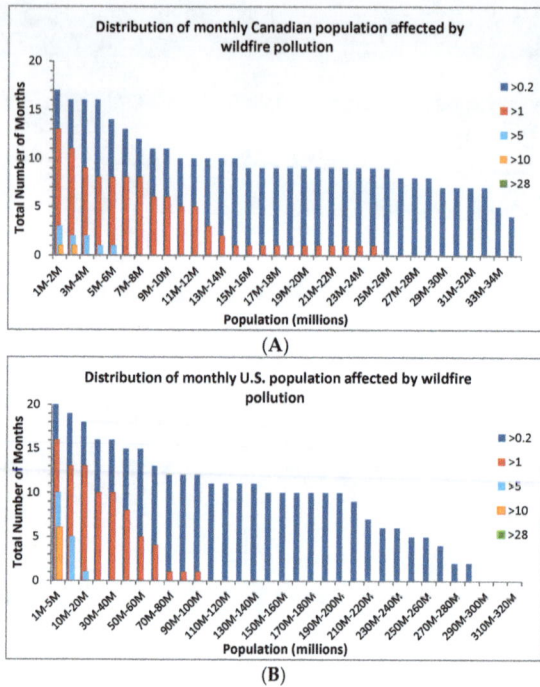

Figure 7. Cumulative distribution of number of months with the portions of the Canadian (**A**) and U.S. (**B**) population affected by monthly fire-PM$_{2.5}$ above five PM$_{2.5}$ concentration thresholds for four five-month wildfire seasons (2013–2016).

Based on the average seasonal statistics (Table S3), over 90% of the Canadian population was affected by seasonal fire-PM$_{2.5}$ > 0.2 µg/m^3 in 2013, 2014 and 2015. The 2016 wildfire season was much milder, with 19% of Canadian population affected by wildfire smoke above this threshold (Figure 8). The corresponding proportions for the 1 µg/m^3 threshold ranged from 0.4% (2016) to 26% (2014). The population affected by concentrations above the 10 µg/m^3 threshold reached its maximum in 2015, with over 100,000 people (Table S3).

In the USA, more than 200 million people (65% of the population) were exposed to average monthly fire-PM$_{2.5}$ > 0.2 µg/m^3 during nine of the 20 months considered (Figure 7). As well, a much greater proportion of the U.S. population was affected by wildfire pollution in 2013, 2014, and 2015 than in 2016, similar to Canada (Figure 8), and for both the USA and Canada, the proportion of the population exposed to wildfire pollution in September was larger than in June.

The total percentages of the U.S. population affected by seasonal fire-PM$_{2.5}$ > 0.2 µg/m^3 ranged from 21% (2016) to 90% (2015) (Figure 8 and Table S4), and the corresponding range above 1 µg/m^3 was 5% (2015) to 15% (2014). Based on the four-year average seasonal statistics for population exposure to fire-PM$_{2.5}$ > 0.2 µg/m^3, a smaller percentage of the U.S. population (69%) than the Canadian population (76%) was exposed to pollution above this threshold (Tables S3 and S4). For concentrations > 1 µg/m^3 the corresponding percentages were 10% in the USA and 12% in Canada. However, for average seasonal fire-PM$_{2.5}$ > 28 µg/m^3, a higher percentage of Americans than Canadian were exposed. The affected U.S. population ranged from 32,549 in 2014 to 56,442 in 2015 (Table S4), giving a four-year average exposure of 0.015% for the U.S. population compared with 0.004% for the Canadian population (Table S3).

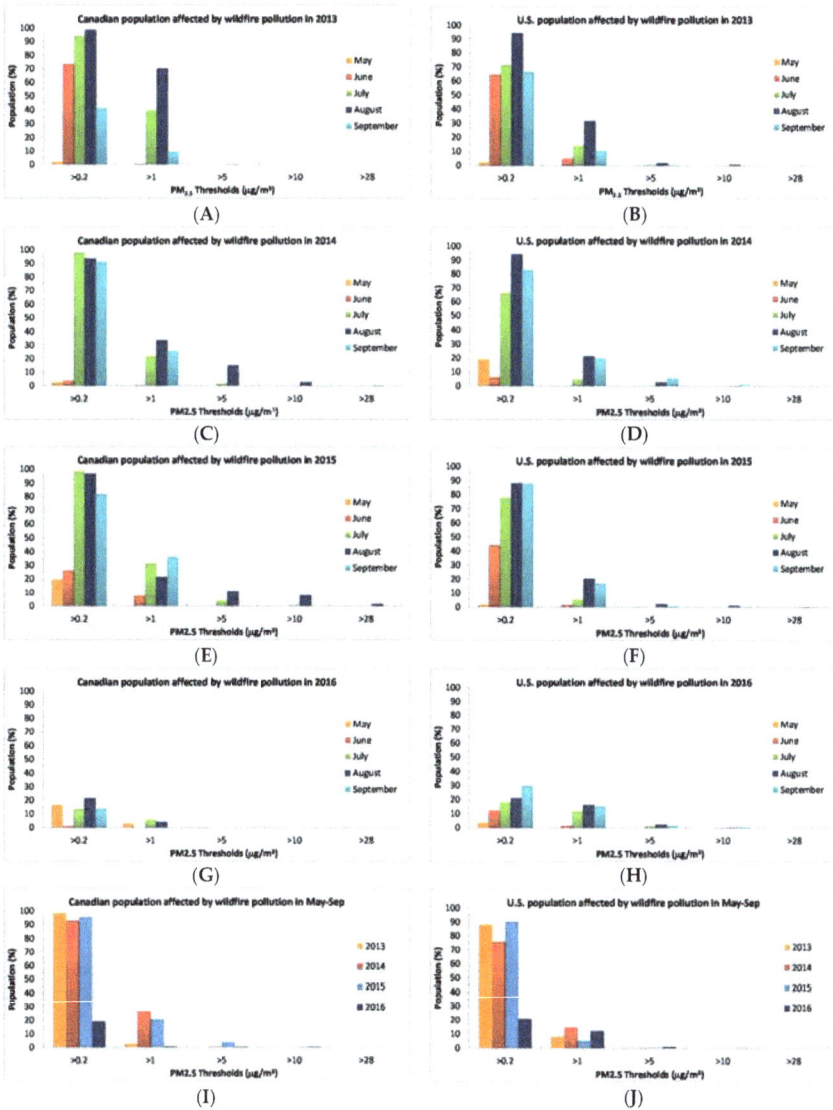

Figure 8. Percentage of population in Canada (left panels: **A**, **C**, **E**, **G**, and **I**) and the USA (right panels: **B**, **D**, **F**, **H**, and **J**), affected by wildfire pollution above five $PM_{2.5}$ concentration thresholds based on the average monthly and seasonal fire-$PM_{2.5}$ contribution to total average monthly and seasonal surface $PM_{2.5}$ concentrations for four wildfire seasons (2013–2016). The percentage of the affected population for Canada and the USA was calculated using the 2016 Canadian and the 2010 U.S. censuses, respectively. See also Tables S3 and S4.

3.3. Frequency of Wildfire-Related Pollution Events

In 2013–2016 most of North America (except Alaska) was affected by wildfire smoke on at least one day (Figure 9). The land-use map for Alaska used by National Resources Canada to determine total fuel consumption was only updated in 2015. Given the direct impact of wildfire emissions estimates from input land-use in FireWork, there may be an underestimation of wildfire emissions for regions of Alaska prior to 2015. The highest daily frequency of wildfire smoke occurred in 2014, where for most of western Canada and the northwestern USA more than 30% of the days from May to September had surface fire-PM$_{2.5}$ greater than 1 µg/m^3 for at least one hour. In parts of Alberta, Saskatchewan and the Northwest Territories, the daily frequency was above 40%, and in part of the southern Northwest Territories it was over 60%. For hourly frequency, in 2013 the majority of Canada and the USA had more than 10% of hourly forecasted fire-PM$_{2.5}$ above 1 µg/m^3. Hourly frequency was slightly lower in 2014 and 2015, but in 2016, only a portion of western North America had hourly frequencies over 10% (Figure 9).

The daily frequency of forecasted fire-PM$_{2.5}$ concentrations greater than 5 µg/m^3 was above 10% for the majority of western Canada in 2013, 2014 and 2015 (Figure 10). The same was true for northern Quebec in 2013. In the case of the USA, regions over 10% were found only in the west over the same period. In 2016, frequencies above 10% were less common when compared with other years and were limited to areas close to the wildfires in northwestern Canada and the northwestern USA and California. Frequencies above 30% were found in 2014 in northwestern Canada, while this percentage was limited to areas very close to the wildfires in the other years. The results of the hourly frequency analysis are similar to those of the daily frequency analysis, with values of 10–20% covering large areas of northwestern Canada and the western USA, especially for 2014. Hourly frequencies over 20% were not present over Canada in 2016 and were limited to areas close to wildfires in 2013, 2014, and 2015 (Figure 10).

The spatial patterns of daily and hourly frequencies for the 10 µg/m^3 concentration threshold were very similar to the patterns observed for the 5 µg/m^3 threshold (Figure 11). However, 2014 was the only year with daily and hourly frequencies above 30%, in northwestern Canada. For a threshold of 28 µg/m^3 (Figure 12), the frequency of days and hours with hourly forecasts above this threshold was generally below 10%.

We can also look at the number of days with elevated fire-PM$_{2.5}$ from a population exposure perspective. Tables S5 and S6 provide this information for Canada and the USA, respectively. In 2014 wildfire season, more than 14% of Canadians were exposed to a daily fire-PM$_{2.5}$ > 5 µg/m^3 on at least 30 days (i.e., 20% or more days) and 12% of Canadians were exposed to a daily fire-PM$_{2.5}$ > 10 µg/m^3 on at least 15 days (Table S5). For the USA, the corresponding values in 2014 were 2% and 3%, but interestingly they were higher (3% and 4%) in 2016 (Table S6), even though in other respects 2016 had a less active fire season (e.g., Table 2).

MAY–SEPTEMBER

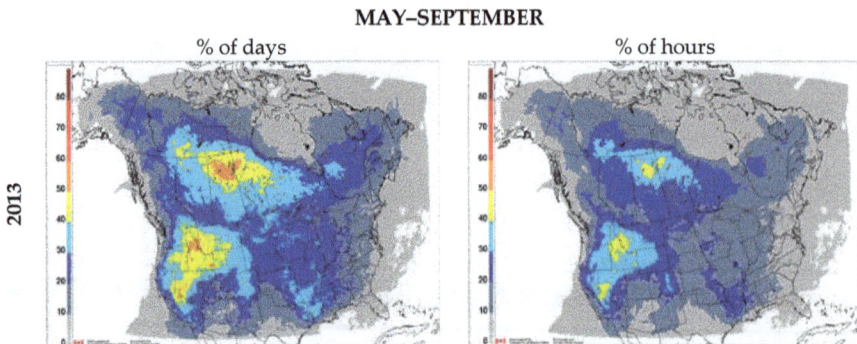

Figure 9. *Cont.*

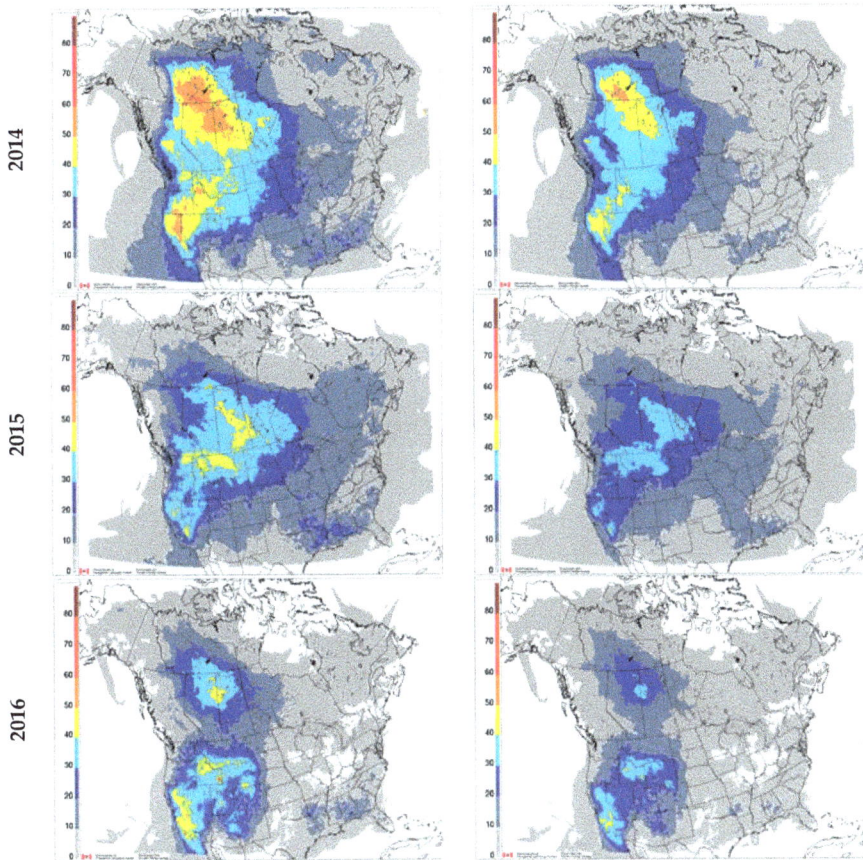

Figure 9. Percentage frequency of the number of days (**left**) and the number of hours (**right**) with forecasted 24-h moving average $PM_{2.5}$ concentration above 1 $\mu g/m^3$ from fire-$PM_{2.5}$ contribution for the period May–September for years 2013, 2014, 2015, and 2016. White areas indicate locations that experienced no days or hours above the threshold during the period.

MAY–SEPTEMBER

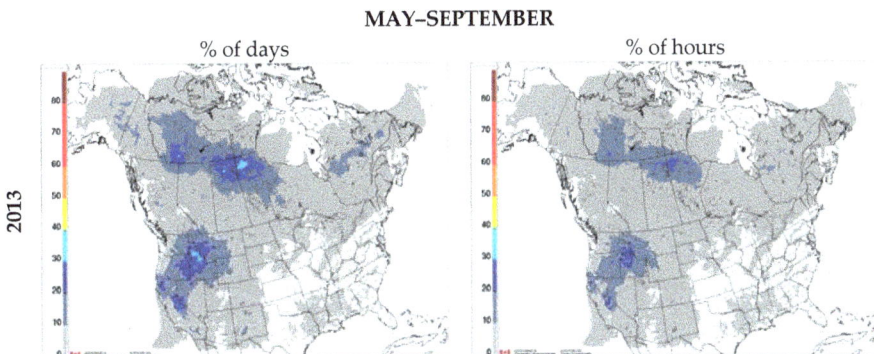

Figure 10. *Cont.*

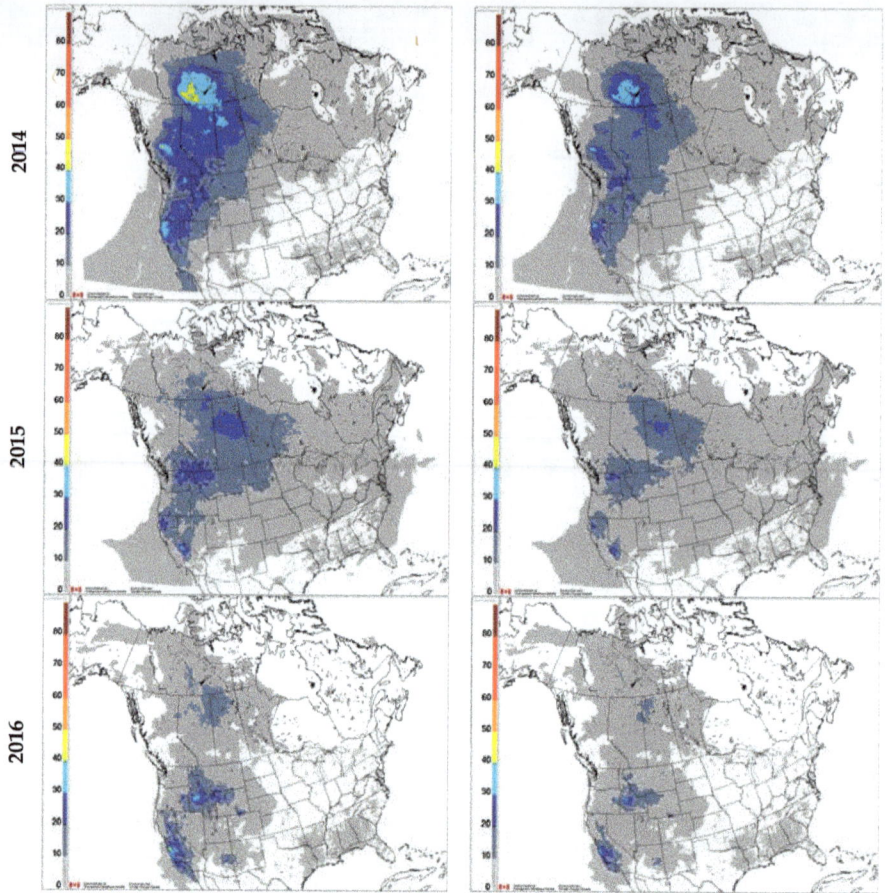

Figure 10. Same as Figure 9 but for 5 µg/m^3.

MAY–SEPTEMBER

% of days   % of hours

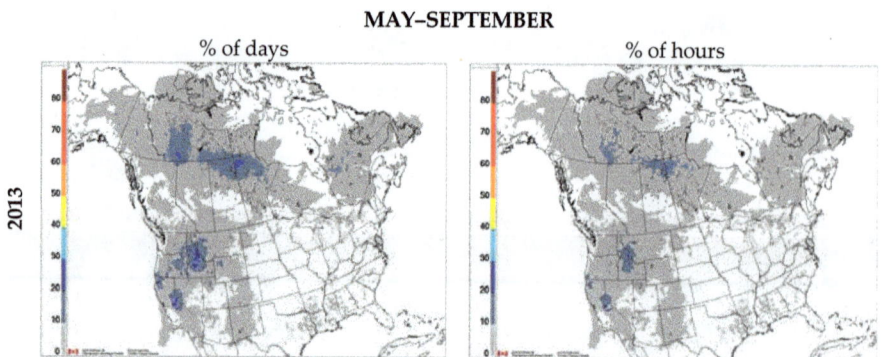

Figure 11. *Cont.*

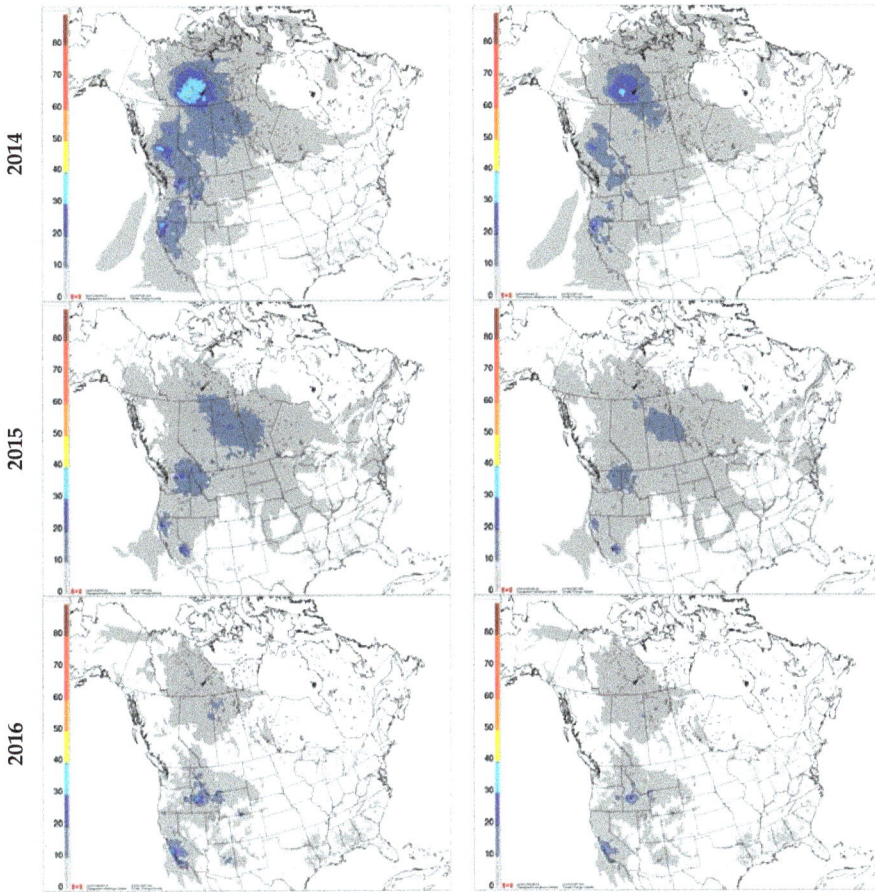

Figure 11. Same as Figure 9 but for 10 μg/m³.

MAY–SEPTEMBER

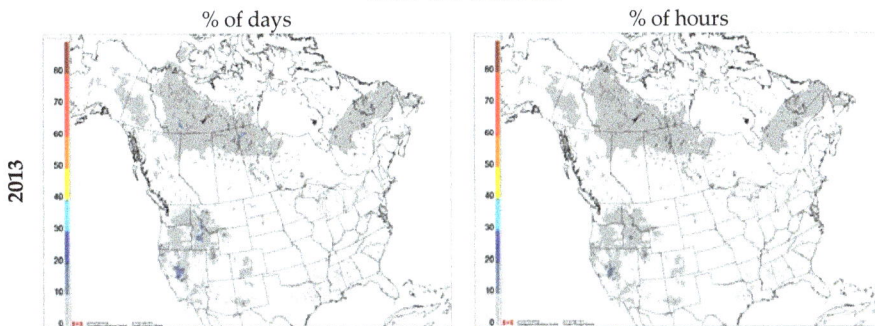

Figure 12. *Cont.*

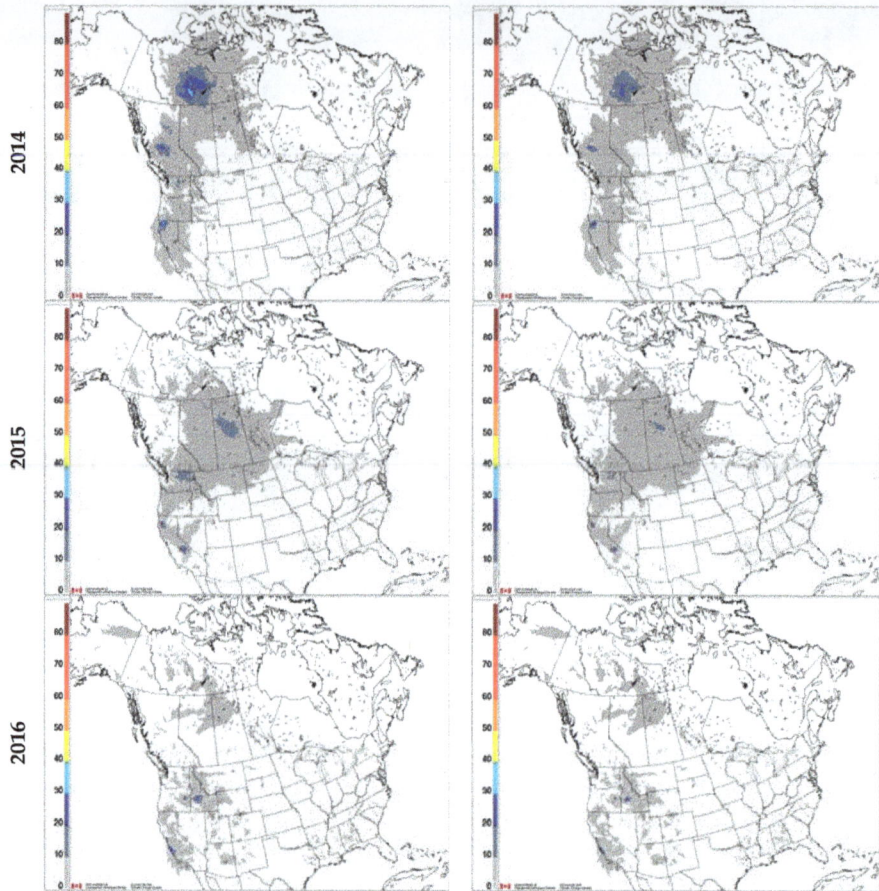

Figure 12. Same as Figure 9 but for 28 μg/m^3.

4. Discussion

The spatial distributions of monthly and seasonal wildfire plumes illustrate the large spatial and temporal variability of wildfire occurrence in North America (Figures 3–6). Our results also highlight how common wildfires are each summer and the large spatial extent of their influence due to long-range transport of wildfire emissions. Analyses of daily elemental and organic carbon measurements from the U.S. IMPROVE speciated PM$_{2.5}$ measurement network support the analysis of the FireWork wildfire smoke forecasts presented here, indicating that smoke from wildfires contributes substantially to PM$_{2.5}$ levels in the western USA [43–46].

A related finding is that wildfires can sometimes impact the same location many times during a single season (Figures 9–12). The frequent presence of fire-PM$_{2.5}$ shown here for 2013–2016, especially in western North America, has implications for regional attainment of PM$_{2.5}$ regulatory objectives. Both Canadian and U.S. standards for PM$_{2.5}$ allow the exclusion of days with concentrations above the national standard due to wildfire smoke. However, to invoke such an exclusion, it is necessary to demonstrate that an unmanaged emission source such as wildfires is the cause of an elevated concentration. FireWork forecasts could provide useful evidence for this purpose.

It is also relevant to compare the distribution of population in North America (Figure 2) with the distribution of wildfire smoke (Figures 3–6). At the continental scale the majority of the North American population lives in the eastern half of the continent whereas the majority of large wildfires occur in the western half. This anticorrelation reduces the degree of population exposure to wildfire smoke. Nevertheless, 32% of the Canadian population lives west of Ontario and 41% of the U.S. population lives west of the Mississippi River, closer to western wildfires. As an example of this difference between wildfire location and population location, consider that for the 2013–2016 period (a) August 2014 was identified as the month with the greatest areal extent of average monthly fire-$PM_{2.5}$ > 28 $\mu g/m^3$, whereas (b) August 2015 was identified as the month with the largest U.S. population exposure to average monthly fire-$PM_{2.5}$ > 28 $\mu g/m^3$.

Short-term exposure to $PM_{2.5}$ from wildfire smoke puts the population at an increased risk of experiencing a wide range of acute health outcomes, particulary those with chronic conditions such as asthma [47] or heart disease [48]. Long-term exposure to $PM_{2.5}$ from all sources, including the contribution from wildfire smoke, increases the risk of developing chronic conditions, such as asthma or heart disease [9]. Global estimates suggest that approximately 340,000 deaths per year can be attributed to smoke from landscape fires, of which only 18% are due to short-term effects while 82% are due to the long-term effects [49]. Our results confirm that fire-$PM_{2.5}$ puts the Canadian and U.S. populations at risk from both short- and long-term exposure. Although our long-term averages covered the five-month fire season rather than the entire year, we found that up to 26.4% of the Canadian population and 14.7% of the U.S. populations were affected by increases of > 1 $\mu g/m^3$ in the extreme fire seasons. This likely indicates that the annual fire-$PM_{2.5}$ averages were > 0.2 $\mu g/m^3$ for these populations, which is defined by the U.S. EPA as a non-negligible impact [38]. Given that wildfires are becoming more frequent and intense across North America [2,42], tools such as FireWork can help to characterize their contribution to the long-term exposure most responsible for the burden of disease attributable to air pollution.

A similar wildfire pollution exposure study was recently published for the USA [23], in which fire-$PM_{2.5}$ contributions were estimated for an earlier five-year period (2008–2012) using paired retrospective simulations performed with another AQ modeling system. As in this study, one simulation considered wildfire emissions and one did not, and then a post-simulation subtraction of predicted paired surface $PM_{2.5}$ fields yielded the fire-$PM_{2.5}$ contribution estimate. One difference between the two studies was the fire seasons sampled: the average annual U.S. area burned during their study period (2008–2012) was 11% higher than that during our study period (2013–2016) [42]. Other differences were the concentrations thresholds (in $\mu g/m^3$) that they considered (0.15, 0.75, 1.5) compared with those that we considered (0.2, 1, 5, 10, 28), and the annual concentration values aggregated from 12 km × 12 km grid cells to the county level that they considered vs. the monthly and five-month values for 10 km × 10 km grid cells that we considered (i.e., higher temporal and spatial resolution). Nevertheless, a limited comparison of the two studies is possible. Based on the 2010 U.S. census (also used in this study), Rappold et al. [23] estimated that 10% of the U.S. population lived in areas where the contribution of fire-$PM_{2.5}$ was >1.5 $\mu g/m^3$. We estimated that between 5.3% (in 2015) and 14.7% (in 2014) of the U.S. population lived in areas where the five-month fire-$PM_{2.5}$ contribution was >1 $\mu g/m^3$. Rappold et al. [23] also estimated that 10.3 million individuals in the U.S. lived in areas having 10 or more days (between 2008 and 2012) with fire-$PM_{2.5}$ contribution > 35 $\mu g/m^3$, and we estimated that 10.6 million individuals in the USA lived in areas having 10 or more days (for all five-month seasons between 2013 and 2016) with fire-$PM_{2.5}$ > 28 $\mu g/m^3$. Considering the differences between these two studies, this basic comparison suggests that the results are comparable and consistent.

This study was an "analysis of opportunity" based on the availability of four years of daily North American wildfire smoke forecasts, and further improvements are likely possible. For example, wildfires in Siberia are known to affect Alaska and western Canada, but FireWork does not currently consider wildfire emissions external to North America. In our operational on-line FireWork performance evaluation, a negative bias in $PM_{2.5}$ concentration forecasts is observed for long-range wildfire pollution advection. As a

Atmosphere **2017**, *8*, 179

consequence, the results presented here are likely to be conservative. Work is ongoing to examine potential improvements to the $PM_{2.5}$ emission factors and the plume-rise parameterization used by FireWork, both of which influence $PM_{2.5}$ concentration predictions. Given the horizontal grid spacing of 10 km used by FireWork, it is expected too that forecasts of near-source concentration will be underestimates due to the limitation of assuming uniform emissions within a grid cell. ECCC also has an AQ objective analysis system that combines AQ model predictions with AQ observations [50]. This system is now used with FireWork, opening the possibility of analyzing optimally-combined fields of model predictions and $PM_{2.5}$ measurements instead of relying solely on model forecast fields. Finally, our results might have differed slightly had we run the same FireWork version for all four years instead of using archived FireWork versions from each year (which ensured that our analyses reflected the information available to ECCC forecasters and stakeholders at the time). However, none of these changes are likely to alter our overall conclusions about the impact of wildfires in North America.

5. Conclusions

The FireWork AQ forecast system with near-real-time wildfire emissions has been run daily for a North American domain by ECCC from 2013 to 2016 during the May–September wildfire season. A multi-year analysis for this period showed the importance of accounting for contributions from wildfire $PM_{2.5}$ emissions to total $PM_{2.5}$ surface concentrations (denoted here as fire-$PM_{2.5}$) during the wildfire season. For both Canada and the USA, the months of July and August usually showed the maximum fire-$PM_{2.5}$, although intense wildfires can also occur in September in the western USA, likely due to a longer summer season [51].

Monthly and seasonal analyses of the mean forecasted fire-$PM_{2.5}$ suggested that, on average, over 76% of Canadians and 69% of Americans were at least minimally affected by wildfire smoke during the four-year study period. Comparison of average monthly fire-$PM_{2.5}$ showed large year-to-year variations in both timing and spatial locations of wildfires between 2013 and 2016. Wildfire impacts are often driven by a few major wildfire events that can lead to poor air quality for several consecutive weeks near the emission sources and beyond. In August 2015 approximately 3 million Canadians and 3 million Americans were exposed to mean monthly fire-$PM_{2.5} > 10 \ \mu g/m^3$.

Calculations of the number of days and hours with forecasted fire-$PM_{2.5}$ above various concentration thresholds ranging from $1 \ \mu g/m^3$ to $28 \ \mu g/m^3$ for 2013–2016 showed that most wildfire events over North America occurred in the western part of the USA and in western, northern, and central Canada. During months of extreme wildfire activity, some areas in northwestern Canada and the western USA had up to 20% of days where the fire-$PM_{2.5}$ was $> 28 \ \mu g/m^3$. The eastern USA and eastern Canada had fewer days with threshold exceedances, but most of North America was affected by fire-$PM_{2.5} > 1 \ \mu g/m^3$ on at least one day per year.

FireWork is a valuable prognostic tool used as guidance by AQ meteorologists to issue forecasts on a daily basis, allowing advance warnings to populations at risk to reduce their exposure. In addition, this study has shown that FireWork is also useful for retrospective analysis of past wildfire events. The statistical analyses of these forecasts over multiple years can be used by public health researchers, AQ regulators and policymakers, and others interested in wildfire impacts to understand and characterize exposure to wildfire smoke and its interannual and geographic variability.

Supplementary Materials: The following are available online at www.mdpi.com/2073-4433/8/9/179/s1.

Acknowledgments: The ECCC FireWork team is very grateful to the Natural Resources Canada team behind the Canadian Wildland Information System (CWFIS), who provided crucial assistance in the development of the FireWork system. We are particularly grateful to Kerry Anderson and Peter Englefield, who provided useful analyses and guidance on the CWFIS system. We also thank the U.S. Forest Service BlueSky team, particularly Susan O'Neill, for analyses and advice on wildfire emission estimates for the USA. We appreciate the assistance of Alexandru Lupu and Cassandra Bolduc for preparing tables presented in this article. Finally, we would like to acknowledge three anonymous reviewers for their valuable and constructive comments, which considerably improved this paper.

Author Contributions: Rodrigo Munoz-Alpizar and Radenko Pavlovic conceived the study; Rodrigo Munoz-Alpizar, Radenko Pavlovic and Sylvain Ménard backfilled missing FireWork forecasts over 2013–2015 five-month periods; Hugo Landry, Samuel Gilbert and Paul-André Beaulieu developed analysis tools; Jacinthe Racine and Annie Duhamel contributed to the results presentation; Rodrigo Munoz-Alpizar, Radenko Pavlovic, Michael D. Moran, Sarah B. Henderson, Jack Chen, Sylvie Gravel and Sylvain Ménard wrote the paper; and Didier Davignon, Sophie Cousineau, and Véronique Bouchet contributed with organizational support and scientific suggestions.

Conflicts of Interest: The authors declare no conflict of interest.

References

1. Beverly, J.L.; Bothwell, P. Wildfire evacuations in Canada 1980–2007. *Nat. Hazards* **2011**, *59*, 571–596. [CrossRef]

2. Canadian Interagency Forest Fire Centre. Canada Report 2016. Available online: http://www.ciffc.ca/images/stories/pdf/2016_canada_report_2017_05_15_v02.pdf (accessed on 15 May 2017).

3. Natural Resources Canada. Cost of Fire Protection. Available online: http://www.nrcan.gc.ca/forests/climate-change/forest-change/17783 (accessed on 15 May 2017).

4. Flannigan, M.; Tymstra, C. The Fort McMurray Wildfire: By the Numbers. Canadian Wildland Fire & Smoke Newsletter. 2016; pp. 1–2. Available online: https://sites.ualberta.ca/~wcwfs/CWFSN/newsletters/CWFSN_Fall_2016.pdf (accessed on 15 September 2017).

5. Urbanski, P.U.; Hao, W.M.; Baker, S. Chemical composition of wildland fire emissions. *Dev. Environ. Sci.* **2009**, *8*, 79–107.

6. Reid, C.E.; Brauer, M.; Johnston, F.H.; Jerrett, M.; Balmes, J.R.; Elliott, C.T. Critical Review of Health Impacts of Wildfire Smoke Exposure. *Environ. Health Perspect.* **2016**, *124*, 1334–1342. [CrossRef] [PubMed]

7. Liu, J.C.; Pereira, G.; Uhl, S.A.; Bravo, M.A.; Bell, M.L. A systematic review of the physical health impacts from non-occupational exposure to wildfire smoke. *Environ. Res.* **2015**, *136*, 120–132. [CrossRef] [PubMed]

8. Liu, J.C.; Mickley, L.J.; Sulprizio, M.P.; Dominici, F.; Yue, X.; Ebisu, K.; Anderson, G.B.; Khan, R.F.; Bravo, M.A.; Bell, M.L. Particulate air pollution from wildfires in the Western US under climate change. *Clim. Chang.* **2016**, *138*, 655–666. [CrossRef] [PubMed]

9. Thurston, G.D.; Kipen, H.; Annesi-Maesano, I.; Balmes, J.; Brook, R.D.; Cromar, K.; de Matteis, S.; Forastiere, F.; Forsberg, B.; Frampton, M.W.; et al. A joint ERS/ATS policy statement: What constitutes an adverse health effect of air pollution? An analytical framework. *Eur. Respir. J.* **2017**, *49*, 1600419. [CrossRef] [PubMed]

10. Canadian Council of Ministers of the Environment (CCME). Guidance Document on Achievement Determination Canadian Ambient Air Quality Standards for Fine Particulate Matter and Ozone. 2012; p. 43. Available online: http://www.ccme.ca/files/Resources/air/aqms/pn_1483_gdad_eng.pdf (accessed on 15 May 2017).

11. Sapkota, A.; Symons, J.; Kleissl, J.; Wang, L.; Parlange, M.; Ondov, J.; Breysse, P.N.; Diette, G.B.; Eggleston, P.A.; Buckley, T.J. Impact of the 2002 Canadian forest fires on particulate matter air quality in Baltimore city. *Environ. Sci. Technol.* **2005**, *39*, 24–32. [CrossRef] [PubMed]

12. Forster, C.; Wandinger, U.; Wotawa, G.; James, P.; Mattis, I.; Althausen, D.; Simmonds, P.; O'Doherty, S.; Jennings, I.M.; Kleefeld, C.; et al. Transport of boreal forest fire emissions from Canada to Europe. *J. Geophys. Res.* **2001**, *106*, 22887–22906. [CrossRef]

13. Adam, M.; Pahlow, M.; Kovalev, V.A.; Ondov, J.M.; Parlange, M.B.; Nair, N. Aerosol optical characterization by nephelometer and lidar: The Baltimore Supersite experiment during the Canadian forest fire smoke intrusion. *J. Geophys. Res.* **2004**, *109*, D16. [CrossRef]

14. Colarco, P.R.; Schoeberl, M.R.; Doddridge, B.G.; Marufu, L.T.; Torres, O.; Welton, E.J. Transport of smoke from Canadian forest fires to the surface near Washington, DC: Injection height, entrainment, and optical properties. *J. Geophys. Res.* **2004**, *109*, 1–12. [CrossRef]

15. Spichtinger, N.; Wenig, M.; James, P.; Wagner, T.; Platt, U.; Stohl, A. Satellite detection of a continental-scale plume of nitrogen oxides from boreal forest fires. *Geophys. Res. Lett.* **2001**, *28*, 4579–4582. [CrossRef]

16. Morris, G.A.; Hershey, S.; Thompson, A.M.; Stohl, A.; Colarco, P.R.; McMillan, W.W.; Warner, J.; Johnson, B.J.; Witte, J.C.; Kucsera, T.L.; et al. Alaskan and Canadian forest fires exacerbate ozone pollution in Houston, Texas, on 19 and 20 July. *J. Geophys. Res.* **2004**, *111*, 1–10. [CrossRef]

17. DeBell, L.J.; Talbot, R.W.; Dibb, J.E.; Munger, J.W.; Fischer, E.V.; Frolking, S.E. A major regional air pollution event in the northeastern United States caused by extensive forest fires in Quebec, Canada. *J. Geophys. Res.* **2004**, *109*, 1–16. [CrossRef]

18. Fiore, A.M.; Pierce, R.B.; Dickerson, R.R.; Lin, M.; Bradley, R. Detecting and attributing episodic high background ozone events. *J. Air Waste Manag. Assoc.* **2014**, *64*, 22–28.

19. Dreessen, J.; Sullivan, J.; Delgado, R. Observations and impacts of transported Canadian wildfire smoke on ozone and aerosol air quality in the Maryland region on June 9–12. *J. Air Waste Manag. Assoc.* **2016**, *66*, 842–862. [CrossRef] [PubMed]

20. Cottle, P.; Strawbridge, K.; McKendry, I. Long-range transport of Siberian wildfire smoke to British Columbia: Lidar observations and air quality impacts. *Atmos. Environ.* **2014**, *90*, 71–77. [CrossRef]

21. Teakles, A.D.; So, R.; Ainslie, B.; Nissen, R.; Schiller, C.; Vingarzan, R.; McKendry, I.; Macdonald, A.M.; Jaffe, D.A.; Bertram, A.K.; et al. Impacts of the July 2012 Siberian fire plume on air quality in the Pacific Northwest. *Atmos. Chem. Phys.* **2017**, *17*, 2593–2611. [CrossRef]

22. Centre for Disease Control. Evidence Review: Exposure Measures for Wildfire Smoke Surveillance. 31 March 2014. Available online: http://www.bccdc.ca/resource-gallery/Documents/Guidelines%20and%20Forms/ Guidelines%20and%20Manuals/Health-Environment/WFSG_EvidenceReview_Smokesurveillance_ FINAL_v2_edstrs.pdf (accessed on 15 May 2017).

23. Rappold, A.G.; Reyes, J.; Pouliot, G.; Cascio, W.E.; Diaz-Sanchez, D. Community Vulnerability to Health Impacts of Wildland Fire Smoke Exposure. *Environ. Sci. Technol.* **2017**, *51*, 6674–6682. [CrossRef] [PubMed]

24. Zhang, Y.; Bocquet, M.; Mallet, V.; Seigneur, C.; Baklanov, A. Real-time air quality forecasting, part I: History, techniques, and current status. *Atmos. Environ.* **2012**, *60*, 632–655. [CrossRef]

25. Zhang, Y.; Bocquet, M.; Mallet, V.; Seigneur, C.; Baklanov, A. Real-time air quality forecasting, part II: State of the science, current research needs, and future prospects. *Atmos. Environ.* **2012**, *60*, 656–676. [CrossRef]

26. Larkin, N.K.; O'Neill, S.M.; Solomon, R.; Raffuse, S.; Strand, T.; Sullivan, D.C.; Krull, C.; Rorig, M.; Peterson, J.; Ferguson, S.A. The BlueSky smoke modeling framework. *Int. J. Wildland Fire* **2010**, *18*, 906–920. [CrossRef]

27. Pavlovic, R.; Chen, J.; Anderson, K.; Moran, M.D.; Beaulieu, P.-A.; Davignon, D.; Cousineau, S. The FireWork air quality forecast system with near-real-time biomass burning emissions: Recent developments and evaluation of performance for the 2015 North American wildfire season. *J. Air Waste Manage. Assoc.* **2016**, *66*, 819–841. [CrossRef] [PubMed]

28. Pavlovic, R.; Chen, J.; Davignon, D.; Moran, M.D.; Beaulieu, P.A.; Landry, H.; Sassi, M.; Gilbert, S.; Munoz-Alpizar, R.; Anderson, K.; et al. FireWork—A Canadian Operational Air Quality Forecast Model With Near-Real-Time Biomass Burning Emissions. Canadian Wildland Fire & Smoke Newsletter. 2016; pp. 18–29. Available online: https://sites.ualberta.ca/~wcwfs/CWFSN/newsletters/CWFSN_Fall_2016.pdf (accessed on 15 September 2017).

29. Yuchi, W.; Yao, J.; McLean, K.E.; Stull, R.; Pavlovic, R.; Davignon, D.; Moran, M.D.; Henderson, S.B. Blending forest fire smoke forecasts with observed data can improve their utility for public health applications. *Atmos. Environ.* **2016**, *145*, 308–317. [CrossRef]

30. Moran, M.D.; Ménard, S.; Pavlovic, R.; Anselmo, D.; Antonopoulos, S.; Makar, P.A.; Gong, W.; Stroud, C.; Zhang, J.; Zheng, Q.; et al. Recent advances in Canada's National Operational AQ Forecasting System. In Proceedings of the 32nd NATO/SPS ITM on Air Pollution Modelling and Its Application, Utrecht, The Netherlands, 7–11 May 2012; pp. 215–220.

31. Moran, M.; Zheng, Q.; Zhang, J.; Pavlovic, R. RAQDPS Version 013: Upgrades to the CMC Operational Regional Air Quality Deterministic Prediction System Released in June 2015. 2015; p. 57. Available online: http://collaboration.cmc.ec.gc.ca/cmc/cmoi/product_guide/docs/lib/op_systems/ doc_opchanges/Technical_Note_GEM-MACH10_v1.5.3+SET2.1.1_Emissions_9Nov2015.pdf (accessed on 15 May 2017).

32. Im, U.; Bianconi, R.; Solazzo, E.; Kioutsioukis, I.; Badia, A.; Balzarini, A.; Baro, R.; Bellasio, R.; Brunner, D.; Chemel, C.; et al. Evaluation of operational online-coupled regional air quality models over Europe and North America in the context of AQMEII phase 2. Part I: Ozone. *Atmos. Environ.* **2015**, *115*, 404–420. [CrossRef]

33. Lee, B.S.; Alexander, M.E.; Hawkes, B.C.; Lynham, T.J.; Stocks, B.J.; Englefield, P. Information systems in support of wildland fire management decision making in Canada. *Comput. Electron. Agric.* **2002**, *37*, 185–198. [CrossRef]

34. Anderson, K.R.; Englefield, P.; Little, J.M.; Reuter, G. An approach to operational forest fire growth predictions for Canada. *Int. J. Wildland Fire* **2009**, *18*, 893–905. [CrossRef]

35. Moran, M.; Pavlovic, R.; Chen, J. FireWork 2016: Release of the Initial Operational Version of the CMC Regional Air Quality Deterministic Prediction System with Near-Real-Time Satellite-Derived Wildfire Emissions. 2016; p. 4. Available online: http://collaboration.cmc.ec.gc.ca/cmc/CMOI/product_guide/docs/tech_notes/technote_raqdps015fw_20160428_e.pdf (accessed on 15 May 2017).

36. Kaiser, J.W.; Heil, A.; Andreae, M.O.; Benedetti, A.; Chubarova, N.; Jones, L.; Morcrette, J.-J.; Razinger, M.; Schultz, M.G.; Suttie, M.; et al. Biomass burning emissions estimated with a global fire assimilation system based on observed fire radiative power. *Biogeosciences* **2012**, *9*, 527–554. [CrossRef]

37. Xing, J.; Pleim, J.; Mathur, R.; Pouliot, G.; Hogrefe Gan, C.-M.; Wei, C. Historical gaseous and primary aerosol emissions in the United States from 1990–2010. *Atmos. Chem. Phys.* **2013**, *13*, 7531–7549. [CrossRef]

38. EPA. Draft Guidance for Comment: Significant Impact Levels for Ozone and Fine Particle in the Prevention of Significant Deterioration Permitting Program. 2016; p. 13. Available online: https://www.epa.gov/sites/production/files/2016-08/documents/pm2_5_sils_and_ozone_draft_guidance.pdf (accessed on 15 May 2017).

39. Statistics Canada. Dissemination Areas, Census 2016. Available online: http://www12.statcan.gc.ca/census-recensement/2016/dp-pd/hlt-fst/pd-pl/Comprehensive.cfm (accessed on 15 May 2017).

40. Statistics Canada. Boundary Files, Census 2016. Available online: http://www12.statcan.gc.ca/census-recensement/2011/geo/bound-limit/bound-limit-eng.cfm (accessed on 15 May 2017).

41. United States Census Bureau. Census 2010. Available online: https://www.census.gov/geo/maps-data/data/tiger-data.html (accessed on 15 May 2017).

42. National Intergancy Fire Center. Available online: https://www.nifc.gov/fireInfo/fireInfo_stats_totalFires.html (accessed on 15 May 2017).

43. Jaffe, D.; Hafner, W.; Chand, D.; Westerling, A.; Spracklen, D. Interannual variations in $PM_{2.5}$ due to wildfires in the Western United States. *Environ. Sci. Technol.* **2008**, *42*, 2812–2818. [CrossRef] [PubMed]

44. Mao, Y.H.; Li, Q.B.; Zhang, L.; Chen, Y.; Randerson, J.T.; Chen, D.; Liou, K.N. Biomass burning contribution to black carbon in the Western United States Mountain Ranges. *Atmos. Chem. Phys.* **2011**, *11*, 11253–11266. [CrossRef]

45. Spracklen, D.V.; Logan, J.A.; Mickley, L.J.; Park, R.J.; Yevich, R.; Westerling, A.L.; Jaffe, D.A. Wildfires drive interannual variability of organic carbon aerosol in the western U.S. in summer. *Geophys. Res. Lett.* **2007**, *34*, L16816. [CrossRef]

46. Zeng, T.; Wang, Y. Nationwide summer peaks of OC/EC ratios in the contiguous United States. *Atmos. Environ.* **2011**, *45*, 578–586. [CrossRef]

47. Henderson, S.B.; Johnston, F.H. Measures of forest fire smoke exposure and their associations with respiratory health outcomes. *Curr. Opin. Allergy Clin. Immunol.* **2012**, *12*, 221–227. [CrossRef] [PubMed]

48. Dennekamp, M.; Straney, L.D.; Erbas, B.; Abramson, M.J.; Keywood, M.; Smith, K.; Sim, M.R.; Glass, D.C.; Monaco, A.D.; Haikerwal, A.; et al. Forest Fire Smoke Exposures and Out-of-Hospital Cardiac Arrests in Melbourne, Australia: A Case-Crossover Study. *Environ. Health Perspect.* **2015**, *123*, 959–964. [CrossRef] [PubMed]

49. Johnston, F.H.; Henderson, S.B.; Chen, Y.; Randerson, J.T.; Marlier, M.; DeFries, R.S.; Kinney, P.; Bowman, D.M.; Brauer, M. Estimated Global Mortality Attributable to Smoke from Landscape Fires. *Environ. Health Perspect.* **2012**, *120*, 695–701. [CrossRef] [PubMed]

50. Robichaud, A.; Ménard, R.; Zaïtseva, Y.; Anselmo, D. Multi-pollutant surface objective analyses and mapping of air quality health index over North America. *Air Qual. Atmos. Health* **2016**, *9*, 743–759. [CrossRef] [PubMed]

51. Jolly, W.M.; Cochrane, M.A.; Freeborn, P.H.; Holden, Z.A.; Brown, T.J.; Williamson, G.J.; Bowman, D.M. Climate-induced variations in global wildfire danger from 1979 to 2013. *Nat. Commun.* **2015**, *6*, 7537. [CrossRef] [PubMed]

atmosphere

MDPI

Review

Interpreting Mobile and Handheld Air Sensor Readings in Relation to Air Quality Standards and Health Effect Reference Values: Tackling the Challenges

George M. Woodall [1,*], Mark D. Hoover [2], Ronald Williams [1], Kristen Benedict [1], Martin Harper [2,†], Jhy-Charm Soo [2], Annie M. Jarabek [1], Michael J. Stewart [1], James S. Brown [1], Janis E. Hulla [3], Motria Caudill [4], Andrea L. Clements [1], Amanda Kaufman [1], Alison J. Parker [5], Martha Keating [1], David Balshaw [6], Kevin Garrahan [7], Laureen Burton [7], Sheila Batka [8], Vijay S. Limaye [8], Pertti J. Hakkinen [9] and Bob Thompson [1]

[1] Environmental Protection Agency, Research Triangle Park, NC 27711, USA; williams.ronald@epa.gov (R.W.); benedict.kristen@epa.gov (K.B.); jarabek.annie@epa.gov (A.M.J.); stewart.michael@epa.gov (M.J.S.); brown.james@epa.gov (J.S.B.); clements.andrea@epa.gov (A.L.C.); kaufman.amanda@epa.gov (A.K.); keating.martha@epa.gov (M.K.); thompson.bob@epa.gov (B.T.)
[2] National Institute for Occupational Safety and Health, Morgantown, WV 26505, USA; mhoover1@cdc.gov (M.D.H.); mharper@zefon.com (M.H.); jsoo@cdc.gov (J.-C.S.)
[3] Army Corps of Engineers, Sacramento, CA 95814, USA; janis.e.hulla@usace.army.mil
[4] Agency for Toxic Substances and Disease Registry, Atlanta, GA 30329, USA; caudill.motria@epa.gov
[5] ORISE Fellow hosted by U.S. Environmental Protection Agency, Washington, DC 20004, USA; parker.alison@epa.gov
[6] National Institute for Environmental Health Sciences, Research Triangle Park, NC 27709, USA; balshaw@niehs.nih.gov
[7] Environmental Protection Agency, Washington, DC 20004, USA; garrahan.kevin@epa.gov (K.G.); burton.laureen@epa.gov (L.B.)
[8] Environmental Protection Agency, Chicago, IL 60605, USA; batka.sheila@epa.gov (S.B.); limaye.vijay@epa.gov (V.S.L.)
[9] National Library of Medicine, Bethesda, MD 20892, USA; hakkinenp@mail.nlm.nih.gov
* Correspondence: woodall.george@epa.gov; Tel.: +1-919-541-3896
† Retired. Current affiliation with Zefon International, Inc., Ocala, FL 34474, USA.

Received: 17 August 2017; Accepted: 14 September 2017; Published: 21 September 2017

Abstract: The US Environmental Protection Agency (EPA) and other federal agencies face a number of challenges in interpreting and reconciling short-duration (seconds to minutes) readings from mobile and handheld air sensors with the longer duration averages (hours to days) associated with the National Ambient Air Quality Standards (NAAQS) for the criteria pollutants-particulate matter (PM), ozone, carbon monoxide, lead, nitrogen oxides, and sulfur oxides. Similar issues are equally relevant to the hazardous air pollutants (HAPs) where chemical-specific health effect reference values are the best indicators of exposure limits; values which are often based on a lifetime of continuous exposure. A multi-agency, staff-level Air Sensors Health Group (ASHG) was convened in 2013. ASHG represents a multi-institutional collaboration of Federal agencies devoted to discovery and discussion of sensor technologies, interpretation of sensor data, defining the state of sensor-related science across each institution, and provides consultation on how sensors might effectively be used to meet a wide range of research and decision support needs. ASHG focuses on several fronts: improving the understanding of what hand-held sensor technologies may be able to deliver; communicating what hand-held sensor readings can provide to a number of audiences; the challenges of how to integrate data generated by multiple entities using new and unproven technologies; and defining best practices in communicating health-related messages to various audiences. This review summarizes the challenges, successes, and promising tools of those initial ASHG efforts and Federal agency progress on crafting similar products for use with other NAAQS pollutants and the HAPs. NOTE: The opinions expressed are those of the authors and do not necessary represent the opinions of their Federal Agencies or the US Government. Mention of product names does not constitute endorsement.

Keywords: air pollutants; ambient air; indoor air; citizen science; toxic chemicals

---

1. Introduction

The *Air Sensors 2013: Data Quality and Applications* workshop, held in Research Triangle Park, N.C., highlighted the substantial advances in the development of portable air sensors capable of providing real-time measurements of ambient air pollution [1]. One anticipated benefit for the use of air sensors is the potential to expand upon the already well-established network of air quality monitors for key air pollutants. Sensor manufacturers introduced a number of new, relatively low-cost, portable air sensors capable of continuously measuring ambient levels of ozone (O_3) and nitrogen dioxide (NO_2). In addition, considerable progress was described regarding the development of air sensors capable of measuring ambient concentrations of particulate matter (PM), and total volatile organic compounds (VOCs). Since that 2013 workshop, the field of portable air quality sensing technology has continued to evolve at a rapid pace, with commercially available products now available for O_3, NO_2, PM, VOCs, as well as for other pollutants, both alone and in complex mixtures. While the potential for these technologies continues to be great, significant challenges in their application remain. Most notably, there is a wide range of data quality differences between different sensor manufactures and models [2], and there is a significant question about how air quality data collected on timescales as short as 1-min should be interpreted in comparison to the available health reference values for air pollutants which are typically based on exposure durations of several hours to many years [3]. For example, commercially available portable air sensors for O_3 can provide the user with minute-by-minute O_3 concentrations; however, interpreting these very short-term measurements with respect to potential adverse health effects is difficult. The health-based National Ambient Air Quality Standard (NAAQS) for O_3 is based on an 8-h average concentration which is backed by thousands of studies from numerous independent researchers and is established through a rigorous process to be scientifically defensible and legally enforceable; no such standards have been developed for these shorter duration exposures. Similar challenges exist when applying Occupational Safety and Health Administration (OSHA) standards. The most comparable OSHA exposure standards include the Permissible Exposure Limit (PEL), which is an 8-h time-weighted average, and the Short Term Exposure Limit (STEL), which is a 15-min time-weighted average.

Detection and monitoring of most non-NAAQS environmental chemicals (including the hazardous air pollutants), toxins and pathogens still largely involves identifying each individual agent, which often requires sending samples to a remote analytical laboratory for analyses. Delivery of laboratory results may take days, weeks or even months. Although the deployment of portable direct-reading instruments, such as photon ionization detectors (PIDs) for total VOCs, can provide some real-time information, these screening instruments lack the sensitivity or selectivity delivered by analytical laboratories and thus, are unable to fully inform a user with a critical need for high-specificity (e.g., rapid response decision-makers). New technologies such as miniaturized light emitting diodes, ultra-violet light detectors and functionalized graphene resistors have enabled the development of chemical detectors capable of delivering laboratory-quality analyses in near real-time. Coupling these types of sensors with global positioning and cell phone technologies may enable the detection, quantification, and visual monitoring of environmental contamination in real-time from remote locations. These promising technologies are expensive, however, and will likely remain out of reach for all but the most dedicated citizen scientist. As described in the following sections, understanding and advancing these types of applications has been a focus of the ASHG: beginning with defining the relevance of air sensors; facing the challenges of communicating across diverse key audiences; dealing with data validity issues; and moving toward a future with more reliable and useful sensor readings.

Formation of the Air Sensors Health Group (ASHG)

Recognizing the potential widespread use of portable air sensors, the likely data interpretation challenges these sensors would present to state and local governments, and the opportunity for collaboration across Federal agencies, the ASHG was formed in 2013. The ASHG is a multi-institutional collaboration of Federal agencies devoted to keeping abreast of new sensor technologies and to assist in the proper interpretation of sensor data as potential indicators of air quality. ASHG consists of experts in a number of areas, including toxicology, public health, engineering, monitoring and sampling, ambient air, indoor air, and occupational health, to name a few.

The ASHG monitors the state of the science at each institution and aims to find common ground on how sensors might effectively be used to meet a wide range of research needs in the occupational, indoor air, and ambient air settings. The ASHG also aims to be a resource for state, regional, and tribal organizations, as well as for citizen scientists and community members considering the use of portable air sensors for air pollution research and decision-making. To date, the ASHG has focused on multiple fronts: assisting EPA Program Offices in developing tools and message statements regarding the potential for adverse health effects from the short-duration air sensor readings for PM and O_3; providing analysis comparing short-term readings to longer-duration averages from existing official monitoring stations used in determining compliance to the NAAQS; and developing prototype visual tools to assist in communicating appropriate interpretive messages. These ASHG contributions have been incorporated into projects and programmatic products which are discussed in later sections of this paper. This review summarizes the challenges, and successes of those initial ASHG efforts, and Federal agency progress on crafting similar products for use with other pollutants.

2. Challenges

2.1. Relevance of Sensor Measurements

A critical concern for interpreting readings from air quality sensors is to have a realistic understanding of what the sensors are actually measuring; how that relates to the expectations of users (including citizen scientists, researchers, regulatory agencies, etc.) for sensitivity, specificity, and robustness of the intended application; and the extent to which those measurements of exposure can appropriately be used to communicate potential hazard and manage risks to public health. The magnitude of this concern varies greatly across the types of pollutants purportedly being measured and is related to a number of factors, such as the chemical and physical nature of the pollutant, the reactivity of the pollutant, concentration and compositional changes over time and location, and the influences with various atmospheric conditions. EPA guidance on achieving high-quality data through systematic planning using the data quality objectives process can be found online [4]. Standards for data quality can differ slightly depending on the context; additional standards are discussed as those contexts are examined in the ensuing sections of this review. The ASHG recognized early in their discussions that this ideal for collecting high-quality data may not be readily obtainable for all potential users of air pollutant sensors.

Low-cost air sensors on the market to date are predominantly of three types; optical, electrochemical, or metal oxide. Generally, these sensor types are not as specific and do not incorporate the front end conditioning or selection that is present in Federal Reference Method (FRM) or Federal Equivalent Method (FEM) measurement devices. Therefore, low-cost sensor measurements may suffer from the influence of co-responsive pollutants, environmental conditions, and even sensor component production variations. The need to understand what the sensor is measuring is of prime importance. In order to better understand the measurement and potential confounding influences, an informed user must understand the physical and chemical nature of a pollutant of interest and how it responds to the environment as well as the sensor measurement technology itself. Field testing may help a user understand the influence of confounding factors and could identify co-responsive pollutants. The interpretation and use of low-cost air sensor data will be enhanced by simultaneous collection of co-responsive pollutant concentrations and environmental data including temperature, relative humidity, wind, and weather (i.e., precipitation, fog), as well as observations about local pollution sources. In this context, chemical sensors must be validated for chemical specificity and sensitivity under the environmental conditions expected in the field prior to any consideration for demonstrating regulatory compliance.

2.1.1. Interpreting Sensor Readings

Managing expectations about what sensor measurement data can and cannot be used for is paramount in communicating with both citizen scientists and the general public at large, who for a variety of reasons (e.g., personal health, general air quality interest) may, in the future, routinely consult a sensor. In anticipation of this interest on the part of the public, the U.S. EPA began a pilot effort in June 2016, to begin addressing the interpretation of short-term sensor readings in the context of air quality [5]. There are several challenges to interpreting these data. Among these challenges are sensor performance and the short-term, sometimes instantaneous output from a sensor. Short-term sensor data come from instruments of unknown performance quality and, importantly, these short-term concentrations cannot be compared to the NAAQS to draw conclusions about what these nearly instantaneous exposures may mean in terms of health impacts. The NAAQS are based on longer exposure durations (e.g., 8-h or 24-h averages) consistent with the health evidence from the reviews of these standards. This health evidence does not support linking 1-min (or shorter) ozone or $PM_{2.5}$ concentrations to adverse health effects, thus a 1-min sensor reading is not directly comparable to the NAAQS, or to the related Air Quality Index (AQI) categories.

To help the public understand the implications of these readings, scientists at EPA have piloted a color-coded Sensor Scale that might be used in conjunction with the AQI to help the public better understand what their short-term sensor data mean in the context of local and regional air quality and consequently make behavioral decisions about outdoor activities. Statistical approaches were used to understand the relationship between short-term ozone and fine particulate matter measurements with longer term averages. The Sensor Scales and explanatory background materials are housed in the Air Sensor Toolbox on the EPA website [6] and have also been described elsewhere. To test the effectiveness of these messages and their visual presentation, EPA conducted focus groups in 2015 and 2017, before and after the deployment of the pilot respectively. Analysis of the outcomes of these focus groups is underway. After considering the input from the focus groups, EPA will refine the messages as appropriate and consider outreach to air quality sensor developers on these focus group findings. Additional work is also on-going to develop similar message schemes for selected HAPs.

In addition to the NAAQS, EPA's Integrated Risk Information System (IRIS) provides inhalation reference concentration (RfC) values for HAPs and other key pollutants important to many EPA Programs with a focus on chronic exposure durations (from years to a lifetime). The Agency for Toxic Substances and Disease Registry (ATSDR), a federal public health agency of the U.S. Department of Health and Human Services, develops similar substance-specific minimal risk level (MRL) values for acute (1–14 days), intermediate (15–364 days), and chronic (365 days and longer) exposure durations [7]. ATSDR's work is focused on Superfund sites, but environmental health specialists apply the MRLs in a wide range of investigations. Many state agencies also develop inhalation health effect reference values in support of their programmatic needs. More generically, the term reference value is used in this text to include all of the various values referred to as standards, guideline values, toxicity values, health benchmarks, etc., and includes values developed for use in emergency response, occupational exposure monitoring, and those protective of the general public. Additional reports are available for a more complete comparison of the available systems of health effect reference values [3,8], and between values for specific pollutants [3,9–13].

2.1.2. Occupational Versus Environmental Exposures

One issue likely to be resolved on a case-by-case basis is the overlap between the environment and the workplace. An employee may be exposed to a chemical in the workplace at a higher concentration than in an outdoor environment. In addition, because the worker is assumed to be fit and healthy and to have periods of recovery from exposure, the levels of concentration deemed tolerable are also higher. A citizen wearing a sensor outdoors, may well carry it into their workplace, and could then experience problems reconciling the levels of pollutants considered acceptable in these different milieus. For example, a sensor designed to help citizens avoid pollution from motor vehicles might measure carbon monoxide. EPA sets a primary NAAQS of 9 ppm averaged over an 8-h period and 35 ppm averaged over a 1-h period, not to be exceeded more than once per year. In the workplace, the situation is not quite as straightforward, but all the limit values which might be applied are higher than those under the NAAQS. The OSHA Permissible Exposure Limit is 50 ppm, while the National Institute for Occupational Safety and Health (NIOSH) Recommended Exposure Limit is 35 ppm, and the American Conference of Governmental Hygienists (ACGIH) Threshold Limit Value is 25 ppm, all averaged over an 8-h period, as with the 9 ppm NAAQS standard. If the monitor alarm is set for the 8-h average NAAQS value, or possibly even at the one-hour average value, the alarm might easily be triggered in the workplace, which might be, for example, a bus garage or foundry. The ensuing discussion with the workplace safety manager or the employer regarding the acceptability of the exposure situation might be difficult for both parties in the absence of well-thought out responses. Nevertheless, in a holistic vision of the exposome, which should consider all exposures over all life-stages, it may not be appropriate to regard ambient, indoor and workplace exposures as somehow "different", to be always measured and assessed separately [14], and so it is to be hoped that it will not be necessary to turn off ambient continuous air monitors "at the factory gates".

2.1.3. Global/International Perspectives

Low cost air quality sensors are of interest worldwide. Multiple studies have distributed sensors in an effort to apply them to local environmental research [15–18]. One of the most notable efforts involving multiple countries and metropolitan areas was the Citi-Sense project [19,20]. This diverse research project involved technology developers, citizen scientists, academics, and professional research organizations using low cost sensor technologies with a purposeful intent [21,22]. The pan-European initiative, EuNetAir, has goals of developing harmony in sensor selection and deployment strategies as well as coordinating sensor evaluation protocols to ensure the timely integration of air quality sensors into monitoring networks [23–25]. The Clean Air Asia consortium, a partnership of multiple Asian-based cities, sensor enthusiasts, academics, and air quality professionals are attempting to improve air quality and improve the overall living conditions in some of the most polluted cities in the world [26].

Even with the apparent world-wide enthusiastic use of low cost sensors to inform public awareness of environmental conditions, there is also a call to ensure the data being collected from these devices across the globe are accurate enough to be used in a purposeful manner [27]. This call for an adequate understanding of sensor performance is not only a reasonable approach but one that must be pursued with the same enthusiasm as those wishing to disseminate low cost sensors to a global population.

A multitude of pseudo air quality index messaging applications are available on the internet from sources around the world [28]. Some of these applications are using low cost sensors to collect air quality data, potentially not accounting for the accuracy of the measurement or the environmental setting in which the measurements are taking place (e.g., indoor, outdoor, near source categories). Furthermore, use of subjective data messaging on health impacts of such sensor measurements without a scientific basis has the potential of confusing the public-at-large and their understanding of sound air quality awareness indices.

The World Health Organization (WHO) describes air pollution as a "major environmental risk to health". The WHO Air Quality Guidelines (AQG) provide an assessment of air pollution health effects and recommend reference values for health-harmful pollution levels of ozone, PM_{10}, $PM_{2.5}$, SO_2, and NO_2 [29]. ATSDR has considered AQGs in the screening process of health evaluations for multiple site investigations in recent years [30–33]. The European Commission (EC) has developed air quality standards for the six US criteria pollutants, as well as benzene, polycyclic aromatic hydrocarbons (PAHs), and three metals: arsenic, cadmium, and nickel.

2.2. Communicating Across Audiences

Portable sensors present a great deal of promise for identifying personal exposure to toxicants present in the environment. However, communication between government or public health agencies, sensor manufacturers, researchers, employers and employees, and citizen scientists is particularly important given the varying degrees of accuracy across sensors, the complexities of environmental exposure, and the difficulties in interpreting potential health risk based on readings from a portable sensor. Thus, the subsections below summarize the various efforts that agencies represented on the ASHG membership have put toward communicating with these groups.

2.2.1. Citizen Scientists and Communities

Over the last decade, members of the public have become increasingly engaged in taking measurements of their environments and otherwise contributing to scientific research. Enabled by the rapid pace of growth of sensor technology, many citizen scientists collect data on air quality in their local environments, both individually and as a part of organized projects. The growth of these technologies and a surge in enthusiasm for these approaches has pushed the boundaries of traditional institution-driven research. Often called "citizen science", these efforts are also referred to as civic or community science, community-based monitoring, crowdmapping, participatory science, open science, or crowdsourcing. In *citizen science*, the public participates voluntarily in the scientific process, addressing problems in ways that may include formulating research questions, conducting scientific experiments, collecting and analyzing data, interpreting results, making new discoveries, developing technologies and applications, and solving complex problems [34]. Of particular relevance to air quality research and air sensors, *community science* or *community citizen science* is "collaboratively led scientific investigation and exploration to address community-defined questions, allowing for engagement in the entirety of the scientific process. Unique in comparison to citizen science, community science may or may not include partnerships with professional scientists, emphasizes the community's ownership of research and access to resulting data, and orients toward community goals and working together in scalable networks to encourage collaborative learning and civic engagement" [35].

The growth of citizen science offers significant opportunities and challenges for federal agencies. Citizen science increases public understanding and community and civic engagement with science and environmental issues, especially locally. Citizen science can connect agencies to the public, provide opportunities for working together towards common goals, support innovation, and make science more accessible and available. It provides data that would otherwise be inaccessible, helps generate a more comprehensive understanding of variation over space and time, and can increase our understanding of social science and human behavior. At the same time, low cost sensor technologies and citizen science introduce challenges to federal agencies, such as communicating risk to project participants, increased public pressure for actions like regulations and enforcement, and communication of data quality. In December 2016, the National Advisory Council for Environmental Policy and Technology provided EPA with advice and recommendations for how to maximize the benefits of citizen science and respond to the corresponding challenges. Recommendations include building technical capacity, providing guidance and communicating data quality needs for different data uses, and integrating citizen science into the full range of EPA's work [34].

EPA is collaborating with citizen scientists to share tools and technology, conduct air monitoring studies, and interpret sensor data. The Air Sensor Toolbox website [6] was created in 2014 to provide resources and tools to citizens interested in learning more about conducting successful air monitoring projects. Topics include how to use sensors, interpreting sensor data, information about EPA air monitoring projects, funding resources, and local air monitoring examples across the United States. The Air Sensor Guidebook [36] provides a comprehensive overview of air pollutants, sensors, study design, data collection, and data interpretation. Other resources include sensor evaluation reports highlighting performance of various sensors on the market, standard operating procedures for sensors, fact sheets, blogs, and training videos from the EPA-sponsored Community Air Monitoring Training in 2015.

Another way EPA engages communities is through collaborations. One such collaboration was conducted in the Ironbound community in Newark, New Jersey through a Regional Applied Research Effort (RARE) grant [37]. EPA scientists trained citizen scientists from the Ironbound Community Corporation (ICC) (The Ironbound section of Newark is a multicultural, multiracial mosaic whose population of 50,000 reflects the diversity and the challenges in urban America. ICC impacts the lives of nearly 1000 people daily and thousands annually. The majority of ICC's 3000 annual clients are from very low to low income households with low literacy and English proficiency and multiple family stressors.) to operate sensor pods developed by the Office of Research and Development (ORD). The collaboration allowed for joint decisions regarding study design, sensor siting, and collection, validation and interpretation of sensor readings. Following the Ironbound project, a similar RARE grant collaborative project was performed in Ponce, Puerto Rico. EPA scientists used lessons learned from the first project to attempt to improve study design topics such as roles and responsibilities and data validation and interpretation. Tasks such as managing, validating, and interpreting large datasets can be an obstacle for community groups involved in air monitoring projects.

ATSDR and NIOSH have partnered with University of Cincinnati, Georgia Tech, and University of Texas Arlington to pilot the use of real-time sensors as sentinels that trigger sample collection via more conventional devices. For example, a low-budget hydrogen sulfide detector can be programmed such that, once a certain threshold is exceeded, a VOC canister will be filled and then sent for laboratory analysis. ATSDR is developing these projects either to characterize peak exposures or as the first stage in assessing the need for a full-blown traditional air monitoring study.

Sensor collocation is an important step to perform before embarking on any monitoring project, regardless of who is conducting the monitoring. This process involves siting sensors in line with regulatory monitors (the gold standard) for a period of time in order to compare the two datasets and establish a regression equation to normalize sensor data and make it more accurate. EPA scientists recently created tools designed to assist citizens in this process, including an easy-to-use Excel™ macro that allows one to compare two datasets, such as sensor and reference data, and with one click provides a regression equation, comparative data graphs, and descriptive statistics. A training document that explains collocation and why it is important, and how to perform a proper collocation, accompanies the macro. EPA is partnering with the Clean Air Carolina community group and the Eastern Band of Cherokee Indians to conduct their own collocation projects, pilot test these resources and provide input on how to improve them for use by the general public.

2.2.2. State, Local and Tribal Agencies

State, tribal and local health and regulatory agencies were identified as a primary target audience early in discussions within the ASHG. The more proximately located agencies and offices are more likely to be called by the public and press, and may be more resource-challenged to interpret data generated using sensors. The EPA Regional Offices were identified to be the most likely first contact point for EPA. Similarly, ATSDR has Regional Offices that are another resource for health agencies.

Health agencies at all levels of government are called upon by the public and press to interpret air monitoring data. City and county health departments, depending on their size and community needs, may have environmental health specialists on staff who are familiar with air monitoring methods. Frequently, local agencies will defer to their respective state health department, many of which have specialists that are supported by the ATSDR Cooperative Agreement Program. State or local agencies may in turn also contact ATSDR or its sister agency, the National Center for Environmental Health (NCEH), which is part of the Centers for Disease Control and Prevention (CDC). As noted earlier in Section 2.1.1, there are several reference value systems which have been applied in interpreting exposures to non-criteria air pollutants, including the RfC and MRL values. Health agencies also frequently make site-specific exposure dose calculations and may evaluate health risks for shorter averaging periods by adapting the dose-response information in toxicology studies on which the reference values are based.

In order to address these needs, EPA Regional Offices have initiated outreach to state, tribal, and local environmental agencies to facilitate information sharing amongst stakeholders and to develop an inventory of sensor-based community air quality investigations. EPA Regional Office staff have highlighted EPA's Air Sensor Toolbox for Citizen Scientists [38] in this outreach, facilitating exchanges on sensor operation and maintenance, funding opportunities, and tools for data interpretation. For example, the Minnesota Pollution Control Agency provided input to the EPA Regional Office in Chicago (Region 5) on its own experience managing an air sensor loan program, and shared documentation it had developed to inform community members about proper sensor operation.

Monitoring for the NAAQS pollutants is delegated to state agencies with oversight from EPA Regional Offices and EPA's Office of Air Quality Planning and Standards (OAQPS); there is a structured regulatory program to determine compliance status with each NAAQS. Emissions of HAP are regulated at the source of the emissions, and monitoring is mostly initiated at the local level (i.e., state, tribal and local agencies), with federal guidance and some federal funding. There are very specific regulatory requirements in monitoring for the NAAQS pollutants but the States and Tribes have more leeway in monitoring for other purposes (e.g., for HAPs). EPA's Superfund Program uses various technologies for emergency response, and new sensors may be useful to inform decisions in these scenarios. Federally recognized Tribes interface with EPA and ATSDR on a government-to-government basis; many receive EPA grants or ATSDR technical assistance to do monitoring and other environmental health projects.

2.2.3. Sensor Manufacturers

Low cost air quality sensors have exploded onto the commercial market in the last decade. Development and manufacturing of these devices is not isolated to a single sector or industry. Some of the earliest developers were design teams associated with academic industrial/design arts programs, telecommunication research groups, and non-profit citizen scientists [39]. In particular, these developers took advantage of inexpensive sensor components and their individual areas of expertise to craft air quality sensors to meet a wide variety of needs [40]. More traditional instrumentation manufacturers have recently started to invest research capital into the low cost sensor area [41]. Their presence is the result of the growing citizen science movement and an obvious market niche.

The existence of a multitude of sensor manufacturers has resulted in both obvious benefits and deficits. On the plus side, there are numerous low cost sensors in a price range typically well under $1000 available for the interested user for both particle as well as gas phase pollutant monitoring. Many offer user friendly features such as immediate data visualization through smart phone applications with wireless data transmission. Some concerns associated with low cost sensor production is that often manufacturers may lack the technical know-how or capabilities to adequately test or calibrate their devices prior to releasing them to the market [27]. Other concerns include reliability issues between replicate copies of the same sensors attributed to either a poor overall manufacturing process associated with the fully assembled sensor or issues with the individual components themselves [42,43].

Recent sensor evaluation efforts by recognized institutions, including the U.S. EPA and others [6,44] would appear to be having a positive impact upon sensor manufacturing. In particular, work conducted in 2012 resulted in some of the first reported performance tests of sensors from a wide variety of manufacturing sources [2]. The evaluations resulted in an almost immediate update of the technology by many of the manufacturers to overcome issues first revealed in the tests (e.g., battery failures, low detection sensitivities, poor telecommunications protocols). Some of those early manufacturers would appear to have left the market while other new ones have joined. These observations suggest that it is likely that more mature instrument manufacturers will continue to invest in this area and capture more of the market share as start-up groups continue to develop unique devices to meet a particular niche.

2.3. Calibration and Validation

As mentioned previously, the calibration procedures conducted by most low-cost sensor manufacturers are often severely inadequate or non-existent and fail to establish the sensor's performance characteristics, especially when used in the ambient environment. End users and research organizations have often taken the lead on establishing the performance of these emerging technologies and sharing their results. Noted groups establishing sensor performance include the European Union's Joint Research Center [45], the U.S. EPA [46], and the South Coast Air Quality Management District's AQ-SPEC laboratory [47]. Results show that while nephelometric (light scattering) devices might perform well in direct chamber-based evaluations, they often reveal significant departures from a true reference grade instrument response under real-world (ambient) conditions [1]. In like manner, while excellent chamber-based response relationships have been observed for select gas phase pollutants [2], ambient test results have been less promising or inconsistent [42]. There are of course, exceptions to these observations and certain devices and sensors appear to show more promise relative to their performance features [48].

Gas detection in the workplace has a long history of being driven by flammability concerns, beginning with the miners' safety lamp, but direct-reading real-time instruments have also been used to determine toxic gases for almost 100 years; colorimetric detector tubes patented for detecting carbon monoxide in the 1930's are examples. More recent developments of infra-red analyzers and gas chromatography detectors (with or without an associated gas chromatograph) have been portable, but not personal, although current research into miniaturization may alter that situation. Electrochemical cells have been developed for specific acute hazards, such as carbon monoxide, hydrogen sulfide and chlorine. These are used in process safety applications, but have also been adapted for use in personal dosimeters. The quality of data used for safety monitoring and personal exposure assessment in the workplace is of paramount importance, particularly in relation to acutely hazardous substances. The European Union countries have developed guidance for the performance, testing, selection, installation, use and maintenance of electrical apparatus used for the direct detection and direct concentration of toxic gases and vapors. Other Standards-setting organizations have published similar products, and a joint Working group of the International Organization for Standardization (ISO) and the International Electrotechnical Commission (IEC) are working on umbrella standards to replace these (these Standards are detailed in Appendix A). In the USA, NIOSH has developed guidance for ensuring data quality for gas and vapor detection, including during emergency response [49,50]. The American Industrial Hygiene Association, through their Gas and Vapor Detection Systems Committee (now Real-Time Detection Systems Committee) has developed guidance for manufactures to report specifications for electronic real-time gas and vapor detection equipment [51].

Recognizing the need to calibrate instrument response under real-world conditions, variable performance based on changing conditions (e.g., aerosol size-distribution), and deteriorating sensor response as due to age, new research efforts have been focused on the idea of auto-calibration methods [52]. Such methods may be able to maintain the accuracy of a node-based distributed sensor system for longer term measurements.

Extending beyond the need for validation in controlled laboratory settings to "real world" conditions and environmental epidemiological and citizen science applications, the National Institute for Environmental Health Sciences (NIEHS) released a phased field validation program in 2013 [53]. Through this effort, teams of engineers, exposure scientists and environmental epidemiologists are working together to test sensor performance at the pilot scale, including iterative improvements in sensor system design and performance and then scaling up the application to a full-sized epidemiological study to demonstrate the added value of temporally and spatially resolved air pollution exposure assessment relative to existing 'citizen' measures.

3. Related Projects and Programs

Although the ASHG has not been a prime mover in development of key products or programs, it has served as an advisory group and has contributed to a number of projects and programs, as noted in the sections below.

3.1. Village Green

The U.S. EPA's Village Green project has provided a test-bed for long-term evaluation and application of emerging air quality sensors in a variety of community settings [54]. While the sensor technologies for particulate matter and ozone are not considered truly low cost (~$6000/each), they do represent mid-tier technologies [55] which are showing good to excellent capabilities for certain attributes. In particular, they are capable of providing extended periods (months) of ambient air quality monitoring using sustainable energy (solar power) with little or no technical support and often with a high degree of agreement with local reference monitoring [56]. Another unique feature of the Village Green is that it was designed to stream data continuously to the public via its web-based data portal. It was in fact, the first real-time public reporting of air quality data for environmental awareness purposes provided to the general public by the US EPA. Since its conception in 2014, Village Green stations have now been deployed in a total of eight (8) U.S. metropolitan areas and involve a variety of air quality sensor and emerging technologies (variable sustainable power supplies, data microprocessing features, etc.).

At the time of its development as a technology test bed and community air quality awareness tool, its capabilities to meet purposeful air quality data analyses were not established. EPA's investigations into short time interval sensor data messaging provided an opportunity to use the Village Green for such an analysis due to its extensive database containing 1-min measurements of both ozone and fine particulate matter. Analysis of that database resulted in a pilot Sensor Scale associated with potential short time interval air quality measurements [5]. EPA is launching a pilot project to test a new tool for making instantaneous outdoor air quality data useful for the public. The new "Sensor Scale" is designed to be used with air quality sensors that provide data in short time increments–often as little as one minute [5]. EPA developed the scale to help people understand the 1-min data the stations provide and how to use those data as an additional tool for planning outdoor activities.

3.2. The E-Enterprise Advanced Monitoring Team (EEAMT)

Under the direction of the E-Enterprise Leadership Council, a joint EPA/State Advanced Monitoring team was formed in April 2015 to address the challenges and opportunities presented by rapidly changing air and water technologies. Data interpretation is one of five priority projects identified in the path forward for EPA, States, and Tribes [57]. In order to provide context and interpretation of advanced monitoring data in formats relevant and understandable to users, the team was charged with advancing (1) statistical analyses to understand the relationship between continuous data and data collected over longer-term averaging times or via discrete (e.g., bi-weekly) sampling;

(2) development of visualization tools (e.g., interactive maps) and websites with appropriate messaging; and (3) development of outreach and communication materials.

Conducting statistical analyses to understand the relationship between short term measurements and longer term standards or discrete measurements is a blanket need across media and pollutants. While messaging already exists to alert the public about air or water quality conditions experienced over specific time periods (e.g., 1-h, 8-h, or 24-h), the same messaging should not be used to translate short term (e.g., 1-min) measurements. Two articles discussing an approach to relate short term ozone and particulate matter ($PM_{2.5}$) measurements with longer term averages have been published [58,59]. Other needs identified by the team include data analysis and messaging for SO_2, NO_2, CO, PM_{10}, benzene, total volatile organic compounds (VOCs), and other specific VOC compounds.

In order to create visualization tools that make continuous and discrete state/federal data more accessible and understandable, EPA and States have determined it is important to present real-time (e.g., hourly concentration values) along with appropriate caveats. Examples of caveats include marking data as "raw", "provisional", or "final" and distinguishing between different forms of data (e.g., regulatory vs. peer reviewed). A need exists to display continuous data along with discrete measurements while giving the data context including geographical and meteorological information. Mockups were drafted as part of a thought process to address these issues and allow for EPA and States to provide guidance or expertise to big data developers desiring to stream, collect, or use regulatory and sensor data. Once short term messaging has been developed, it is important to develop appropriate outreach and communication materials that EPA and States can use for consistent messaging including frequently asked questions (FAQs) and a standardized, centralized repository of metrics that break down toxicity/health information by pollutant for both short term and long term effects (e.g., gaseous, metals, PM, pH, toxins, dissolved oxygen, etc.). Outreach material should describe the limitations of sensors and sensor networks including information on what we do not know and cannot measure with confidence.

3.3. Homeland Security Applications

Air sensors have many important applications relating to homeland security, but two roles are especially important: (1) serving as sentinels to detect the release of dangerous substances into the air, and (2) providing measurements of the amount of dangerous substances present in the air so that health risks can then be assessed and appropriate decisions made to protect public health.

Serving as sentinels to provide early warning of the release of dangerous substances is a very critical function for real-time air sensors. In this capacity, air sensors are designed to monitor background levels of various substances (dangerous chemicals or/or their indicators) and detect any significant increases in their presence. Air sensor networks have already been deployed extensively as sentinels in several major cities, chemical plants, military installations, and during major public events. These networks have been deployed on roof tops, in underground transportation systems, and in various types of mobile ground vehicles and aircraft.

In addition to their role as sentinels to detect the release of dangerous chemicals, air sensors may be used to measure concentrations of harmful chemicals which can potentially provide an estimate of health risks to persons breathing the air. Estimating health risks from acute or short-term exposure durations is a challenging task, as has already been noted in this paper. Most of the tools used to assess health risks are more centered on long-term, chronic, or lifetime exposure to chemicals. For example, as EPA's premier program for assessing health risks from chemicals, the Integrated Risk Information System (IRIS) has focused on dose-response data for chronic exposures. Health effect reference values derived to assess long-term health risks have very little, if any, value for assessing health risks from acute, short-term, or intermittent exposure scenarios, and few have been developed to assess those health risks to the general public.

A number of chemical release accidents, including the December 1984 accidental release of methyl isocyanate from a Union Carbide chemical plant in Bhopal, India, led to the formation of the Acute

Exposure Guideline Level (AEGL) program in 1995 [60]. Standing Operating Procedures for deriving AEGL values were developed by the National Research Council (NRC) [61] for use in both planning for and during a catastrophic chemical release event, using values covering inhalation exposures from 10-min to 8-h.

Shortly after the events of 11 September 2001, the need for additional sets of values for use for remediation in the aftermath of such events led to the development of another set of health metrics, called Provisional Advisory Levels (PALs). PALs were derived by building upon the AEGL methodology developed by the NRC [61], to create values for use at 1-day, 30-days, 90-days, and 2-years durations [62]. Under the PALs program, EPA has drafted over 3000 numeric values for over one hundred priority chemical agents, two routes of exposure (ingestion and inhalation), three levels of harmful effects, and four relatively short exposure durations. These values, in addition to those already published by many other organizations [3], can provide a context on which to compare sensor-derived exposure levels.

3.4. Indoor Air–Non-Industrial

Indoor air quality (IAQ), which is the air quality within buildings (i.e., homes, schools, offices and other non-industrial buildings) and other enclosed spaces, can affect the health, comfort and ability to perform for occupants, from infants to senior citizens. IAQ involves many factors including: outdoor air contributions; chemical, micro-biological, and particulate contaminants; and, characteristics of the indoor climate such as temperature, humidity and airflow. Americans spend about 90 percent of their time indoors [63–66], where pollutant levels, like some VOCs, may be two to five times higher—and occasionally 100 times higher than outdoors [67–69]. EPA traditionally has focused its IAQ efforts to identify sources and developing guidance to reduce human exposure to unhealthy indoor air and to provide low-cost mitigation strategies consistent with public health practices. In addition, many international professional, trade, and standards organizations have dedicated committees addressing a broad range of indoor air quality issues. With the increased availability of low-cost sensor technology, however, the resources we use to assess indoor air have been expanding.

One challenge for sensor work within indoor environments is the fact that the several pollutants important in the indoor environment cannot be detected accurately by current low-cost sensor technology. Indoor air sensors have the same challenges that ambient air sensors encounter when attempting to evaluate short-term sensor readings. Most of the indoor air pollutants also share the challenge associated with HAPs, the lack of enforceable or agreed upon health based standards to with which to compare sensor readings. As with HAPs, the available health effect reference values may be inappropriate for direct use with indoor air sensor readings. There are several other challenges unique to using sensors indoors—for example, unlike the ability of ambient air sensors to be collocated with regulatory monitors to help assess accuracy, there is no similar availability for indoor air.

While guidance development around IAQ best practices will continue within EPA, the potential for increased use of sensor technology for indoor air quality continues to evolve. Sensors that work with occupants' health-assessment technologies and those that can actuate building systems or individual appliances automatically may help enhance IAQ efforts in the built environment. There is a great potential that, as low-cost sensor technology improves, it may be a viable complement to comprehensive building system approaches currently used by the IAQ community and would help inform and improve IAQ stakeholders' ability (both professional and consumer level) to use IAQ management tools, interpret air quality data for their space, and assess the benefits of IAQ-related actions.

It is important that we not only continue the engineering "fixes" that are constantly improving source control, as well as ventilation and filtration/air cleaning systems, but also work to improve sensor technology as a tool to help optimize building performance and increase the accessibility of these tools as a complement to integrated strategies for improving IAQ for all parts of society. Within the context of IAQ, this may involve creating future systems that can be automated, improving the integration of these low-cost tools with building and health-assessment systems, establishing guidance that creates IAQ best practices that integrates these tools into existing public health practices and provides appropriate messaging for the public to better understand their health in their indoor environments.

4. Maximizing the Usefulness of Sensor Readings

As discussions within the ASHG have progressed over the years, approaches have been considered on how sensors may improve our understanding of real-world dosimetry, and how we might better organize the process of developing and using sensors through life-cycle analysis. Both of these approaches are more immediately relevant to research-grade sensor technologies; however, as more improvements in sensitivity and specificity are made to the low-cost sensors available to citizen scientists, these approaches will apply to a more universal set of sensor users.

4.1. Transitioning Sensors into Dosimeters and Future of Exposure Science

Radiation biology provides a paradigm for transitioning sensor-derived data into estimates of internal dose. Biological effect data from early animal studies modeled radiation exposures to define internal dose and algorithms were developed so that the readout from portable detector could be translated into the dose delivered to internal target organs.

It is the internal dose (the concentration of the agent at the cell/organ/tissue being affected) and not the environmental exposure level that is more closely related to toxicological effects. Health-based standard setting and development of health effect reference values are moving to apply emerging advances in exposure and toxicological sciences aimed at a more seamless integration of exposure and internal dose metrics, together with the incorporation, coherent alignment, and coordination of novel data streams [70–72]. Elucidation of the linkages between exposures and adverse effects in humans and the ecosystem will result in a better understanding on which to develop effective management strategies.

Constructing a robust context for this integration calls for coordinated research with human-health and ecologic-health scientists to identify, collect, and evaluate data that capture internal and external markers of exposure in a format that improves the analysis and modeling of exposure–response relationships and links to emerging methods for hazard identification such as high-throughput test systems (HTS). Figure 1 depicts selected scientific and technologic advances for measuring and monitoring considered in relation to a conceptual integration of exposure and dose as espoused by the National Research Council [71].

Mattingly et al. [73] developed an exposure ontology (ExO) designed to address the lack of exposure information required to elucidate environmental contributions to diseases, translate molecular insights from new technologies such as HTS, and aid assessment of human and ecological risks. The ExO formalized definition such as "exposure receptor" and centralized the role of exposure science with the intent to extend its ability to integrate and analyze exposure information within the broader context of environmental health. The exposure receptor can be an organ, tissue or cell, and the exposure stressor can be a biological, physical, or psychosocial agent, so that exposure assessment may include estimating the magnitude, frequency, and duration of an exposure, along with characteristics of the specific receptor. This concept of the exposome is consistent with the life-cycle approach articulated above regarding sensors. Such integration will expand the impact of exposure data and inform existing environmental health data by providing associated real-world exposure context.

On the response side of the exposure-dose-response linkage, adverse outcome pathways (AOP) are emerging as an important construct for the integration of toxicological effects across various levels of biological organization (e.g., genomic, cellular, target tissue, individual, population) based on HTS and identification of molecular initiating events [74,75]. This construct is also completely consistent with the ExO, and more recently Teeguarden et al. [76] have proposed formal linkage of exposure science with the AOP by means of the aggregate exposure pathway (AEP).

Therefore, emerging technologies now provide a method whereby measures of environmental exposure can be translated to an internal dose which can then be used in applications such as biomonitoring of populations or in cell systems. These approaches essentially provide a scalable platform with which to depict both exposure and internal dose. Alignment of exposures across experimental toxicity-testing systems can be achieved by understanding, measuring, and applying this information on the processes that control the time course of concentrations and delivery of chemicals and particles to target cells in various test systems at different scales (e.g., scale of ecological epidemiological studies, target tissue dose for in vivo animal studies, delivered concentration in HTS) that may be used as the basis for developing health-based values to which sensor data may be compared. Such exposure and internal dosimetry considerations can provide a context for the interpretation of emerging sensor data and may also inform future considerations of the form of various health-based reference values.

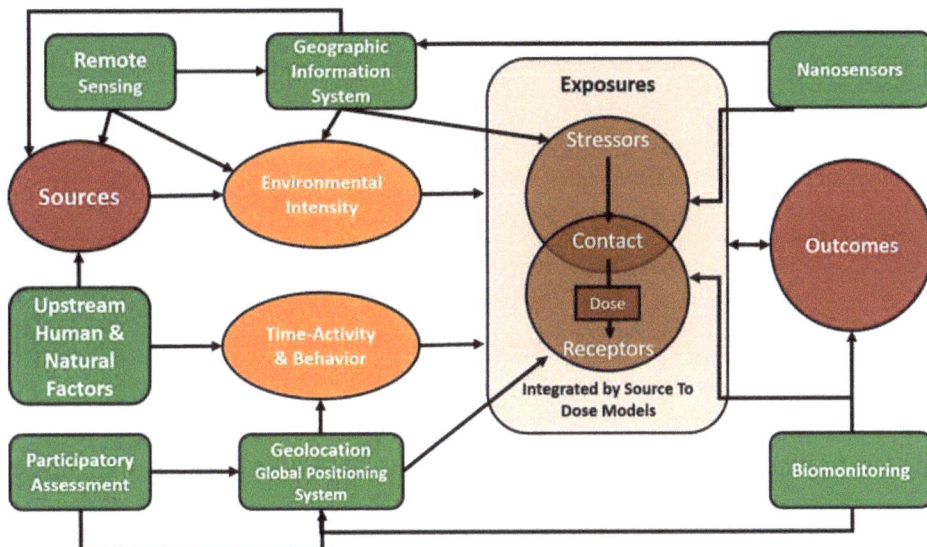

Figure 1. Conceptual integration of exposure and dose (from National Research Council).

4.1.1. Dosimeters in the Workplace

The Army projects aimed at developing a naphthalene dosimeter are an example of how biomarkers of exposure are being integrated with chemical sensor-derived data [77]. The first in the series of naphthalene dosimeter projects was funded through the Army's SBIR program. It developed an instrument capable of measuring naphthalene from air every three minutes [78]. A second project, underway at NIOSH, is an independent evaluation of the performance of the instrument. A third project lead by the U.S. Army Research Institute of Environmental Medicine is deploying the prototype instrument on military fuel-handlers to measure the concentration of naphthalene in their breathing zone. Concurrent to collecting the personal real-time exposure data, biomarkers of naphthalene exposure are being collected from the exhaled breath, urine and skin of the individual wearing the air sensor. Following the radiation dosimetry paradigm, the concurrently collected exposure and biomarker data will be used in a fourth project that will inform development of a model to estimate internal dose [79].

4.1.2. Dosimetry for NAAQS Air Pollutants

Portable sensors available for some NAAQS pollutants (namely, PM, O_3, and NO_2) have been evaluated by EPA. There are not specific metabolites that can be used to quantify the general relationship between exposure and dose. However, unlike naphthalene which enters the body by both dermal absorption and inhalation, at least these three NAAQS pollutants enter the body predominately by inhalation. In general, greater than 80% of inhaled ozone is absorbed in the respiratory tract (see Table 5-1 of U.S. EPA, 2013 [80]). Thus, dose can be approximated as the product of O_3 concentration, minute ventilation, and duration of exposure. Conceivably, an estimate of minute ventilation could be derived using an accelerometer in combination with data on the mass and gender of an individual. A more exact linkage to predicted decrements in lung function due to O_3 exposure, at least in healthy individuals, could be calculated by linking dosimeter data with an existing model [81]. This type of approach could be applied to both occupational and ambient ozone exposures or adapted for other compounds where health endpoints are closely related to inhaled dose.

Development of a dosimeter for PM is complicated due to the dependence of particle deposition on inhaled particle size, route of breathing, tidal volume, breathing rate, and lung size [82]. A couple of PM samplers that mimic the deposition in adults during nasal breathing and light exercise based on the International Commission on Radiological Protection (ICRP) model [83] have been developed. First, Koehler et al. [84] employed the use of a foam plug in which particle deposition efficiencies are similar to the ICRP [83] predicted total respiratory tract deposition (average of adult males and females) for particles between 0.05 and 2 μm. Total deposited dose in the respiratory tract can be determined either gravimetrically or by digesting the foam and extracting metals or organs for quantification by other means. Second, TSI has developed a Nanoparticle Surface Area Monitor (Model 3550) which can be used to determine particle surface area depositing in the tracheobronchial and alveolar regions of the lung for particles between 0.01 and 1 μm. The Model 3550 is designed to match predicted deposition in an adult male. Its operation is based on the diffusion charging of particles followed by detection of the aerosol using an electrometer. Particle doses can be assessed by the second, computed as a time-weighted average, or cumulative total deposited particle surface area. Available portable PM sensors typically use light scattering to estimate PM mass or number concentration in air for micron-sized or larger particle fractions such as $PM_{2.5}$. Unless an underlying PM size distribution is known or assumed, estimates of dose cannot be derived based on data from these portable PM sensors.

The utility of a dosimeter for ambient NO_2 is questionable. For NO_2, one of the critical endpoints is asthma exacerbation via an increase in airways responsiveness, as discussed in Section 5.2.2 of the Integrated Science Assessment (ISA) for NO_2 [85]. Since the increase in airway responsiveness does not appear to be associated with NO_2 dose for NO_2 concentrations between 100 and 600 ppb [86], it may be sufficient to monitor NO_2 concentration. Additionally, ambient NO_2 concentrations are generally only elevated near roadways, as discussed in Section 2.5.3 of the 2016 ISA [85]. An elevated NO_2 sensor reading could serve as a warning for a person with asthma to take actions to reduce exposure, e.g., by using recirculation of air in an automobile or avoiding outdoor activities in close proximity to major roadways or during periods of increased traffic.

4.1.3. Dosimeters for Hazardous Air Pollutants (HAPs)

There are 187 air pollutants designated as HAPs, including several VOC, that are emitted by point sources and under the purview of Section 112 of the Clean Air Act [87]. These pollutants are known or suspected to cause cancer or other serious health effects, such as reproductive effects or birth defects, or adverse environmental effects and are often also encountered in the workplace. As sensor technologies emerge to address these pollutants, our recommendations for life-cycle characterization and dosimetry as discussed above will be essential to create proper context for interpretation of sample measurements with comparisons to appropriate health effect reference values, when such values are available [3].

4.2. Adopting a Life-Cycle Approach

An overarching life-cycle framework and decision-making process that the ASHG has encouraged as an ideal for air quality sensor applications is illustrated in Figure 2. The life-cycle concept was originally developed for use in emergency response situations [88], adapted and applied for radioactive air sampling and instrumentation [89], adopted as a systematic way to organize the framework of the White House's signature initiative on Nanotechnology for Sensors and Sensors for Nanotechnology [90], and most recently expanded to meet all manner of the emerging sensor needs for safety, health, well-being and productivity [91]. The lifecycle begins with a clear and complete identification of the purpose of the measurement, including what needs to be measured, under which conditions it needs to be measured, and how well it needs to be measured. The lifecycle guides the research and development, prototype testing, qualification type testing, production control testing, and training needs for the sensor system. The lifecycle further defines procedures for acceptance testing, initial calibration, functional checks, conduct and evaluation of operational experience, maintenance and recalibration, and periodic performance testing to confirm continued successful use of the sensor system. Effective following of the life-cycle process ensures that the sensor methods and instrumentation will work as intended under realistic conditions. Documentation and continuous improvement are essential at each step.

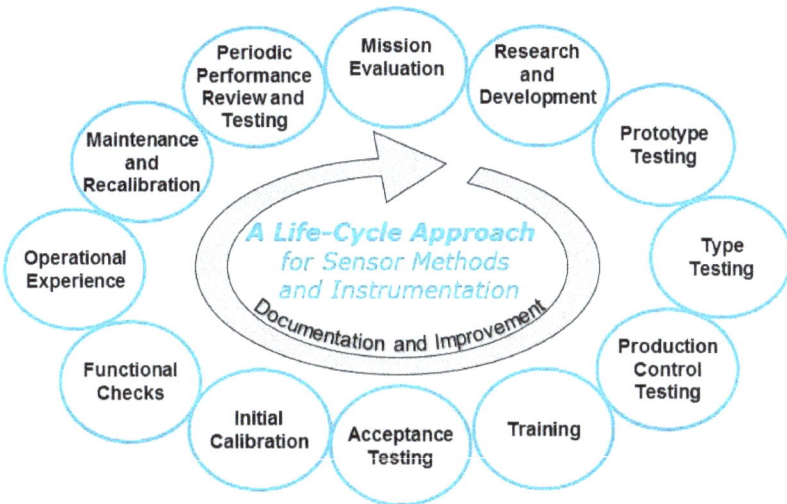

Figure 2. The life-cycle approach for sensor methods and instrumentation [88–91].

Use of the lifecycle supports an approach to sensor methods and instrumentation that is consistent with the roles served by resources such as *My Air, My Health: An HHS/EPA Challenge* [92] to develop and validate new methods, and the online *AirNow* [93] maps and forecast data that are collected and shared using federal reference or equivalent monitoring techniques or techniques approved by the state, local or tribal monitoring agencies.

4.2.1. A Working Definition of Air Quality Sensors and Health Informatics

A proposed working definition of *Air Quality Sensors and Health Informatics* might be "the science and practice of determining which information is relevant to meeting air sensor and health objectives; developing and implementing effective mechanisms for collecting, validating, storing, sharing, analyzing, modeling, and applying the information; and then confirming that appropriate decisions were made and that desired mission outcomes were achieved" [94]. The additional steps in the informatics lifecycle include "conveying experience to the broader community, contributing to generalized knowledge, and updating standards and training" [95]. Successful informatics endeavors will apply all of the steps in the process. ASHG members have also assisted with the development of an attempt to define "data readiness levels", which could help with relevance and reliability issues for the collection, sharing, and application of air sensor data [96].

4.2.2. Roles and Responsibilities of Sensor and Data Customers, Creators, Curators, and Analysts

In the context of our working definition of informatics for air sensors and health, the roles and responsibilities of the myriad individuals who are engaged in the development and application of air quality sensors can be viewed as fitting into four categories: sensor and data customers (who specify the sensors and data needs for their intended purposes), sensor and data creators (who will develop relevant and reliable sensors and data to meet the customer needs), sensor and data curators (who will maintain and ensure the quality of the sensors and sensor data), and sensor data analysts (who will develop and apply models for data analysis and interpretation that are consistent with the quality and quantity of the data and that those data meet the customers' needs). In some instances, the same individuals may perform all roles, and in the larger global reality the individuals and their roles may extend over significant distances, organizations, and time periods. As shown in Figure 3, effective communication across the many customer, creator, curator, and analyst interfaces is essential, and that communication across each of the six interfaces must work effectively in both directions [95]. This vision follows the views of Hendren et al. [97] on a collaborative approach to assessing, evaluating, and advancing the state of the field for data curation in the emerging field of nanomaterials and nanotechnology.

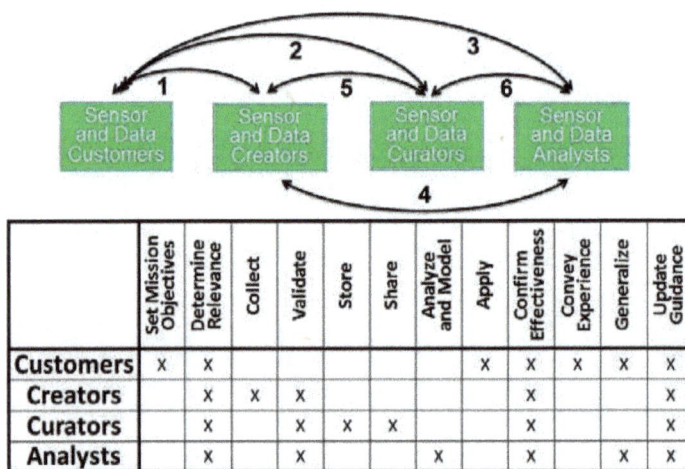

	Set Mission Objectives	Determine Relevance	Collect	Validate	Store	Share	Analyze and Model	Apply	Confirm Effectiveness	Convey Experience	Generalize	Update Guidance
Customers	X	X						X	X	X	X	X
Creators		X	X	X					X			X
Curators		X		X	X	X			X			X
Analysts		X		X			X		X		X	X

Figure 3. Communication interfaces, roles, and responsibilities for air quality sensor and health data customers, creators, curators, and analysts [95].

5. Forecast of Advancing Technologies

Major areas of opportunity to advance the state of the art and application of sensor technologies include: strengthening and sustaining infrastructures for investments and collaborations, improving the sharing of information, integrating data and a coherent interpretation across various venues to construct a cumulative accounting of exposures, and advancing breakthroughs in miniaturization of sensor systems. At this writing, there are efforts across multiple organizations (public and private; large and small; formal and informal) with additional involvement by independent inventors. Entrepreneurial organizations such as Aclima have become involved in collecting data from mobile sensors attached to Google street-view cars [98] and are leading the way to making those data accessible.

Equally important are opportunities to increase the linkage of measurements of exposure via sensor readings to additional parameters useful to estimate dosimetry: biometric data (e.g., pulse rate, breath rate, and the like, which are now available from devices such as FitBit); geospatial measurements of location and daily movement patterns; and of the physical environment (temperature, humidity, etc.). These capabilities are most actively being developed within the occupational exposure arena but migration of these technologies to the low-cost sensor market is anticipated as value-added features. Properly informing and providing useful guidance to ensure these technologies are appropriately engaged will be another challenge for groups like the ASHG.

5.1. Infrastructure Needs

5.1.1. Forecast and Statement of Needed Investment

When real-time sensing is combined with concurrent GPS coordinates, the derived data are multidimensional. Today, sensor-derived data include streaming of latitude, longitude, altitude in addition to the chemical data. However, the hardware and software to fully capture the potential usefulness is lacking. For example, visualization of multidimensional data might be imaged as a video of a multi-colored topographical map that flexes as an individual moves through different levels of the positional and chemical data. Depth of color might represent various concentrations of chemicals detected by the sensors. If the full potential is to be captured, investment is essential. Infrastructure is needed to not only capture multidimensional data streams but also analyze it as it is being produced. This is especially relevant to homeland security and for rapid response following a catastrophic event. The application of sensor technology to the field of epidemiology can also be cited as an investment opportunity. New evidence of cause-and-effect will become evident by developing the ability to retrieve archived multidimensional sensor-derived data and overlaying it with geographical disease prevalence information. Challenges of scalability across dimensions of time and geography must be addressed to properly interface with available health-based standards and reference values.

As described in Section 3.1, there is great interest, and investment, in transforming toxicity testing with the application of high throughput analyses using cell cultures [70]. Generating evidence to demonstrate the validity of these testing methods is an ongoing challenge. In this context, we have an opportunity to transform systematic hazard identification and risk assessment processes by combining real-time exposure monitoring with real-time image analysis. Again, investment in the data capture, analytics and archiving capabilities is essential if this opportunity is to be realized.

5.1.2. Scientific Literature Collection and Coordination

Another essential investment is needed to network the fields of Health Science, Electrical Engineering and Computer Science. A first step might be to encourage sensor-related journals to petition the National Library of Medicine for incorporation into the database. The National Library of Medicine's PubMed database [99] provides the ability to search for peer-reviewed publications. This includes research about the use of sensors to assess indoor and outdoor air. As an example, PubMed includes air sensor-related publications from a journal called "Environmental Monitoring and Assessment" such as this 2017 one about "Public engagement on urban air pollution: an exploratory study of two interventions" [100]. However, not all scientific journals are indexed in PubMed and this is especially true for journals specializing in emerging areas and technologies Editors and publishers can submit their journals for inclusion in PubMed; however, not all journals will be accepted based on NLM's selection criteria [101]. The relatively new PubMed Commons allows for the sharing of opinions and information about PubMed citations [102]. This could include comments about the techniques used in a publication or alerting readers to consider looking at publications with more current approaches and findings.

5.2. Miniaturization

Miniaturization has been essential to recent successes in moving laboratory-scale technologies to the field. For example, to meet the demands for personal exposure monitoring of the size characteristics and spatial distribution of ultrafine particles, Fierz et al. [103] have developed a compact, real-time instrument called the diffusion size classifier (DiSC) which provides particle size information that is in good agreement with the much larger and more expensive laboratory-based aerosol spectrometers. Another personal monitoring device recently released into the market is the portable aerosol mobility spectrometer (PAMS) which simultaneously measures the number-weighted size distribution of submicrometer aerosol, including the nanoparticle fraction [104]. In developing the PAMS, NIOSH investigators used miniaturization to overcome the prohibitive size, weight, and cost limitations of the previous technology, and they also eliminated burdensome regulatory and administrative limitations for record-keeping and transportation of the instrument by replacing the traditionally used radioactive source for particle charge conditioning with a nonradioactive bipolar diffusion charger. Compared to traditionally used instruments, these innovations provided reductions in size (by a factor of 20), weight (by a factor of 10–15), and cost (by a factor of 4), along with improvements in analytical performance. However, further improvements in instrument size, weight, cost, and performance are still needed to make these technologies more widely affordable, deployable, and able to operate in the wide range of particle number concentrations that can be encountered in both workplaces and ambient environments. Needed improvements include smaller, more reliable power sources, including the possibility of body-heat or motion-driven power sources, and more compact data collection, processing, and memory capabilities. As described in the perspective article by Fadel et al. [105], advances in nanotechnologies could enable and accelerate the development of inexpensive, portable devices for the broad detection, identification, and quantification of biological and chemical substances, including sensors for air quality and health applications.

6. Summary and Conclusions

The goals of the ASHG are to provide an open dialogue from the multiple disciplines and agencies represented on a number of related issues: (1) improving the understanding of what sensor technologies may be able to deliver for these agencies to meet their missions; (2) communication of what small-scale sensor readings can provide to a number of audiences; and most centrally, (3) best practices in communicating health-related messages to numerous audiences.

In this paper, we have attempted to cross-reference other related projects and provide additional resources to an interested reader to pursue additional information from credible sources. Additional aims were to provide an update on the advances made to date under the auspices of these various programs, to forecast potential applications of rapidly emerging sensor technologies and to foster a collaborative response to challenges involved in their application to research and support of decisions related to air quality management.

Acknowledgments: Funding for publication costs were provided by the U.S. Environmental Protection Agency, Washington, DC USA, This project was supported in part by an appointment to the Internship/Research Participation Program at the U.S. Environmental Protection Agency, administered by the Oak Ridge Institute for Science and Education through an interagency agreement between the U.S. Department of Energy and EPA, with additional support provided by the National Institutes of Health, National Library of Medicine.

Author Contributions: George Woodall, Mark Hoover, Ronald Williams and Janis Hulla were the primary architects of the outline for this review paper; all authors contributed in writing the paper.

Conflicts of Interest: The authors declare no conflict of interest.

Appendix A

Standards for equipment used to detect and determine toxic gases and vapors in a workplace or similar situation.

A.1. Published

Europe:

1. EN 45544:2000 Workplace atmospheres. Electrical apparatus used for the direct detection and direct concentration measurement of toxic gases and vapours. Part 1: General requirements and test methods; Part 2: Performance requirements for apparatus used for measuring concentrations in the region of limit values; Part 3: Performance requirements for apparatus used for measuring concentrations well above limit values; Part 4: Guide for selection, installation, use and maintenance.

USA:

1. Underwriters Laboratory: UL 2075 Gas and Vapor Detectors and Sensors
2. American National Standards Institute/International Safety Association: ANSI/ISA-92.00.01-2010 Performance Requirements for Toxic Gas Detectors; ANSI/ISA 92.00.02-2013 Installation, Operation, and Maintenance of Toxic Gas-Detection Instruments
3. American Society for Testing and Materials: ASTM E2885-13 Standard Specification for Handheld Point Chemical Vapor Detectors (HPCVD) for Homeland Security Application

International Electrotechnical Commission:

1. IEC 60079-29-1:2007 Explosive atmospheres—Part 29-1: Gas detectors—Performance requirements of detectors for flammable gases
2. IEC 60079-29-2:2007 Explosive atmospheres—Part 29-2: Gas detectors—Selection, installation, use and maintenance of detectors for flammable gases and oxygen

Other:

1. Australian/New Zealand Standard: AS/NZS 4641:2007 Electrical apparatus for the detection of oxygen and other gases and vapours at toxic levels—General requirements and test methods.
2. ISO database: https://www.iso.org/committee/52702/x/catalogue/p/0/u/1/w/0/d/0
3. CEN Database: https://standards.cen.eu/dyn/www/f?p=204:7:0::::FSP_ORG_ID:6245&cs= 178094E67E1897102F190938A48C7A285

A.2. Standards Proceeding through Process

ISO/IEC (IEC 62990-1) Workplace Atmospheres—Part 1: Gas detectors—Performance requirements of detectors for toxic gases

ISO/IEC (IEC 62990-2) Work-place Atmospheres—Part 2: Gas detectors—Selection, installation, use and maintenance of detectors for toxic gases and vapours and oxygen sensors.

References

1. Williams, R. Findings from the 2013 EPA Sensors Workshop. Available online: https://www.epa.gov/air-research/findings-2013-epa-air-sensors-workshop (accessed on 5 May 2017).
2. Williams, R.; Long, R.; Beaver, M.; Kaufman, A.; Zeiger, F.; Heimbinder, M.; Hang, I.; Yap, R.; Acharya, B.; Ginwald, B.; et al. *Sensor Evaluation Report*; U.S. Environmental Protection Agency: Washington, DC, USA, 2014.

3. U.S. Environmental Protection Agency. *Graphical Arrays of Chemical-Specific Health Effect Reference Values for Inhalation Exposures*; U.S. Environmental Protection Agency: Research Triangle Park, NC, USA, 2009; EPA/600/R-09/061. Available online: http://cfpub.epa.gov/ncea/cfm/recordisplay.cfm?deid=211003 (accessed on 5 May 2017).
4. U.S. Environmental Protection Agency. Guidance on Systematic Planning Using the Data Quality Objectives Process. Available online: https://www.epa.gov/fedfac/guidance-systematic-planning-using-data-quality-objectives-process (accessed on 5 May 2017).
5. U.S. Environmental Protection Agency. Air Sensor Toolbox: What Do My Sensor Readings Mean? Sensor Scale Pilot Project. Available online: https://www.epa.gov/air-sensor-toolbox/what-do-my-sensor-readings-mean-sensor-scale-pilot-project (accessed on 2 May 2017).
6. U.S. Environmental Protection Agency. Air Sensor Toolbox for Citizen Scientists, Researchers and Developers. Available online: https://www.epa.gov/air-sensor-toolbox (accessed on 2 May 2017).
7. ATSDR. Agency for Toxic Substances and Disease Registry Minimal Risk Levels (MRLs) 2016. Available online: https://www.atsdr.cdc.gov/mrls/index.asp (accessed on 12 June 2017).
8. Woodall, G.; Lipscomb, J.; Taylor, M. Review of health-based reference values for inhalation exposures. 2017; in preparation.
9. U.S. Environmental Protection Agency. *Inhalation Health Effect Reference Values for Toluene (CASRN 108-88-3)*; U.S. Environmental Protection Agency: Research Triangle Park, NC, USA, 2012. Available online: http://oaspub.epa.gov/eims/eims-comm.getfile?p_download_id=512650 (accessed on 2 May 2017).
10. U.S. Environmental Protection Agency. *Chemical-Specific Reference Values for Benzene (CASRN 71-43-2)*; U.S. Environmental Protection Agency: Research Triangle Park, NC, USA, 2012; EPA/600/R-12/047F1. Available online: http://oaspub.epa.gov/eims/eimscomm.getfile?p_download_id=512648 (accessed on 2 May 2017).
11. U.S. Environmental Protection Agency. *Inhalation Health Effect Reference Values for Ethylbenzene (CASRN 100-41-4)*; U.S. Environmental Protection Agency: Research Triangle Park, NC, USA, 2012; EPA/600/R-12/047F2. Available online: http://oaspub.epa.gov/eims/eimscomm.getfile?p_download_id=512649 (accessed on 2 May 2017).
12. U.S. Environmental Protection Agency. *Inhalation Health Effect Reference Values for Xylene—All Isomers (CASRNs Mixed Isomers—1330-20-7; m-xylene—95-47-6; o-xylene—108-38-3; p-xylene—106-42-3)*; U.S. Environmental Protection Agency: Research Triangle Park, NC, USA, 2012; EPA/600/R-12/047F4. Available online: http://oaspub.epa.gov/eims/eimscomm.getfile?p_download_id=512651 (accessed on 2 May 2017).
13. U.S. Environmental Protection Agency. *Inhalation Health Effect Reference Values for Manganese (CASRN 7439-96-5—Manganese) and Compounds (CASRN 1344-43-0; 1317-35-7; and 1129-60-5)*; U.S. Environmental Protection Agency: Research Triangle Park, NC, USA, 2012; EPA/600/R-12/047F5. Available online: http://oaspub.epa.gov/eims/eimscomm.getfile?p_download_id=512652 (accessed on 2 May 2017).
14. Harper, M.; Weis, C.; Pleil, J.D.; Blount, B.C.; Miller, A.; Hoover, M.D.; Jahn, S. Commentary on the contributions and future role of occupational exposure science in a vision and strategy for the discipline of exposure science. *J. Expo. Sci. Environ. Epidemiol.* **2015**, *25*, 381–387. [CrossRef] [PubMed]
15. Fishbain, B.; Lerner, U.; Castell, N.; Cole-Hunter, T.; Popoola, O.; Broday, D.M.; Iñiguez, T.M.; Nieuwenhuijsen, M.; Jovasevic-Stojanovic, M.; Topalovic, D.; et al. An evaluation tool kit of air quality micro-sensing units. *Sci. Total Environ.* **2017**, *575*, 639–648. [CrossRef] [PubMed]
16. Bart, M.; Williams, D.E.; Ainslie, B.; McKendry, I.; Salmond, J.; Grange, S.K.; Alavi-Shoshtari, M.; Steyn, D.; Henshaw, G.S. High density ozone monitoring using gas sensitive semi-conductor sensors in the lower Fraser Valley, British Columbia. *Environ. Sci. Technol.* **2014**, *48*, 3970–3977. [CrossRef] [PubMed]
17. Gao, M.L.; Cao, J.J.; Seto, E. A distributed network of low-cost continuous reading sensors to measure spatiotemporal variations of PM2.5 in Xi'an, China. *Environ. Pollut.* **2015**, *199*, 56–65. [CrossRef] [PubMed]
18. Mead, M.I.; Popoola, O.A.M.; Stewart, G.B.; Landshoff, P.; Calleja, M.; Hayes, M.; Baldovi, J.J.; McLeod, M.W.; Hodgson, T.F.; Dicks, J.; et al. The use of electrochemical sensors for monitoring urban air quality in low-cost, high-density networks. *Atmos. Environ.* **2013**, *70*, 186–203. [CrossRef]
19. Jovasevic-Stojanovic, M. CITI-SENSE Development of Sensor-Based Citizens' Observatory Community for Improving Quality of Life in Cities. Available online: http://www.citi-sense.eu/Portals/106/Documents/Dissemination%20material/CITI-SENSE_Newsletter%20No4.pdf (accessed on 19 September 2017).

20. CITI-SENSE. Deliverable D9.16 Project Overview for the Lay People. 2016. Available online: http://co.citi-sense.eu/TheProject/Publications/Deliverables.aspx (accessed on 2 May 2017).

21. CITI-SENSE. Development of Sensor-Based Citizens' Observatory Community for Improving Quality of Life in Cities. Available online: http://www.citi-sense.eu/ (accessed on 2 May 2017).

22. Aspuru, I.; García, I.; Herranz, K.; Santander, A. Citi-Sense: Methods and tools for empowering citizens to observe acoustic comfort in outdoor public spaces. *Noise Mapp.* **2016**, *3*, 37–48. [CrossRef]

23. European Commission. Air Quality Standards. 8/6/16. Available online: http://ec.europa.eu/environment/air/quality/standards.htm (accessed on 2 May 2017).

24. Borrego, C.; Costa, A.M.; Ginja, J.; Amorim, M.; Coutinho, M.; Karatzas, K.; Sioumis, T.; Katsifarakis, N.; Konstantinidis, K.; De Vito, S.; et al. Assessment of air quality microsensors versus reference methods: The EuNetAir joint exercise. *Atmos. Environ.* **2016**, *147*, 246–263. [CrossRef]

25. Penza, M.; Consortium, E. COST Action TD1105: Overview of sensor-systems for air-quality monitoring. *Procedia Eng.* **2014**, *87*, 1370–1377. [CrossRef]

26. Clean Air Asia. Clean Air Asia India. Available online: http://cleanairasia.org/india/ (accessed on 2 May 2017).

27. Lewis, A.; Edwards, P. Validate personal air-pollution sensors. *Nature* **2016**, *535*, 29–31. [CrossRef] [PubMed]

28. BreezoMeter. BreezoMeter: Hyperlocal Air Quality Data, Available Globally. Available online: https://breezometer.com/ (accessed on 2 May 2017).

29. WHO. *Ambient (Outdoor) Air Quality and Health. Fact Sheet*; WHO: Geneva, Switzerland, 2016. Available online: http://www.who.int/mediacentre/factsheets/fs313/en/ (accessed on 2 May 2017).

30. ATSDR. *Health Consultation: Evaluation of Particulate Matter, Bennett Landfill Fire, Chester, South Carolina*; ATSDR: Atlanta, GA, USA, 2015.

31. ATSDR. *Health Consultation: Assessing the Public Health Implications of the Criteria (NAAQS) Air Pollutants and Hydrogen Sulfide as Part of the Midlothian Area Air Quality Petition Response, Midlothian, Ellis County, Texas*; ATSDR: Atlanta, GA, USA, 2016. Available online: https://www.atsdr.cdc.gov/sites/midlothian/health_consultations.html (accessed on 2 May 2017).

32. ATSDR. *Health consultation: Brooklyn Township PM2.5, Brooklyn Township, Susquehanna County, Pennsylvania*; ATSDR: Atlanta, GA, USA, 2016; Cost Recovery Number: 3A4K00. Available online: https://www.atsdr.cdc.gov/HAC/pha/BrooklynTownship/BrooklynTwnsp_pm2-5_HC_Final_04-22-2016_508.pdf (accessed on 2 May 2017).

33. ATSDR. *Health Consultation: Review of Analysis of Particulate Matter and Metal Exposures in Air, KCBX (AKA, Chicago Petroleum Coke sites), Cook County, IL, USA*; ATSDR: Atlanta, GA, USA, 2016.

34. NACEPT. *NACEPT 2016 Report: Environmental Protection belongs to the Public, a Vision for Citizen Science at EPA*; U.S. Environmental Protection Agency: Washington, DC, USA, 2016; EPA 219-R-16-001. Available online: https://www.epa.gov/faca/nacept-2016-report-environmental-protection-belongs-public-vision-citizen-science-epa (accessed on 2 May 2017).

35. Dosemagen, S.; Gehrke, G. Civic Technology and Community Science: A New Model for Public Participation in Environmental Decisions in Confronting the Challenges of Public Participation: Issues in Environmental, Planning, and Health Decision-Making. In Proceedings of the Iowa State University Summer Symposia on Science Communication, Ames, IA, USA, 30 May–1 June 2013.

36. Williams, R.; Kilaru, V.; Snyder, E.; Kaufman, A.; Dye, T.; Rutter, A.; Russell, A.; Hafner, H. *Air Sensor Guidebook*; U.S. Enviromental Protection Agency: Research Triangle Park, NC, USA, 2016; EPA/600/R-14/159. Available online: https://cfpub.epa.gov/si/si_public_record_report.cfm?direntryid=277996 (accessed on 2 May 2017).

37. U.S. Environmental Protection Agency. Citizen Science in Newark, New Jersey. 2015. Available online: https://www.epa.gov/sciencematters/citizen-science-newark-new-jersey (accessed on 12 July 2017).

38. U.S. Environmental Protection Agency. Air sensor toolbox: Resources and funding. 2016. Available online: https://www.epa.gov/air-sensor-toolbox/air-sensor-toolbox-resources-and-funding#TER (accessed on 5 June 2017).

39. Heimbinder, M. Apps & Sensors for Air Pollution. Taking Space: The HabitatMap & AirCasting Blog 2012 4/9/2012. Available online: http://www.takingspace.org/epa-apps-sensors-for-air-pollution-workshop/ (accessed on 2 May 2017).

40. Snyder, E.G.; Watkins, T.H.; Solomon, P.A.; Thoma, E.D.; Williams, R.W.; Hagler, G.S.; Shelow, D.; Hindin, D.A.; Kilaru, V.J.; Preuss, P.W. The changing paradigm of air pollution monitoring. *Environ. Sci. Technol.* **2013**, *47*, 11369–11377. [CrossRef] [PubMed]

41. MacDonnell, M.; Raymond, M.; Wyker, D.; Finster, M.; Chang, Y.; Raymond, T.; Temple, B.; Scofield, M.; Vallano, D. *Mobile Sensors and Applications for Air Pollutants*; U.S. Environmental Protection Agency: Washington, DC, USA, 2013.

42. Jiao, W.; Hagler, G.; Williams, R.; Sharpe, R.; Brown, R.; Garver, D.; Judge, R.; Caudill, M.; Rickard, J.; Davis, M.; et al. Community Air Sensor Network (CAIRSENSE) project: Evaluation of low-cost sensor performance in a suburban environment in the southeastern United States. *Atmos. Meas. Tech.* **2016**, *9*, 5281–5292. [CrossRef]

43. Smith, K.; Edwards, P.M.; Evans, M.J.J.; Lee, J.D.; Shaw, M.D.; Squires, F.; Wilde, S.; Lewis, A.C. Clustering approaches that improve the reproducibility of low-cost air pollution sensors. *Faraday Discuss.* **2017**, *200*, 621–637. [CrossRef] [PubMed]

44. Kaufman, A.; Brown, A.; Barzyk, T.; Williams, R. The Citizen Science Toolbox: A One-Stop Resources for Air Sensor Technology. In *EM: Air and Waste Management Association's Magazine for Environmental Managers*; Air & Waste Management Association: Pittsburgh, PA, USA, 2014.

45. Spinelle, L.; Gerboles, M.; Aleixandre, M.; Bonavitacola, F. Evaluation of metal oxides sensors for the monitoring of O3 in ambient air at Ppb level. *Chem. Eng. Trans.* **2016**, *54*, 319–324.

46. Williams, R. *Evaluation of Elm and Speck Sensors*; U.S. Environmental Protection Agency: Washington, DC, USA, 2015.

47. Air Quality Sensor Performance Evaluation Center (AQ-SPEC). South Coast Air Quality Management District. Available online: http://www.aqmd.gov/aq-spec/ (accessed on 2 May 2017).

48. Lin, C.; Gillespie, J.; Schuder, M.D.; Duberstein, W.; Beverland, I.J.; Heal, M.R. Evaluation and calibration of Aeroqual series 500 portable gas sensors for accurate measurement of ambient ozone and nitrogen dioxide. *Atmos. Environ.* **2015**, *100*, 111–116. [CrossRef]

49. NIOSH. *Components for Evaluation of Direct-Reading Monitors for Gases and Vapors*; NIOSH: Cincinnati, OH, USA, 2012; Publication No. 2012-12-162. Available online: https://www.cdc.gov/niosh/docket/archive/pdfs/niosh-220/2012-162.pdf (accessed on 2 May 2017).

50. NIOSH. *Addendum to Components for Evaluation of Direct-Reading Monitors for Gases and Vapors: Hazard Detection in First Responder Environments*; NIOSH: Cincinnati, OH, USA, 2012; Publication No. 2012-163. Available online: https://www.cdc.gov/niosh/docket/archive/pdfs/niosh-220/2012-163.pdf (accessed on 2 May 2017).

51. AIHA. Reporting Specifications for Electronic Real Time Gas and Vapor Detection Equipment Fact Sheet. Available online: https://www.aiha.org/governmentaffairs/PositionStatements/Reporting%20Specifications%20for%20Real%20Time%20Detection_Final.pdf (accessed on 2 May 2017).

52. Moltchanov, S.; Levy, I.; Etzion, Y.; Lerner, U.; Broday, D.M.; Fishbain, B. On the feasibility of measuring urban air pollution by wireless distributed sensor networks. *Sci. Total Environ.* **2015**, *502*, 537–547. [CrossRef] [PubMed]

53. National Institute of Environmental Health Science. Validation and Demonstration of Devices for Environmental Exposure Assessment (R21/R33). 2013. Available online: https://grants.nih.gov/grants/guide/rfa-files/RFA-ES-13-013.html (accessed on 2 May 2017).

54. U.S. Environmental Protection Agency. Village Green Project. 4/12/17. Available online: https://www.epa.gov/air-research/village-green-project (accessed on 2 May 2017).

55. U.S. Environmental Protection Agency. (2013) DRAFT Roadmap for Next Generation Air Monitoring. Available online: https://www.epa.gov/sites/production/files/2014-09/documents/roadmap-20130308.pdf (accessed on 12 July 2017).

56. Wan, J.; Hagler, G.; Williams, R.; Sharpe, B.; Weinstock, L.; Rice, J. Field assessment of the Village Green Project: An autonomous community air quality monitoring system. *Environ. Sci. Technol.* **2015**. [CrossRef]

57. Hindin, D.; Grumbles, B.; Wyeth, G.; Beneditc, K.; Watkins, T.; Aburn, G., Jr.; Ulrich, M.; Lang, S.; Poole, K.; Dapolito Dunn, A. Advanced monitoring technology: Opportunities and challenges a path forward for EPA, states, and tribes. *EM Environ. Manag.* **2016**, *11*, 16–21.

58. Keating, M.; Benedict, K.; Evans, R.; Jenkins, S.; Mannshardt, E.; Stone, S.L. Interpreting and communicating short-term air sensor data. *EM Environ. Manag.* **2016**, *11*, 22–25.

59. Mannshardt, E.; Benedict, K.; Jenkins, S.; Keating, M.; Mintz, D.; Stone, S. Analysis of short-term ozone and PM2.5 measurements: Characteristics and relationships for air sensor messaging. *J. Air Waste Manag. Assoc.* **2016**, *67*, 462–474. [CrossRef] [PubMed]

60. U.S. Environmental Protection Agency. History of Acute Exposure Guideline Levels (AEGLs). Available online: https://www.epa.gov/aegl/history-acute-exposure-guideline-levels-aegls (accessed on 12 July 2017).

61. NRC. *Standing Operating Procedures for Developing Acute Exposure Guideline Levels (Aegls) for Hazardous Chemicals*; National Academy Press: Washington, DC, USA, 2001.

62. Young, R.A.; Bast, C.B.; Wood, C.S.; Adeshina, F. Overview of the Standing Operating Procedure (SOP) for the development of Provisional Advisory Levels (PALs). *Inhal. Toxicol.* **2009**, *21*, 1–11. [CrossRef] [PubMed]

63. Klepeis, N.E.; Nelson, W.C.; Ott, W.R.; Robinson, J.P.; Tsang, A.M.; Switzer, P.; Behar, J.V.; Hern, S.C.; Engelmann, W.H. The National Human Activity Pattern Survey (NHAPS): A resource for assessing exposure to environmental pollutants. *J. Expo. Anal. Environ. Epidemiol.* **2001**, *11*, 231–252. [CrossRef] [PubMed]

64. Ott, W. *Human Activity Patterns: A Review of the Literature for Estimating Time Spent Indoors, Outdoors, and In-Transit*; U.S. Environmental Protection Agency, Office of Research and Development: Las Vegas, NV, USA, 1989.

65. Robinson, J.P.; Thomas, J. *Time Spent in Activities, Locations, and Microenvironments: A California National Comparison*; U.S. Environmental Protection Agency: Las Vegas, NV, USA, 1992.

66. U.S. Environmental Protection Agency. *Child-Specific Exposure Factors Handbook (Final Report) 2008*; U.S. Environmental Protection Agency: Washington, DC, USA, 2008; EPA/600/R-06/096F. Available online: http://cfpub.epa.gov/ncea/cfm/recordisplay.cfm?deid=199243 (accessed on 12 July 2017).

67. Wallace, L.A. *Project Summary: The Total Exposure Assessment Methodology (TEAM) Study*; U.S. Environmental Protection Agency: Washington, DC, USA, 1987.

68. SAB. *Integrated Human Exposure Committee Commentary on Indoor Air Strategy*; U.S. Environmental Protection Agency: Washington, DC, USA, 1998.

69. Nehr, S.; Hosen, E.; Tanabe, S.I. Emerging developments in the standardized chemical characterization of indoor air quality. *Environ. Int.* **2017**, *98*, 233–237. [CrossRef] [PubMed]

70. NRC. *Toxicity Testing in the 21st Century: A Vision and a Strategy*; National Academies Press: Washington, DC, USA, 2007.

71. NRC. *Exposure Science in the 21st Century: A Vision and a Strategy*; National Academies Press: Washington, DC, USA, 2012.

72. National Academies of Sciences, Engineering, and Medicine. *Using 21st Century Science to Improve Risk-Related Evaluations*; National Academies Press: Washington, DC, USA, 2017.

73. Mattingly, C.J.; Mckone, T.E.; Callahan, M.A.; Blake, J.A.; Hubal, E.A. Providing the missing link: The exposure science ontology ExO. *Environ. Sci. Technol.* **2012**, *46*, 3046–3053. [CrossRef] [PubMed]

74. Villeneuve, D.L.; Crump, D.; Garcia-Reyero, N.; Hecker, M.; Hutchinson, T.H.; Lalone, C.A.; Landesmann, B.; Lettieri, T.; Munn, S.; Nepelska, M.; et al. Adverse outcome pathway (AOP) development I: Strategies and principles. *Toxicol. Sci.* **2014**, *142*, 312–320. [CrossRef] [PubMed]

75. Villeneuve, D.L.; Crump, D.; Garcia-Reyero, N.; Hecker, M.; Hutchinson, T.H.; Lalone, C.A.; Landesmann, B.; Lettieri, T.; Munn, S.; Nepelska, M.; et al. Adverse outcome pathway development II: Best practices. *Toxicol. Sci.* **2014**, *142*, 321–330. [CrossRef] [PubMed]

76. Teeguarden, J.G.; Tan, Y.M.; Edwards, S.W.; Leonard, J.A.; Anderson, K.A.; Corley, R.A.; Kile, M.L.; Simonich, S.M.; Stone, D.; Tanguay, R.L.; et al. Completing the link between exposure science and toxicology for improved environmental health decision making: The aggregate exposure pathway framework. *Environ. Sci. Technol.* **2016**, *50*, 4579–4586. [CrossRef] [PubMed]

77. Hulla, J.; Snawder, J.E.; Proctor, S.P.; Chapman, G.D. DOD impact assessment and management of naphthalene-related risks. *Toxicologist* **2010**, *114*, 400.

78. Hug, W.F.; Bhartia, R.; Reid, R.D.; Reid, M.R.; Oswal, P.; Lane, A.L.; Sijapati, K.; Sullivan, K.; Hulla, J.E.; Snawder, J.; et al. Advanced environmental, chemical, and biological sensing technologies IX. In *Wearable Real-Time Direct-Reading Naphthalene and VOC Personal Exposure Monitor*; Vo-Dinh, T., Lieberman, R.A., Gauglitz, G., Eds.; SPIE: Bellingham, WA, USA, 2012; Volume 8366.

79. Hulla, J.; Proctor, S.; Snawder, J.E. The naphthalene dosimeter—Vanguard technology for improved health protection. *Toxicologist* **2015**, *144*, 110.

80. U.S. Environmental Protection Agency. *Integrated Science Assessment for Ozone and Related Photochemical Oxidants*; U.S. Environmental Protection Agency, Office of Research and Development, National Center for Environmental Assessment-RTP Division: Research Triangle Park, NC, USA, 2013.

81. Mcdonnell, W.F.; Stewart, P.W.; Smith, M.V.; Kim, C.S.; Schelegle, E.S. Prediction of lung function response for populations exposed to a wide range of ozone conditions. *Inhal. Toxicol.* **2012**, *24*, 619–633. [CrossRef] [PubMed]

82. Brown, J.S.; Gordon, T.; Price, O.; Asgharian, B. Thoracic and respirable particle definitions for human health risk assessment. *Part. Fibre Toxicol.* **2013**, *10*, 12. [CrossRef] [PubMed]

83. ICRP. *Human Respiratory Tract Model for Radiological Protection: A Report of a Task Group of the International Commission on Radiological Protection*; Pergamon Press: New York, NY, USA, 1994; Volume 24, pp. 1–482.

84. Koehler, K.A.; Clark, P.; Volckens, J. Development of a sampler for total aerosol deposition in the human respiratory tract. *Ann. Occup. Hyg.* **2009**, *53*, 731–738. [CrossRef] [PubMed]

85. U.S. Environmental Protection Agency. *Integrated Science Assessment for Oxides of Nitrogen-Health Criteria (Final Report)*; U.S. Environmental Protection Agency, Office of Research and Development, National Center for Environmental Assessment: Research Triangle Park, NC, USA, 2016.

86. Brown, J.S. Nitrogen dioxide exposure and airway responsiveness in individuals with asthma. *Inhal. Toxicol.* **2015**, *27*, 1–14. [CrossRef] [PubMed]

87. U.S. Congress. Clean Air Act as Amended in 1990, Section 112 (b) (1) Hazardous Air Pollutants. 1990. Available online: https://www.epa.gov/history/epa-history-clean-air-act-amendments-1990 (accessed on 12 July 2017).

88. Hoover, M.D.; Cox, M. Public protection from nuclear, chemical, and biological terrorism: Health Physics Society 2004 summer school. In *A Life-Cycle Approach for Development and Use of Emergency Response and Health Protection Instrumentation*; Brodsky, A., Johnson, R.H., Groans, R.E., Eds.; Medical Physics: Madison, WI, USA, 2004; pp. 317–324.

89. Hoover, M.D.; Cox, M. Radioactive air sampling methods. In *A Life-Cycle Approach to Development and Application of Air Sampling Methods and Instrumentation*; Maiello, M.L., Hoover, M.D., Eds.; CRC Press/Taylor & Francis: Boca Raton, FL, USA, 2011; pp. 43–52.

90. NNI. *Nanotechnology for Sensors and Sensors for Nanotechnology: Improving and Protecting Health, Safety, and Environment*; NNI: Washington, DC, USA, 2012.

91. Hoover, M.D.; Debord, D.G. Turning numbers into knowledge: Sensors for safety, health, well-being, and productivity. *Synergist* **2015**, *26*, 22–26.

92. InnoCentive. My Air, My Health: An HHS/EPA Challenge. Available online: https://www.innocentive.com/ar/challenge/9932947 (accessed on 12 July 2017).

93. U.S. Environmental Protection Agency. AirNow. Available online: https://www.airnow.gov/ (accessed on 12 July 2017).

94. De la Iglesia, D.; Harper, S.; Hoover, M.D.; Klaessig, F.; Lippelli, P.; Maddux, B.; Morse, J.; Nel, A.; Rajan, K.; Reznik-Zellen, R.; et al. Nanoinformatics 2020 Roadmap. National Nanomanufacturing Network 2011. Available online: http://eprints.internano.org/607/ (accessed on 20 June 2017).

95. Hoover, M.D.; Myers, D.S.; Cash, L.J.; Guilmette, R.A.; Kreyling, W.G.; Oberdörster, G.; Smith, R.; Cassata, J.R.; Boecker, B.B.; Grissom, M.P. Application of an informatics-based decision-making framework and process to the assessment of radiation safety in nanotechnology. *Health Phys.* **2015**, *108*, 179–194. [CrossRef] [PubMed]

96. NNI. *Nanotechnology Signature Initiative: Nanotechnology Knowledge Infrastructure (NKI) Data Readiness Levels Discussion Draft*; NNI: Washington, DC, USA, 2013. Available online: https://www.nano.gov/node/1015 (accessed on 20 June 2017).

97. Hendren, C.O.; Powers, C.M.; Hoover, M.D.; Harper, S.L. The Nanomaterial Data Curation Initiative: A collaborative approach to assessing, evaluating, and advancing the state of the field. *Beilstein J. Nanotechnol.* **2015**, *6*, 1752–1762. [CrossRef] [PubMed]

98. Google. Mapping the Invisible: Street View Cars Add Air Pollution Sensors. Available online: https://environment.google/projects/airview/ (accessed on 12 July 2017).

99. U.S. National Library of Medicine. PubMed. Available online: https://www.ncbi.nlm.nih.gov/pubmed (accessed on 17 August 2017).

100. Oltra, C.; Sala, R.; Boso, A.; Asensio, S.L. Public engagement on urban air pollution: An exploratory study of two interventions. *Environ. Monit. Assess.* **2017**, *189*, 296. [CrossRef] [PubMed]

Atmosphere **2017**, *8*, 182

101. U.S. National Library of Medicine. Fact Sheet: MEDLINE® Journal Selection. 30 June 2017. Available online: https://www.nlm.nih.gov/pubs/factsheets/jsel.html (accessed on 17 August 2017).

102. U.S. National Library of Medicine. PubMed Commons. Available online: https://www.ncbi.nlm.nih.gov/pubmedcommons/ (accessed on 17 August 2017).

103. Fierz, M.; Houle, C.; Steigmeier, P.; Burtscher, H. Design, Calibration, and field performance of a miniature diffusion size classifier. *Aerosol Sci. Technol.* **2011**, *45*, 1–10. [CrossRef]

104. Kulkarni, P.; Qi, C.; Fukushima, N. Miniature differential mobility analyzer for compact field-portable spectrometers. *Aerosol Sci. Technol.* **2016**, *50*, 1167–1179. [CrossRef] [PubMed]

105. Fadel, T.R.; Farrell, D.F.; Friedersdorf, L.E.; Griep, M.H.; Hoover, M.D.; Meador, M.A.; Meyyappan, M. Toward the responsible development and commercialization of sensor nanotechnologies. *ACS Sens.* **2016**, *1*, 207–216. [CrossRef] [PubMed]

atmosphere

MDPI

Article

Mobile DOAS Observations of Tropospheric NO$_2$ Using an UltraLight Trike and Flux Calculation

Daniel-Eduard Constantin [1,*], Alexis Merlaud [2], Mirela Voiculescu [1], Carmelia Dragomir [1], Lucian Georgescu [1], Francois Hendrick [2], Gaia Pinardi [2] and Michel Van Roozendael [2]

[1] European Center of Excellence for the Environment, Faculty of Sciences and Environment, "Dunarea de Jos" University of Galati, Str. Domneasca 111, Galati 800008, Romania; mirela.voiculescu@ugal.ro (M.V.); carmelia.dragomir@ugal.ro (C.D.); lucian.georgescu@ugal.ro (L.G.)

[2] Royal Belgian Institute for Space Aeronomy, Ringlaan-3-Avenue Circulaire, B-1180 Brussels, Belgium; alexis.merlaud@aeronomie.be (A.M.); francois.hendrick@aeronomie.be (F.H.); gaia.pinardi@aeronomie.be (G.P.); michel.vanroozendael@aeronomie.be (M.V.R.)

* Correspondence: daniel.constantin@ugal.ro

Academic Editors: Pius Lee, Rick Saylor and Jeff McQueen
Received: 24 February 2017; Accepted: 19 April 2017; Published: 22 April 2017

Abstract: In this study, we report on airborne Differential Optical Absorption Spectroscopy (DOAS) observations of tropospheric NO$_2$ using an Ultralight Trike (ULT) and associated flux calculations. The instrument onboard the ULT was developed for measuring the tropospheric NO$_2$ Vertical Column Density (VCD) and it was operated for several days between 2011 and 2014, in the South-East of Romania. Collocated measurements were performed using a car-DOAS instrument. Most of the airborne and mobile ground-based measurements were performed close to an industrial platform located nearby Galati city (45.43° N, 28.03° E). We found a correlation of R = 0.71 between tropospheric NO$_2$ VCDs deduced from airborne DOAS observations and mobile ground-based DOAS observations. We also present a comparison between stratospheric NO$_2$ Slant Column Density (SCD) derived from the Dutch OMI NO$_2$ (DOMINO) satellite data product and stratospheric SCDs obtained from ground and airborne measurements. The airborne DOAS observations performed on 13 August 2014 were used to quantify the NO$_2$ flux originating from an industrial platform located nearby Galati city. Measurements during a flight above the industrial plume showed a maximum tropospheric NO$_2$ VCD of $(1.41 \pm 0.27) \times 10^{16}$ molecules/cm^2 and an associated NO$_2$ flux of $(3.45 \pm 0.89) \times 10^{-3}$ kg/s.

Keywords: mobile DOAS; airborne observations; nitrogen dioxide; emission flux

---

1. Introduction

Nitrogen dioxide (NO$_2$) is a chemical gaseous compound with an important role in the Earth's atmosphere. NO$_2$ is a key trace element in the chemistry of ozone, since it is involved in the catalytic destruction of ozone in the stratosphere [1], while in the troposphere its photolysis leads directly to the formation of ozone (O$_3$) in the presence of VOCs (volatile organic compounds). NO$_2$ is released in the atmosphere from natural sources (soil, lightning, solar cosmic rays) and anthropogenic emissions (fossil fuels and biomass burning, industrial activities). Long-term exposure to NO$_2$ may affect the respiratory system and lead to coronary diseases. NO$_2$ can lead to acidification of the aquatic ecosystem following the oxidation to HNO$_3$.

The Differential Optical Absorption Spectroscopy (DOAS) technique [2] has been used for NO$_2$ atmospheric measurements since the early 1970s [3,4]. Nowadays, besides ground-based zenith sky measurements, DOAS techniques have developed into Multi-Axis Differential Optical Absorption Spectroscopy (MAX-DOAS) observations [5]. The mobile DOAS technique was recently used on several platforms such as: cars [6,7], airplanes [8–10], Unmanned Aerial Vehicles (UAVs) [11] or satellites [12–16].

Airborne observations have a number of important advantages for atmospheric research such as: the flexibility during the flights and the possibility to access remote areas such as oceans, deserts, rural or areas without roads.

The ULMs (Ultralight Motorized) are airborne platforms with an important scientific potential for atmospheric research. So far, ULMs have been used to study the ultraviolet actinic radiation flux [17], formaldehyde distribution [18], aerosol profiles [19], SO_2, NO_2 and ozone distribution [20–23].

This work highlights the capability of a low-cost system (ULT-DOAS) used for measurements of tropospheric NO_2 VCD and associated flux calculations. This study presents airborne DOAS observations of tropospheric NO_2 using an Ultralight Trike (ULT) and associated flux calculations. The work presented here was motivated by the need to further assess the intrapixel variability of NO_2 detected by UV-VIS DOAS instruments onboard satellites. This work comes in the context of a validation programme of the future Atmospheric Sentinels, starting with the Sentinel-5 Precursor to be launched in summer 2017. A similar system based on DOAS onboard the ULM was used during the Airborne Romanian Measurements of Aerosols and Trace gases (AROMAT) campaign, held in Romania in August 2015 [24]. The AROMAT campaign was conducted under the aegis of the European Space Agency (ESA) in the framework of a series of ESA field campaigns.

2. Methodology

2.1. Experimental and Instrumental Descriptions

The mobile DOAS observations were performed onboard of an Ultralight Trike (ULT), in the South-East of Romania (Figure 1) during several days between 2011 and 2014. All measurements were performed under clear-sky conditions (see Table 1). The Mobile DOAS system used for measurements will be described in the following as the ULT-DOAS system. The measurements took place in an area around Galati (located at 45.43° N, 28.03° E), Braila (45.26° N, 27.95° E), and close to the industrial areas of Slobozia (44.56° N, 27.35° E). Note that an operational steel mill factory is located in the vicinity of Galati, while Slobozia was chosen due to the presence of a fertilizer factory. Due to security concerns, direct flights above the cities or industrial platforms were not performed. Most of the airborne DOAS observations were performed in nadir geometry. Details about the ULT-DOAS measurements are presented in Table 1.

Figure 1. The main locations where airborne and/or ground-based mobile Differential Optical Absorption Spectroscopy (DOAS) observations were performed.

Table 1. Coordinates and temporal coverage of the mobile DOAS measurements.

Day	Time Interval UTC *	Route of ULT-DOAS ** Measurements	NO$_2$ Source Target
1 September 2011	8.31–9.45	Galati–Braila	Braila urban area
25 August 2012	7.53–8.89	Galati–Braila	Braila urban area
21 July 2014	9.51–10.96	Galati–Slobozia	Slobozia industrial area
13 August 2014	7.32–8.19	Galati	Galati industrial area

* Coordinated Universal Time. ** ULT-DOAS = Ultralight Trike-Differential Optical Absorption Spectroscopy.

Airborne-DOAS measurements were accompanied by car-DOAS measurements and static twilight observations on 21 July 2014. Static twilight DOAS measurements, used for determination of the NO$_2$ content from the reference spectra, were performed in a rural area close to Galati city. The car-DOAS measurements were performed right before or after the experimental flights.

The aircraft used for all the flight experiments presented in this paper is a double-seated, open-cockpit ultralight aircraft, trike type (model Fanagoria 21, produced by Plovdiv Air Bulgaria). The flexible wing (Atlant-21, produced by Plovdiv Air Bulgaria) has an area of 16 m^2. The ULT is powered by a Subaru EA81 engine with 75 HP. The cruise speed is 75 km/h and the maximum speed is 100 km/h relative to the ground. The aircraft has a maximum total weight at take-off of 450 kg.

The ULT-DOAS instrument consists of a compact Czerny-Turner spectrometer (AvaSpec-ULS2048XL-USB2, of 175 × 110 × 44 mm dimensions and 855 g weight) placed in the Ultralight Trike. Figure 2 presents the instrumental DOAS set-up. The spectral range of the spectrometer is 280–550 nm with 0.7 nm resolution (FWHM—Full Width at Half Maximum) with a focal length of 75 mm. The entrance slit is 50 μm and the grating is 1200 L/mm, blazed at 250 nm. The spectrometer is connected to the telescope through a 400 μm chrome-plated brass optical fiber. The telescope achieves a 2.3° field-of-view with fused silica collimating lenses. Each spectrum is recorded by a laptop and georeferenced by a GPS receiver. The spectrometer and the GPS receiver are powered by the laptop USB ports. The entire set-up is powered by 12 V of the ULT through an inverter. Each measurement is a 10-second average of 10 scans accumulations at an integration time between 50 and 150 ms.

This work is mostly based on nadir-DOAS observations but we also present zenith-sky observations onboard ULT for stratospheric NO$_2$ measurements. The same DOAS system was used in the case of the zenith sky car-DOAS observations.

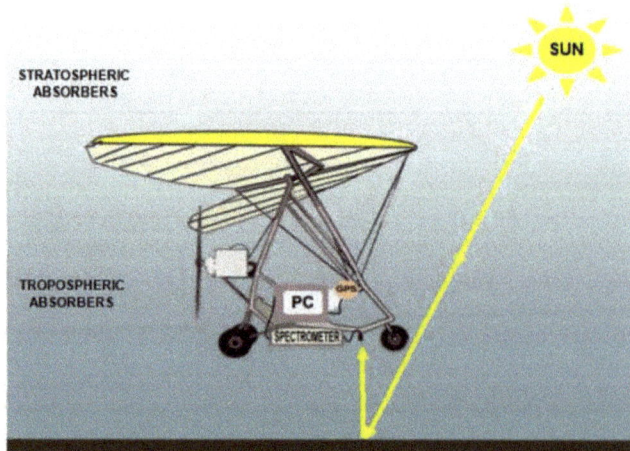

Figure 2. Schematic of the ULT-DOAS measurement principle.

2.2. Determination of the NO$_2$ Tropospheric Vertical Column and Flux Calculation

2.2.1. Retrieval of NO$_2$ Slant Column

The analysis of the measured spectra was performed using the QDOAS software [25]. For the NO$_2$ fit, the spectral window of 425–490 nm was used. The NO$_2$ spectral analysis included five absorption cross sections: the NO$_2$ cross sections at 298 K and 220K [26], the O$_3$ cross section at 223 K [27], the O$_4$ cross section [28] and a Ring spectrum [29]. A fifth-degree polynomial to account for scattering processes and broad-band absorption in the atmosphere was used in the DOAS analysis. The direct result of the spectral analysis is a differential slant column density (DSCD), which is the integrated trace gas concentration along the light path through the atmosphere. The DSCD is the difference between the slant column densities in the measured spectra (SCD$_{meas}$) and the Fraunhofer reference spectrum (SCD$_{ref}$). The NO$_2$ amount in the Fraunhofer reference spectrum is unknown and its retrieval is important for the determination of the SCD$_{meas}$ (Equation (1)).

$$SCD_{meas} = DSCD + SCD_{ref} \tag{1}$$

Figure 3 presents a typical DOAS fit using a spectrum recorded during the experiment close to the Galati steel factory, on 13 August 2014.

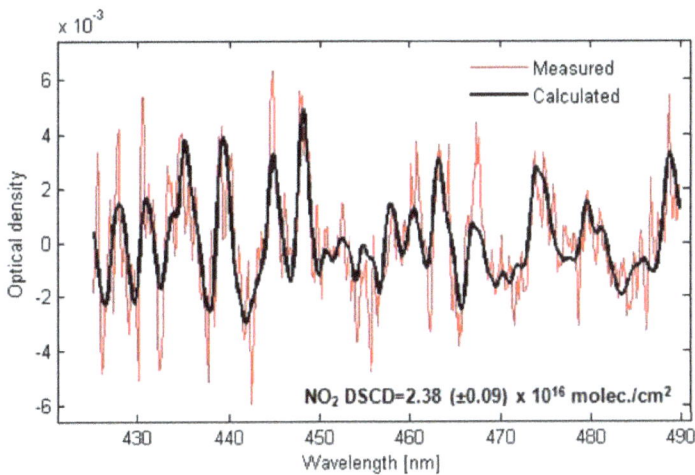

Figure 3. Example of a DOAS fit realized with the QDOAS software; the analyzed spectrum was recorded close to Galati, on 13 August 2014. Black line corresponds to molecular cross sections scaled to the detected absorptions in the measured spectrum (red line).

The Slant Column Density (SCD) is converted to a Vertical Column Density (VCD) by means of an Air Mass Factor (AMF), which is defined as the ratio between SCD and VCD (Equation (2))

$$AMF = \frac{SCD}{VCD} = \frac{\tau_{SCD}}{\tau_{VCD}} \tag{2}$$

where τ_{SCD} and τ_{VCD} are the optical thickness for the slant column (SCD) and vertical column (VCD), respectively.

Since the measured spectra contain information about both stratospheric and tropospheric NO$_2$ content, the SCD$_{meas}$ can be written as:

$$SCD_{meas} = AMF_{tropo} \cdot VCD_{tropo} + AMF_{strato} \cdot VCD_{strato} - SCD_{ref} \tag{3}$$

The above equations can be further simplified assuming that SCD_{ref} is dominated by stratospheric NO_2. Using this assumption, the stratospheric contributions can be canceled by the NO_2 amount in the reference spectrum (Equation (4)) [8].

$$VCD_{tropo} = DSCD/AMF_{tropo} \tag{4}$$

The assumption presented above is valid if the reference spectrum (needed for the spectral evaluation) is recorded at noon, in an area with a very low NO_2 content and if SCD_{strato} does not vary in time. The Fraunhofer reference spectrum could also be a zenith-sky spectrum recorded at high altitude over the boundary layer [30]. However, in this work the tropospheric NO_2 VCD is based on calculations using Equation (3).

2.2.2. Deduction of the SCD_{ref} and VCD_{strato}

To avoid systematic errors due to the use of multiple reference spectra, only one spectrum will be used for the spectral analysis of all DOAS observations presented in this paper.

The NO_2 amount in the reference spectrum was calculated using a photo-chemically modified Langley plot [6,31]. The SCD_{ref} corresponds to a zenith spectrum recorded at noon, in a clean rural area close to Galati city. The spectrum was recorded on 13 August 2014 (9.70 UTC and solar zenith angle (SZA) = 31.55°).

The photo-chemically modified Langley plot was applied for the twilight sunrise observations performed on 21 July 2014. By applying the Langley plot method, we calculated the SCD_{ref} as 4.1×10^{15} molecules/cm^2 of NO_2.

The stratospheric contribution used for the retrieval of the VCD_{tropo} is derived from the assimilated vertical stratospheric columns simulated by Dutch OMI NO_2 (DOMINO). Table 2 shows the satellite overpass data sets that were used for the retrieval algorithm presented in this work.

Table 2. OMI satellite overpasses data sets.

Day	Orbit Nr.	Overpass Time UTC	Stratospheric VCD [$\times 10^{15}$ molecules/cm^2]
1 September 2011	37,923	11:04:28	3.76
25 August 2012	43,151	11:11:07	3.75
21 July 2014	53,272	11:17:36	4.14
13 August 2014	53,607	11:23:47	3.74

VCD = Vertical Column Density.

Figure 4A presents the SCD determined at twilight sunrise on 21 July 2014 compared with the SCD_{strato} derived from the DOMINO Level 2 product [32] scaled with a chemically modified AMF calculated by PSCBOX [33,34]. More details about the retrieval of the SCD_{strato} using twilight observations and model simulations are presented in [6]. A good agreement between the two types of SCD_{strato} determinations is obtained, which gives confidence in the stratospheric SCD measured by our static DOAS observations.

Figure 4B shows the SCDstrato derived from DOMINO compared with the SCDs determined by car-DOAS zenith-sky observations and ULT-DOAS measurements performed in the zenith geometry, for the same day of 21 July 2014. From this plot, one can see that the car-DOAS measurements are dominated by tropospheric NO_2 while the zenith-sky ULT-DOAS observation presents a low amount of NO_2 close to the stratospheric NO_2 SCD derived from OMI. This is due to the fact that zenith-sky ULT-DOAS observations are performed above the NO_2 plume or above the planetary boundary layer.

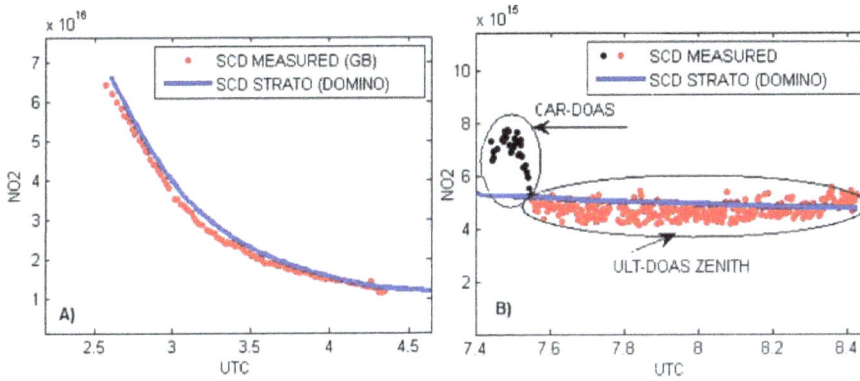

Figure 4. Comparisons between measured SCD using various methods of determination and stratospheric SCD derived from OMI (21 July 2014); (**A**) Comparison between SCD determined from ground-based (GB) observations and stratospheric SCD derived from DOMINO; (**B**) Comparison between SCD determined from CAR-DOAS (black dots) and ULT-DOAS (red dots) and stratospheric SCD derived from DOMINO.

2.2.3. Radiative Transfer Calculation

In order to determine the VCD, the SCD retrieved with the DOAS method has to be converted using an appropriate AMF. The geometric approximation for the airborne DOAS observations assumes a simple reflection of the sunlight on the earth's surface. In this case (neglecting the earth's sphericity) the nadir AMF can be described as a function of the solar zenith angle (SZA) as:

$$AMF_{geo} = 1 + 1/\cos(SZA) \tag{5}$$

Since the ULT-DOAS measurements were performed in the open atmosphere using scattered sunlight radiation, the radiative transfer during the observations needs to be modeled to interpret the retrieved data. In this work, the AMF was calculated using the radiative transfer model (RTM) UVspec/DISORT [35], which is a fully spherical model. This RTM has been validated using six other different codes [34].

The general assumptions made for the radiative transfer calculation using the RTM UVspec/DISORT are introduced in Table 3.

Table 3. Input parameters used for the radiative transfer model (RTM) calculations.

Trace Gas	Nitrogen Dioxide		
wavelength		440 nm	
flight altitude		1000 m	1500 m
albedo		0.1	
visibility	1 km	5 km	20 km
line of sight		0°	

Results of AMF simulations for the nadir flight performed on 13 August 2014, using the input parameters presented in Table 3, are displayed in Figure 5. The geometric AMF for the nadir view is also shown. The visibility parameter accounts for the effect of aerosols.

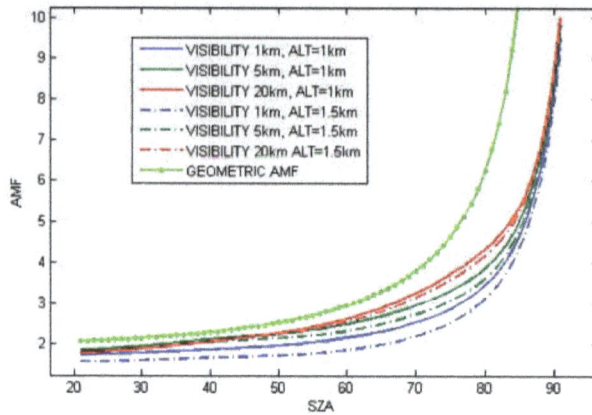

Figure 5. AMF simulations obtained from RTM calculations using UVspec/DISORT for various input parameters, for 13 August 2014. AMF = Air Mass Factor.

3. Results and Discussions

The airborne DOAS observations were designed to determine the distribution of tropospheric NO_2 from the South-East of Romania from urban, industrial and rural areas and associated flux.

The first flights were performed on 1 September 2011 and 25 August 2012, and aimed at measuring the NO_2 around the industrial area of Galati city and from Braila city. The flight performed on 21 July 2014 aimed to measure the NO_2 emitted by a fertilizer factory located nearby Slobozia city; unfortunately, during the DOAS flight the fertilizer factory was not operational. Figure 6 presents the horizontal distribution of the tropospheric NO_2 determined in nadir geometry for 1 September 2011, while Figure 7 depicts the results during a similar flight, but on 25 August 2012. During this experiment, the plume from the industrial platform was not fully crossed by the optical instrument onboard the ULT. The wind was northerly resulting in local increases of the NO_2 amount detected by the spectrometer. The maximum tropospheric NO_2 VCD detected during this experiment was $(1.1 \pm 0.24) \times 10^{16}$ molecules/cm^2 while the minimum tropospheric NO_2 VCD was $(2.1 \pm 0.81) \times 10^{15}$ molecules/cm^2.

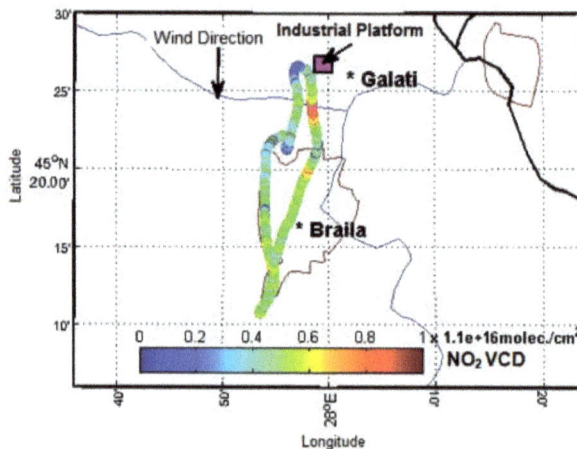

Figure 6. Map of tropospheric NO_2 VCD determined on 1 September 2011 using ULT-DOAS observations.

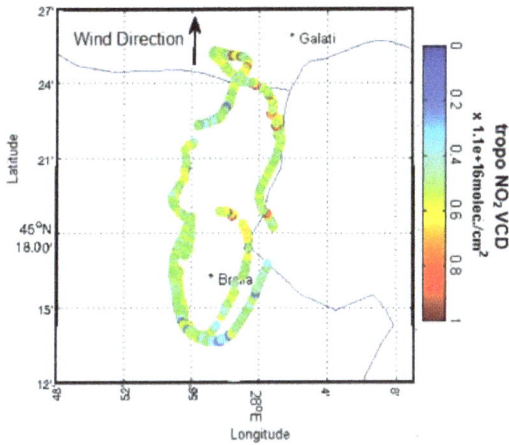

Figure 7. Map of tropospheric NO$_2$ VCD determined on 25 August 2012 using ULT-DOAS observations.

The trajectory of the flight on 13 August 2014 gave us the opportunity of calculating the NO$_2$ flux emissions around the industrial area of Galati city. This was not possible for the other flights because encircling the NO$_2$ source was not authorized.

On 1 September 2011, the NO$_2$ amount was low relative to the other day. The flight comprised almost two complete circles around Braila; however, the NO$_2$ displayed no clear variation. The horizontal distribution of NO$_2$ was quite homogenous over Braila city on this day.

A double experiment was performed on 13 August 2014, using both a ULT-DOAS and a car-based DOAS system. The mobile ground-based DOAS observations were performed using the same equipment during 9.75–10 UTC, while the ULT-DOAS observations were performed during 7.30–8.15 UTC.

Figure 8 shows the tropospheric NO$_2$ VCD derived along the trajectory of the ULT-DOAS measurements. The right plot shows a photograph of the plume crossed by the ULT flights. During the same day, approximately 1 h after the acquisition of the ULT-DOAS measurements, a zenith-sky car-DOAS system was used to sample the NO$_2$ plume at the same location as the ULT-DOAS observations. We assume that the quantity of the NO$_2$ emitted by the steel factory was almost constant during the airborne and car-DOAS system.

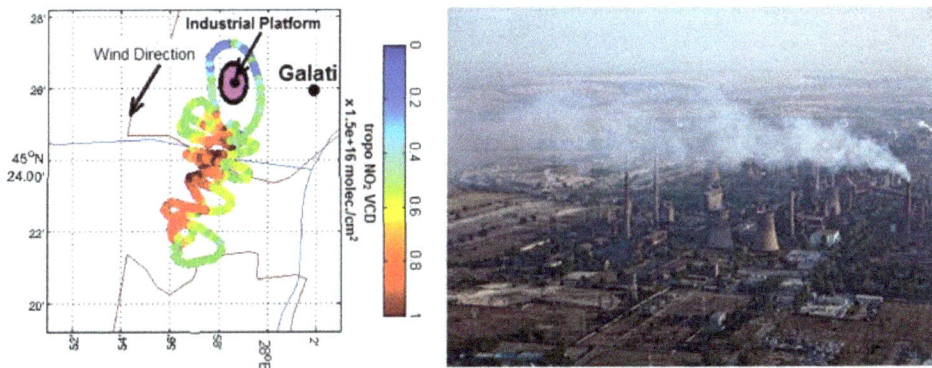

Figure 8. The tropospheric NO$_2$ VCD along the flight trajectory using the ULT-DOAS system on 13 August 2014 (**left**); Photography of the NO$_2$ plume determined on the same day (**right**).

Figure 9 presents the NO$_2$ VCD derived from the nadir airborne DOAS observations performed over the industrial area of Galati city compared with zenith-sky ground-based mobile DOAS measurements performed over the same area in the same day (13 August 2014). In these figures, we show the original SCDs (A) and the retrieved tropospheric NO$_2$ VCD (B). Figure 9A shows that the DSCDs determined from the ULT-DOAS system are ~30% higher than the DSCDs determined using car-DOAS observations. This difference is attributed to the different observation geometries. After appropriate AMF calculation (see Section 2.2.2), both observations show close results.

The ULT flight above the industrial plume led to the detection of a maximum tropospheric NO$_2$ VCD of (1.41 ± 0.27) × 10^{16} molecules/cm^2 while the car-DOAS observation shows a maximum tropospheric NO$_2$ VCD of (1.36 ± 0.21) × 10^{16} molecules/cm^2. Figure 10 displays the correlation between the tropospheric NO$_2$ VCD retrieved by the ULT-DOAS and the car-DOAS instrument, where closest spatially coincident data were selected. A Pearson correlation coefficient R = 0.71 was obtained between ground mobile DOAS observations and airborne DOAS measurements.

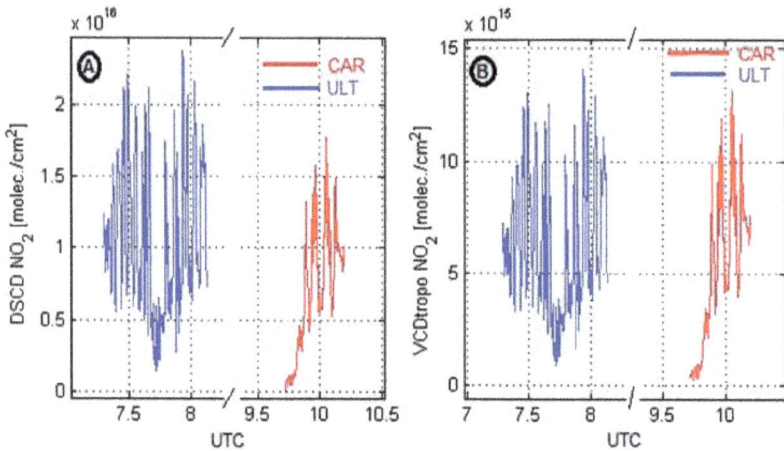

Figure 9. Comparisons between ULT-DOAS and car-DOAS observations performed on 13 August 2014. (**A**) The results of the preliminary DOAS analysis (DSCDs) and (**B**) after determination of the vertical columns (VCDs).

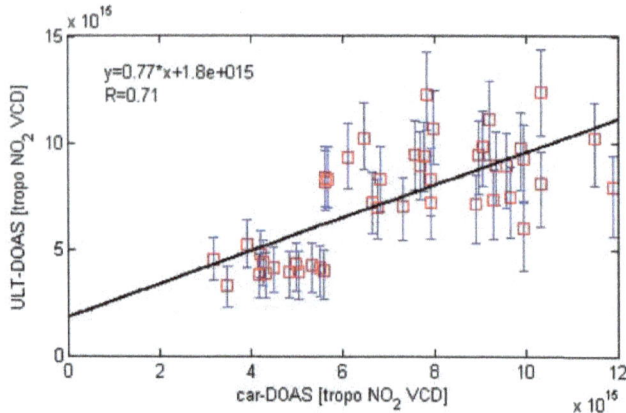

Figure 10. Correlation between tropospheric VCDs measured by the ULT-DOAS and the car-DOAS instrument on 13 August 2014.

NO₂ Flux Calculation

The NO₂ flux above the industrial platform located nearby Galati city was calculated by performing upwind and downwind measurements around the point source using ULT-DOAS observations on 13 August 2014. The calculation of the emission flux is based on the following parameters: the NO₂ VCD determined from the transect over the plume, the wind speed and the wind factor correction, taking into account the angle between the flight direction and wind direction (Equation (6)), [20,36,37]:

$$\text{Flux}_{\text{NO}_2} = \sum_i \text{VCD}_{\text{NO}_2}(s_i) \cdot v \cdot \sin(\alpha_i) \cdot \Delta s_i \tag{6}$$

where VCD_{NO2} is the NO₂ tropospheric vertical column, v is the wind speed, α the angle between wind direction and driving route, i is the observations index, and Δs_i is the distance between two successive spectra.

The wind data used for the NO₂ flux calculation rely on measurements of the automatic weather station (Davis Vantage Pro2) located in the campus university of Galati city (45.44° N, 28.05° E), while vertical wind profiles come from the Hybrid Single Particle Lagrangian Integrated Trajectory Model (HYSPLIT) [38] model using archived dataset GDAS 0.5° × 0.5°. The weather station is located at 30 m height and acquires data every 30 min. Since no atmospheric sounding was possible during the experiments, the wind measured on the ground was scaled to the output of HYSPLIT model simulation at 1000 m altitude. During the period of the ULT-DOAS measurements, the mean NO₂ emission flux was determined to be $(3.45 \pm 0.89) \times 10^{-3}$ kg/s. A local environmental report indicates ~600 tons/year NO_x emissions emitted by the steel factory [39]. Using a Leighton ratio (L = [NO]/[NO₂]) of 0.3, we calculated that the steel factory has emitted a mean of ~10×10^{-3} kg/s of NO₂. The difference between the two types of estimation may be attributed to the fact that the NO_x emissions from the steel factory are dependent on the quantity of the steel produced, which can vary from one day to another. Also, the derived NO₂ fluxes must be dealt with some care because of the probably incomplete NO to NO₂ conversion [40].

4. Conclusions

Ultralight-trike DOAS observations were performed in the South-East of Romania during four days between 2011 and 2014. The first two flights were focused over Braila city, the third aimed at measurements of the NO₂ plume emitted by a fertilizer factory near Slobozia city. Unfortunately, during the DOAS flight the fertilizer factory was not operational. The last flight, performed on 13 August 2014, was focused over the industrial area of Galati city. Nadir observations were performed around the industrial platform of Galati city aiming at measuring the tropospheric NO₂ VCD around the source and at evaluating the associated NO₂ flux. To retrieve the tropospheric NO₂ VCD from ULT-DOAS observations, complementary ground- and space-based measurements were used.

We showed that the tropospheric NO₂ VCD deduced from the ULT-DOAS observations are consistent with measurements performed from the ground using a zenith-sky car-DOAS system. Although two hours separated the two experiments, a correlation coefficient of R = 0.71 was found between the two results, a tropospheric NO₂ VCD of $(1.41 \pm 0.27) \times 10^{16}$ molecules/cm² and an estimated associated flux of $(3.45 \pm 0.89) \times 10^{-3}$ kg/s was measured close to the industrial area of Galati city on 13 August 2014, the only day that it was possible to determine the NO₂ flux.

Also, we showed that the stratospheric SCD derived from ground-based and airborne measurements correlates well with stratospheric NO₂ derived from observations by the OMI satellite sensor.

Based on this study, we conclude that the ULT is an efficient tool which allows determining with a high resolution the NO₂ distribution around urban or industrial sources. Also, the ULT-DOAS system is a very efficient tool for measuring fluxes due to its flexibility during the flights and the possibility to access remote areas. The ULT-DOAS system might also constitute a promising tool for satellite validation and calibration under clear-sky conditions, especially for upcoming high-resolution sensors such as the TROPOMI/Sentinel-5 Precursor instrument to be launched in summer 2017.

Acknowledgments: The work of D.E. Constantin was supported by Project PN-II-RU-TE-2014-4-2584, a grant of the Romanian National Authority for Scientific Research and Innovation, CNCS UEFISCDI. The DOMINO data product was taken from the ESA TEMIS archive (www.temis.nl) maintained at KNMI, The Netherlands.

Author Contributions: Daniel-Eduard Constantin and Alexis Merlaud conceived and designed the study; Daniel-Eduard Constantin analyzed the data; and Mirela Voiculescu, Carmelia Dragomir, Lucian Georgescu, Gaia Pinardi, Francois Hendrick and Michel Van Roozendael improved the paper.

Conflicts of Interest: The authors declare no conflict of interest.

References

1. Solomon, S.; Portmann, R.W.; Sanders, R.W.; Daniel, J.S.; Madsen, W.; Bartram, B.; Dutton, E.G. On the role of nitrogen dioxide in the absorption of solar radiation. *J. Geophys. Res.* **1999**, *104*, 12047–12058. [CrossRef]
2. Platt, U. Differential optical absorption spectroscopy (DOAS). *Chem. Anal. Ser.* **1994**, *127*, 27–83.
3. Brewer, A.W.; McElroy, C.T.; Kerr, J.B. Nitrogen dioxide concentration in the atmosphere. *Nature* **1973**, *246*, 129–133. [CrossRef]
4. Noxon, J.F. Nitrogen dioxide in the stratosphere and troposphere measured by ground-based absorption spectroscopy. *Science* **1975**, *189*, 547–549. [CrossRef] [PubMed]
5. Hendrick, F.; Müller, J.-F.; Clémer, K.; Wang, P.; De Mazière, M.; Fayt, C.; Gielen, C.; Hermans, C.; Ma, J.Z.; Pinardi, G.; et al. Four years of ground-based MAX-DOAS observations of HONO and NO_2 in the Beijing area. *Atmos. Chem. Phys.* **2014**, *14*, 765–781. [CrossRef]
6. Constantin, D.-E.; Merlaud, A.; van Roozendael, M.; Voiculescu, M.; Fayt, C.; Hendrick, F.; Pinardi, G.; Georgescu, L. Measurements of Tropospheric NO_2 in Romania Using a Zenith-Sky Mobile DOAS System and Comparisons with Satellite Observations. *Sensors* **2013**, *13*, 3922–3940. [CrossRef]
7. Dragomir, C.; Constantin, D.-E.; Voiculescu, M.; Georgescu, L.; Merlaud, A.; Roozendael, M.V. Modeling results of atmospheric dispersion of NO_2 in an urban area using METI–LIS and comparison with coincident mobile DOAS measurements. *Atmos. Pollut. Res.* **2015**, *6*, 503–510. [CrossRef]
8. Merlaud, A.; van Roozendael, M.; van Gent, J.; Fayt, C.; Maes, J.; Toledo-Fuentes, X.; Ronveaux, O.; de Mazière, M. DOAS measurements of NO_2 from an ultralight aircraft during the Earth Challenge expedition. *Atmos. Meas. Tech.* **2012**, *5*, 2057–2068. [CrossRef]
9. Meier, A.C.; Schönhardt, A.; Bösch, T.; Richter, A.; Seyler, A.; Ruhtz, T.; Constantin, D.-E.; Shaiganfar, R.; Wagner, T.; Merlaud, A.; et al. High-resolution airborne imaging DOAS-measurements of NO_2 above Bucharest during AROMAT. *Atmos. Meas. Tech. Discuss.* **2016**. [CrossRef]
10. Tack, F.; Merlaud, A.; Iordache, M.-D.; Danckaert, T.; Yu, H.; Fayt, C.; Meuleman, K.; Deutsch, F.; Fierens, F.; van Roozendael, M. High resolution mapping of the NO_2 spatial distribution over Belgian urban areas based on airborne APEX remote sensing. *Atmos. Meas. Tech. Discuss.* **2017**. [CrossRef]
11. Merlaud, A.; Constantin, D.; Fayt, C.; Maes, J.; Mingireanu, F.; Mocanu, I.; Georgescu, L.; Roozendael, M. Small whiskbroom imager for atmospheric composition monitoring (SWING) from an unmanned areal vehicle (UAV). In Proceedings of the 21st ESA Symposium on European Rocket and Balloon Programmes and related Research, Thun, Switzerland, 9–13 June 2014; pp. 1–7.
12. Bovensmann, H.; Burrows, J.P.; Buchwitz, M.; Frerick, J.; Nöel, S.; Rozanov, V.V.; Chance, K.V.; Goede, A.H.P. SCIAMACHY—Mission objectives and measurement modes. *J. Atmos. Sci.* **1999**, *56*, 127–150. [CrossRef]
13. Levelt, P.; van den Oord, G.; Dobber, M.; Malkki, A.; Visser, H.; de Vries, J.; Stammes, P.; Lundell, J.; Saari, H. The ozone monitoring instrument. *IEEE T. Geosci. Remote.* **2006**, *44*, 1093–1101. [CrossRef]
14. Munro, R.; Eisinger, M.; Anderson, C.; Callies, J.; Corpaccioli, E.; Lang, R.; Lefebvre, A.; Livschitz, Y.; Albinana, A.P. GOME-2 on MetOp. In Proceedings of the 2006 EUMETSAT Meteorological Satellite Conference, Helsinki, Finland, 12–16 June 2006; p. 48.
15. van der A, R.J.; Eskes, H.J.; Boersma, K.F.; van Noije, T.P.C.; van Roozendael, M.; de Smedt, I.; Peters, D.H.M.U.; Meijer, E.W. Trends, seasonal variability and dominant NO_x source derived from a ten year record of NO_2 measured from space. *J. Geophys. Res.* **2008**, *113*, D04302. [CrossRef]

16. Varotsos, C.; Christodoulakis, J.; Tzanis, C.; Cracknell, A.P. Signature of tropospheric ozone and nitrogen dioxide from space: A case study for Athens, Greece. *Atmos. Environ.* **2014**, *89*, 721–730. [CrossRef]

17. Junkermann, W. An ultralight aircraft as platform for research in the lower troposphere: System performance and first results from radiation transfer studies in stratiform aerosol layers and broken cloud conditions. *J. Atmos. Ocean. Technol.* **2001**, *18*, 934. [CrossRef]

18. Junkermann, W. On the distribution of formaldehyde in the western Po-Valley, Italy, during FORMAT 2002/2003. *Atmos. Chem. Phys.* **2009**, *9*, 9187–9196. [CrossRef]

19. Chazette, P.; Sanak, J.; Dulac, F. New Approach for Aerosol Profiling with a Lidar Onboard an Ultralight Aircraft: Application to the African Monsoon Multidisciplinary Analysis. *Environ. Sci. Technol.* **2007**, *41*, 8335–8341. [CrossRef] [PubMed]

20. Wang, P.; Richter, A.; Bruns, M.; Burrows, J.P.; Scheele, R.; Junkermann, W.; Heue, K.-P.; Wagner, T.; Platt, U.; Pundt, I. Airborne multi-axis DOAS measurements of tropospheric SO_2 plumes in the Po-valley, Italy. *Atmos. Chem. Phys.* **2006**, *6*, 329–338. [CrossRef]

21. Grutter, M.; Basaldud, R.; Rivera, C.; Harig, R.; Junkerman, W.; Caetano, E.; Delgado-Granados, H. SO_2 emissions from Popocatépetl volcano: Emission rates and plume imaging using optical remote sensing techniques. *Atmos. Chem. Phys.* **2008**, *8*, 6655–6663. [CrossRef]

22. General, S.; Pöhler, D.; Sihler, H.; Bobrowski, N.; Frieß, U.; Zielcke, J.; Horbanski, M.; Shepson, P.B.; Stirm, B.H.; Simpson, W.R.; et al. The Heidelberg Airborne Imaging DOAS Instrument (HAIDI)—A novel imaging DOAS device for 2-D and 3-D imaging of trace gases and aerosols. *Atmos. Meas. Tech.* **2014**, *7*, 3459–3485. [CrossRef]

23. Liu, L.; Flatøy, F.; Ordóñez, C.; Braathen, G.O.; Hak, C.; Junkermann, W.; Andreani-Aksoyoglu, S.; Mellqvist, J.; Galle, B.; Prévôt, A.S.H.; et al. Photochemical modelling in the Po basin with focus on formaldehyde and ozone. *Atmos. Chem. Phys.* **2007**, *7*, 121–137. [CrossRef]

24. Airborne Romanian Measurements of Aerosols and Trace Gases (AROMAT). Available online: http://uv-vis.aeronomie.be/aromat/index.php (accessed on 15 January 2017).

25. Danckaert, T.; Fayt, C.; van Roozendael, M.; de Smedt, I.; Letocart, V.; Merlaud, A.; Pinardi, G. QDOAS Software user manual Version 2.111-April 2016, UV-Visible DOAS Research Group of the Royal Belgian Institute for Space Aeronomy Web Site. Available online: http://uv-vis.aeronomie.be/software/QDOAS/QDOAS_manual.png (accessed on 15 December 2016).

26. Vandaele, A.; Hermans, C.; Simon, P.; Carleer, M.; Colin, R.; Fally, S.; Mérienne, F.; Jenouvrier, A.; Coquart, B. Measurements of the NO_2 absorption cross-section from 42000 cm^{-1} to 10000 cm^{-1} (238–1000 nm) at 220 K and 294 K (220 K). *J. Quant. Spectrosc. Radiat. Transf.* **1998**, *59*, 171–184. [CrossRef]

27. Bogumil, K.; Orphal, J.; Homann, T.; Voigt, S.; Spietz, P.; Fleischmann, O.C.; Vogel, A.; Hartmann, M.; Kromminga, H.; Bovensmann, H.; et al. Measurements of molecular absorption spectra with the SCIAMACHY pre-flight model: Instrument characterization and reference data for atmospheric remote-sensing in the 230–2380 nm region. *J. Photochem. Photobiol. A Chem.* **2003**, *157*, 167–184. [CrossRef]

28. Thalman, R.; Volkamer, R. Temperature dependent absorption cross-sections of O_2-O_2 collision pairs between 340 and 630 nm and at atmospherically relevant pressure. *Phys. Chem. Chem. Phys.* **2013**, *15*, 15371–15381. [CrossRef] [PubMed]

29. Chance, K.V.; Spurr, R.J.D. Ring effect studies: Rayleigh scattering, including molecular parameters for rotational Raman scattering, and the Fraunhofer spectrum. *Appl. Opt.* **1997**, *36*, 5224–5230. [CrossRef] [PubMed]

30. Baidar, S.; Oetjen, H.; Coburn, S.; Dix, B.; Ortega, I.; Sinreich, R.; Volkamer, R. The CU Airborne MAX-DOAS instrument: Vertical profiling of aerosol extinction and trace gases. *Atmos. Meas. Tech.* **2013**, *6*, 719–739. [CrossRef]

31. Vaughan, G.; Quinn, P.T.; Green, A.C.; Bean, J.; Roscoe, H.K.; van Roozendael, M.; Goutail, F. SAOZ measurements of stratospheric NO_2 at Aberystwyth. *J. Environ. Monit.* **2006**, *8*, 353–361. [CrossRef] [PubMed]

32. Dirksen, R.J.; Boersma, K.F.; Eskes, H.J.; Ionov, D.V.; Bucsela, E.J.; Levelt, P.F.; Kelder, H.M. Evaluation of stratospheric NO_2 retrieved from the Ozone Monitoring Instrument: Intercomparison, diurnal cycle and trending. *J. Geophys. Res.* **2011**, *116*, D08305. [CrossRef]

33. Hendrick, F.; Barret, B.; van Roozendael, M.; Boesch, H.; Butz, A.; De Mazière, M.; Goutail, F.; Hermans, C.; Lambert, J.-C.; Pfeilsticker, K.; et al. Retrieval of nitrogen dioxide stratospheric profiles from ground-based zenith-sky UV-visible observations: Validation of the technique through correlative comparisons. *Atmos. Chem. Phys.* **2004**, *4*, 2091–2106. [CrossRef]

34. Hendrick, F.; van Roozendael, M.; Kylling, A.; Petritoli, A.; Rozanov, A.; Sanghavi, S.; Schofield, R.; von Friedeburg, C.; Wagner, T.; Wittrock, F.; et al. Intercomparison exercise between different radiative transfer models used for the interpretation of ground-based zenith-sky and multi-axis DOAS observations. *Atmos. Chem. Phys.* **2006**, *6*, 93–108. [CrossRef]

35. Mayer, B.; Kylling, A. Technical note: The libRadtran software package for radiative transfer calculations—Description and examples of use. *Atmos. Chem. Phys.* **2005**, *5*, 1855–1877. [CrossRef]

36. Rivera, C.; Sosa, G.; Wöhrnschimmel, H.; de Foy, B.; Johansson, M.; Galle, B. Tula industrial complex (Mexico) emissions of SO_2 and NO_2 during the MCMA 2006 field campaign using a mobile mini-DOAS system. *Atmos. Chem. Phys.* **2009**, *9*, 6351–6361. [CrossRef]

37. Johansson, M.; Rivera, C.; de Foy, B.; Lei, W.; Song, J.; Zhang, Y.; Galle, B.; Molina, L. Mobile mini-DOAS measurement of the outflow of NO_2 and HCHO from Mexico City. *Atmos. Chem. Phys.* **2009**, *9*, 5647–5653. [CrossRef]

38. Stein, A.F.; Draxler, R.R.; Rolph, G.D.; Stunder, B.J.B.; Cohen, M.D.; Ngan, F. NOAA's HYSPLIT atmospheric transport and dispersion modeling system. *Bull. Am. Meteorol. Soc.* **2015**, *96*, 2059–2077. [CrossRef]

39. *Studiu Privind Calitatea Aerului in Municipiul Galati*; Municipiul Galati: Galati, Romania, 2016.

40. Frins, E.; Bobrowski, N.; Osorio, M.; Casaballe, N.; Belsterli, G.; Wagner, T.; Platt, U. Scanning and mobile multi-axis DOAS measurements of SO_2 and NO_2 emissions from an electric power plant in Montevideo, Uruguay. *Atmos. Environ.* **2014**, *98*, 347–356. [CrossRef]

Article

Evaluation of Analysis by Cross-Validation. Part I: Using Verification Metrics

Richard Ménard *and Martin Deshaies-Jacques

Air Quality Research Division, Environment and Climate Change Canada, 2121 Transcanada Highway, Dorval, QC H9P 1J3, Canada; martin.deshaies-jacques@canada.ca
* Correspondence: richard.menard@canada.ca; Tel.: +1-514-421-4613

Received: 5 September 2017; Accepted: 24 February 2018; Published: 27 February 2018

Abstract: We examine how passive and active observations are useful to evaluate an air quality analysis. By leaving out observations from the analysis, we form passive observations, and the observations used in the analysis are called active observations. We evaluated the surface air quality analysis of O_3 and $PM_{2.5}$ against passive and active observations using standard model verification metrics such as bias, fractional bias, fraction of correct within a factor of 2, correlation and variance. The results show that verification of analyses against active observations always give an overestimation of the correlation and an underestimation of the variance. Evaluation against passive or any independent observations display a minimum of variance and maximum of correlation as we vary the observation weight, thus providing a mean to obtain the optimal observation weight. For the time and dates considered, the correlation between (independent) observations and the model is 0.55 for O_3 and 0.3 for $PM_{2.5}$ and for the analysis, with optimal observation weight, increases to 0.74 for O_3 and 0.54 for $PM_{2.5}$. We show that bias can be a misleading measure of evaluation and recommend the use of a fractional bias such as the modified normalized mean bias (MNMB). An evaluation of the model bias and variance as a function of model values also show a clear linear dependence with the model values for both O_3 and $PM_{2.5}$.

Keywords: chemical data assimilation; air quality model diagnostics; cross-validation

1. Introduction

Since 2003, Environment and Climate Change Canada (ECCC) has been producing hourly surface analyses of pollutants covering North America [1,2] which became operational products in February 2013 [3]. The analyses are produced using an optimum interpolation scheme that combines the operational air quality forecast model GEM-MACH output [4] (CHRONOS model output was used prior to 2010 [5]) with real-time hourly observations of O_3, $PM_{2.5}$, PM_{10}, NO_2, and SO_2 from the AirNow gateway with additional observations from Canada. As those surface analyses are not used to initialize an air quality model, it raises the issue on how to evaluate them. We conduct routine evaluations using the same set of observations as those used to produce the analysis. Once in a while, when there is a change in the system, a more thorough evaluation is conducted where we leave out a certain fraction of the observations and use them as independent observations, a process known as cross-validation. Observations used in producing the analysis are called *active observations* while those not used for evaluation are *passive observations*. Cross-validation is used to validate any model that depends on data. In air quality applications it has been used, for example, for mapping and exposure models [6–8]. The purpose of this two-part paper is to examine the relative merit of using active or passive observations (or independent observations in general) viewed from different evaluation metrics, but also to develop, in the second part, a mathematical framework to estimate the analysis error, and in doing so, to improve the analysis.

The evaluation of an analysis is important, even in the case where it is used to initialize an air quality forecast model, since the evaluation of the resulting air quality forecast may not be a good measure of the quality of the analysis. In air quality forecasting, the forecast error growth is small, depicts little sensitivity to initial conditions and is in fact more sensitive to numerous modeling errors such as: photochemistry, clouds, meteorology, boundary conditions and emissions just to name a few [9–13]. Furthermore, chemical species that are observed are incomplete compared to species needed to initialize an air quality model, incomplete in terms of the number of species observed as well as in their kind [9,11,13,14]. Only a fraction of the observed species (either of secondary or primary pollutants) are usable for data assimilation; important chemical mechanisms are left completely unobserved and for aerosols, information on size distribution is quite limited and almost nonexistent when it comes to speciation [9,13]. In addition, the observational coverage is limited to the surface or to total column measurements which, up until now, were available at one or two local times per day. There are thus many assumptions to be made from an analysis to a proper 3D initial chemical condition and surface emission correction and its subsequent impact on the air quality forecast. These considerations warrant an independent evaluation of the quality of the analysis on its own [15].

Evaluating an analysis with observations is quite different from evaluating a model with observations, since analyses are created from observations. From a statistical point of view, the observation and analysis cannot be considered independent. However, let us assume that observation errors are spatially uncorrelated. Then, since the passive and active observation sites are never collocated, then the errors from passive observations are uncorrelated with errors of active observations (i.e., observations that are used for the analysis). Furthermore, since the modelling errors are usually assumed to be uncorrelated with observation errors, then it is also uncorrelated with the analysis errors. Cross-validation thus offers a means to evaluate analyses with statistically independent (passive) observations [16].

In part one of this paper, we evaluate the relative merit of passive and active observations in the evaluation of analyses using standard metrics used for model evaluation. We show how and when the use of active observations can be misleading and that passive observations can provide a means to identify optimal analyses. Our examples show that optimal analyses, at the independent observation sites, have much smaller biases than the model biases and increase the correlation coefficient by nearly a factor of 2.

The paper is thus organized as follows. First we present the analysis scheme we will be using, as well as the cross-validation design, the evaluation metrics and the configuration of the experiments. Then in Section 3, we assess the quality of the analyses in both active and passive observation spaces using standard air quality evaluation metrics, identify some pitfalls of some metrics and advocate using active observations. Conclusions are presented in Section 4.

2. Experimental Design

2.1. Design of the Objective Analysis Solver

In optimum interpolation there is no use of an explicit interpolation observation operator. The correlation between a pair of locations, either from two observation sites or from an observation site to a model grid point, is computed as a function of distance using a prescribed correlation function. The observation operator is in effect a delta function applied over a continuous spatial domain [17].

In this study we interpolate the gridded analysis field to observation locations, using bilinear interpolation, to compute residuals such as observation-minus-analysis. Thus there can be a discrepancy between the observation operator used to generate the analysis, i.e., delta functions, and the observation operator used to interpolate the analysis field at the observation location, i.e., bilinear interpolation. To eliminate this discrepancy in observation operators we have revised the optimum interpolation scheme to use explicitly the same bilinear interpolation in handling the error covariance. We will give details below.

As in the operational optimum interpolation, the inversion of the innovation covariance matrix for the analysis solver is done using Choleski decomposition on the full matrix. The number of observations to be processed per analysis being of the order of a thousand or less, there was no need for computational simplification for large number of observations by using either data selection [18] or compact support correlation functions [17,19]. Thus, the analysis scheme used in this study computes explicitly the gain matrix $\widetilde{\mathbf{K}}$ as,

$$\widetilde{\mathbf{K}} = \widetilde{\mathbf{B}}\mathbf{H}^T \left(\mathbf{H}\widetilde{\mathbf{B}}\mathbf{H}^T + \widetilde{\mathbf{R}}\right)^{-1} \tag{1}$$

where $\mathbf{H}$ is a bilinear interpolation operator, $\widetilde{\mathbf{B}}$ is the prescribed background error covariance and $\widetilde{\mathbf{R}}$ is the prescribed observation error covariance. The tilde $(\widetilde{\cdot})$ emphasizes that these are prescribed, potentially suboptimal, quantities.

The computational demand of the Kalman gain was kept low by computing the background error correlation function only at model grid points needed for the bilinear interpolation. For example, to calculate the correlation between a pair of observations requires the computation of correlation between four points surrounding observation 1 (needed for the bilinear interpolation) and the other four points surrounding observation 2, thus forming a 4×4 correlation matrix $\mathbf{C}$ between the target model grid points. Then we calculate $\mathbf{H}\mathbf{C}\mathbf{H}^T$ which gives the correlation between two observation sites. This procedure is generalized for the N observations needed for the analysis. Equation (1) also involves the computation of $\widetilde{\mathbf{B}}\mathbf{H}^T$ that we compute as a set of N representers (i.e., columns of $\widetilde{\mathbf{B}}\mathbf{H}^T$), each being a 2D field that maps the background error covariance in model space with a single observation location, using again the bilinear interpolation approach to get a single interpolated representor for each observation location. By doing so we keep the consistency between the observation operators used for interpolation of a field and the observation operator used to manipulate matrices.

2.2. Cross-Validation

Cross-validation is a technique to evaluate an analysis (or in general any model that depends on observations) by partitioning the original observation data set into a training set, used to create the analysis, and an independent (or passive) set, used to evaluate the analysis. The most common cross-validation designs are: the k-fold cross-validation, where the original observation data set is partitioned into k equal size subsamples and the leave-one-out cross-validation, where N subsamples are created, each with one different observation set aside for the evaluation while the other $N - 1$ observations are used in producing the analysis. The cross-validation is then repeated with all the different sets until all observations have been used for evaluation. Clearly, there are k analyses computed in the k-fold cross-validation and N in the leave-one-out cross-validation, which is being computationally demanding when N is large. The main disadvantage of the k-fold cross-validation is that the analyses being evaluated uses a smaller number of observations (actually $(k-1)N/k$) than the original observation data set, whereas the leave-one-out cross-validation evaluates analysis that uses nearly the same number of observations (actually $N - 1$) as the original observation data set. This actually matters with the k-fold cross-validation if we need an estimate of the analysis error variance (or any other second moments) as the analysis error variance depends on the number of observations used.

Let $\mathbf{O}_j$ be a vector that contains the jth set of observations used for evaluation, and let $\mathbf{A}_{(j)}$ be a vector of analysis value interpolated at the verification observation locations of $\mathbf{O}_j$ and where the analysis used all observations except those in $\mathbf{O}_j$ (the index in parenthesis, i.e., (j), indicates all sets except the set j). It is customary in cross-validation literature (e.g., [20]) to construct a mean square error cost function, often denoted by CV,

$$\text{CV} = \sum_j (\mathbf{O}_j - \mathbf{A}_{(j)})^T (\mathbf{O}_j - \mathbf{A}_{(j)}) \tag{2}$$

that represents a misfit quadratic error of the model $\mathbf{A}$- in our case the analysis. Different model $\mathbf{A}$ can be compared and selected from which the CV value is smallest. Likewise, a tunable parameter in $\mathbf{A}$ can be obtained by minimizing the cost function CV with respect to that parameter. As we shall discuss later in this paper, in Section 4 and onwards, the bias of $(\mathbf{O}_j - \mathbf{A}_{(j)})$ needs to be removed from the cost function in order to estimate the input error covariance parameters.

In applications and thus in all experiments that follow, the analyses and verification against passive observations are made only with a set of observations that have passed a quality control. The quality control is nearly identical to the quality control used for the operational implementation of the analysis of surface pollutants at ECCC (see supplementary material in Robichaud et al. [3]). It consists in discarding observations that report a negative value, or whose value exceeds a certain unrealistic threshold set to 300 ppbv for ozone (300 µg/m^3 for PM$_{2.5}$). Observations are also discarded based on innovations (or observed-minus-background values) when, for ozone, they exceed 50 ppbv (100 µg/m^3 for PM$_{2.5}$) in absolute value. The quality-controlled observations are then separated into three sets of observations of equal numbers, i.e., a 3-fold cross-validation procedure, as illustrated in Figure 1.

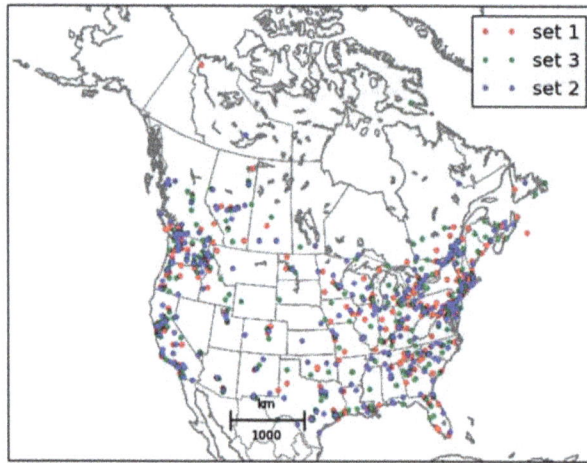

Figure 1. Spatial distribution of the three subsets of PM$_{2.5}$ observations used for cross-validation. The selection algorithm is based on regular picking of station by ID number.

The selection into three sets is made by station ID number, selecting on a regular basis each fourth station, starting with station 1 for the first set, station 2 for the second set and station 3 for the third set, and resulting in locally spatially random distribution of each sets of stations. The cross-validation is then made by leaving one set out of the three sets, and using the remaining two sets to produce the analysis.

2.3. Verification Metrics

We will evaluate the analyses against passive and active observations with the following standard evaluation metrics used for air quality models [21–24]; the bias, the modified normalized mean bias (MNMB), the fraction of correct within a factor of 2 (FC2), the variance (var($O - A$)) and the correlation coefficient ($cor(O, A)$), where the statistics is computed over time t for each station, and then the resulting metric is averaged over all the verifying station i,

$$\text{bias} = \frac{1}{N_i}\sum_i \left\{ \frac{1}{N_k}\sum_k (O_i(t_k) - A_i(t_k)) \right\} \tag{3}$$

$$\text{MNMB} = \frac{1}{N_i}\sum_i \left\{ \frac{2}{N_k}\sum_k \left(\frac{O_i(t_k) - A_i(t_k)}{O_i(t_k) + A_i(t_k)} \right) \right\} \tag{4}$$

$$\text{FC2} = \frac{1}{N_i}\sum_i \left\{ \frac{1}{N_k} count \left\{ 0.5 \leq \frac{A_i(t_k)}{O_i(t_k)} \leq 2 \right\} \right\} \tag{5}$$

$$\text{var}(O - A) = \frac{1}{N_i}\sum_i \left\{ \frac{1}{N_k - 1}\sum_k \left[(O_i(t_k) - A_i(t_k)) - \overline{(O_i - A_i)} \right]^2 \right\} \tag{6}$$

$$\text{cor}(O, A) = \frac{1}{N_i}\sum_i \left\{ \frac{1}{N_k - 1} \frac{\sum_k (O_i(t_k) - \overline{O_i})(A_i(t_k) - \overline{A_i})}{\sqrt{\sum_k (O_i(t_k) - \overline{O_i})^2 \sum_k (A_i(t_k) - \overline{A_i})^2}} \right\} \tag{7}$$

where $O_i(t_k)$ is the observed value at time t_k at the station i, $A_i(t_k)$ is the analysis at time t_k interpolated at the location of the station i, N_k is the total number of time sample per station, N_s is the total number of stations (in the sample or over the domain), and the overbar $\overline{(\ )}$ denotes the time average. The bias and the MNMB are metrics of the first moment that have distinctive properties. The bias gives a representative measure of the systematic discrepancy between analyzed and observed values over the whole set of observations used for verification. However, since atmospheric constituents exhibit a range of values that can vary in time and space, and different constituents have different range of values and may as well be expressed with different units, a relative error measure such as the MNMB is often preferred [24]. The MNMB is a dimensionless quantity that falls in the range $[-2, +2]$. The factor of 2 is introduced so to give a % error interpretation to the MNMB. This metric has the additional advantage of treating over- and under-estimation in a symmetric way [24]. However, the MNMB is relatively insensitive to relatively large discrepancies between analysis (or model) values and observed values, that is when its values are close to +2 (200%) or −2 (−200%) [23].

The fraction of correction within a factor of 2 (FC2) is a measure of reliability. It is based on counts and has the distinctive advantage that it is insensitive to outliers. It is worth mentioning that it accounts both high values outliers and also low values outliers that is a unique property of this metric [22]. The FC2 metric is also symmetric with respect to permutation of A and O, it is also dimensionless and its values must lie between 0 and 1. Our experience with this metric indicates that it is relatively insentive for relatively good agreement between analysis and observed values.

The variance, $\text{var}(O - A)$, and the correlation coefficient $\text{cor}(O, A)$ are metrics that depend on the spread of the discrepancy between analysis and observed values. The variance is not a dimensionless metric. It gives a representative measure of the spread of the discrepancy between analyses and observations and is not sensitive to systematic errors. As we will show in Section 4 and also shown in Marseille et al. [16], $\text{var}(O - A)$ with passive observations has the distinct advantage of providing a measure of the true analysis error variance (i.e., the error with respect to the truth) and $\text{var}(O - A)$ can be considered as a cross-validation cost function CV, Equation (2), with debiased $(O - A)$ increments. As for any second moment metric, $\text{var}(O - A)$ is sensitive to outliers; they must be removed, and this is done by gross check of the $(O - B)$, as explained in the previous subsection Section 2.2. Finally, the correlation coefficient is a dimensionless quantity that lies in the range $[-1, +1]$. It is also invariant to shifts in the mean (i.e., not sensitive to systematic errors), and multiplicative rescaling of either analysis or observations. The correlation is also relatively insensitive to improvement when the correlation is close to 1 or −1.

2.4. Description of the Ensemble of Analyses and Their Verification Statistics

A series of hourly analyses of O_3 and $PM_{2.5}$ at 21 UTC for a period of 60 days (14 June to 12 August 2014) were performed with given input error statistics using the operational model GEM-MACH and the real-time AirNow observations as described in the introduction and with quality controlled observations (see Section 2.2 above). In all experiments, the observation and background error variances, σ_o^2 and σ_b^2, used in the analysis are uniform. The prescribed observation error and background error covariances are given as $\tilde{\mathbf{R}} = \sigma_o^2\mathbf{I}$, $\tilde{\mathbf{B}} = \sigma_b^2\mathbf{C}$, where the correlation model $\mathbf{C}$ is a homogeneous isotropic second-order autoregressive model with a correlation length obtained by maximum likelihood, as in Ménard et al. [17]. Note that aside from quality control, that ends up rejecting some observations, the analysis uses the observation values and model realizations as is, with no bias correction.

We repeat the series of 60 day analyses for different observation and background error variances chosen in such a way that their sum $\sigma_o^2 + \sigma_b^2$ is equal to var$(O - B)$ but with different ratios of error variances $\gamma = \sigma_o^2/\sigma_b^2$. We perform the series of analyses over a wide range of γ ratios in the interval $[10^{-2}, 10^2]$, thus creating on one end analyses with very large observation weights, i.e., $\gamma \ll 1$, such that the analysis interpolated at the active observation sites tend to match the observed value, and on the other end, with $\gamma \gg 1$, creating analyses with very small observation weight producing analyses that are very close to the background (model) state.

The condition $\sigma_o^2 + \sigma_b^2 = \text{var}(O - B)$, called the *innovation variance consistency*, is an important constraint that is useful for the estimation of the *true* error statistics [25]. Indeed, the stronger condition for the full covariance matrices, the *innovation covariance consistency* criterion, takes the form: $< (O - B)$ $(O - B)^T >= \tilde{R} + H\tilde{B}H^T$, where $<>$ represents the mean over an ensemble of realizations, H is the interpolation from model grid to the observation location (or observation operator). It is one of the two necessary and sufficient condition to obtain the *true* error covariance statistics (in observation space) [25,26].

As explained in Section 2.3 above, the verification metrics are first calculated over 60 days for a given hour and for each station. Then the metric is averaged over all the verifying stations, resulting in one value of the metric for each hour of the day. Here, however, we computed the metrics for 21 UTC only. If N_s is the total number of stations, the statistics over one of the 3-fold subset then involves an average of the metric over $N_s/3$ passive stations. Doing this for all three subsets, and taking the average of the subsets' results, is equivalent to taking the average of the metric over all stations. In the results that will be presented in the following sections, we always present the average metric over the three passive subsets so that, in the end, the sample size of the passive observation experiments and of the active observation experiments are equal and thus can be presented side by side on the same graphic.

3. Verification against Passive and Active Observations

In this series of experiments, analyses of O_3 and $PM_{2.5}$ were produced using a fixed homogeneous isotropic correlation function, where the correlation length was obtained by maximum likelihood using a second-order auto-regressive model and error variances computed using a local Hollingsworth-Lönnberg fit [17]. A correlation length of 124 km was obtained for O_3 and of 196 km for $PM_{2.5}$. Our correlation length is defined from the curvature at the origin as in Daley [27] and is different from the length-scale parameter of the correlation model (see Ménard et al. [17] for a discussion of these issues). We did a series of 60-days analyses for different values of σ_o^2 and σ_b^2 but such that their sum respects the innovation variance consistency, $\sigma_o^2 + \sigma_b^2 = \text{var}(O - B)$, an important condition for an optimal analysis [25], as explained in Section 2.4. This is the experimental procedure that has been used to generate the Figures 2–7. The results are shown for a wide range of variance ratios $\gamma = \sigma_o^2/\sigma_b^2$ from 10^{-2} to 10^2 in Figures 2–5 and 7 in particular. Note that $\gamma \ll 1$ corresponds to a very large observation weight while $\gamma \gg 1$ correspond to very small observation weight.

The var$(O - A)$ using passive observations (red curve with circles) and active observations (black curve with squares) is presented in Figure 2 for O_3 (left panel) and $PM_{2.5}$ (right panel). The solid blue line represents var$(O - B)$, the variance of observation-minus-model, i.e., prior to an analysis. As mentioned in Section 2.4, in the cross-validation experiments we averaged the verification metric over the 3-fold subsets so that, in effect, the total number of observations that end up being used for verification is N_s, the total number of stations. We thus argue that the verification sampling error for the cross-validation experiments (red curve) is the same as for the active observations using the full analysis (i.e., analysis using the total number of stations; black curve). In addition, note that from Figure 1 the station sampling strategy gives rise to spatially random selection of stations, so that the individual metric on each set should be comparable. Furthermore, there is roughly 1300 quality controlled O_3 observations over the domain and 750 $PM_{2.5}$ quality controlled observations, each with 60 time samples or less. To give some qualitative idea of the sampling error, the different metric values for the individual 3-fold sets are presented in the Supplementary Material Figures S1 and S2, where we can see that for $var(O − A)$ and $cor(O, A)$ the metric values for the individual sets are nearly indistinguishable from the means of the 3-subset.

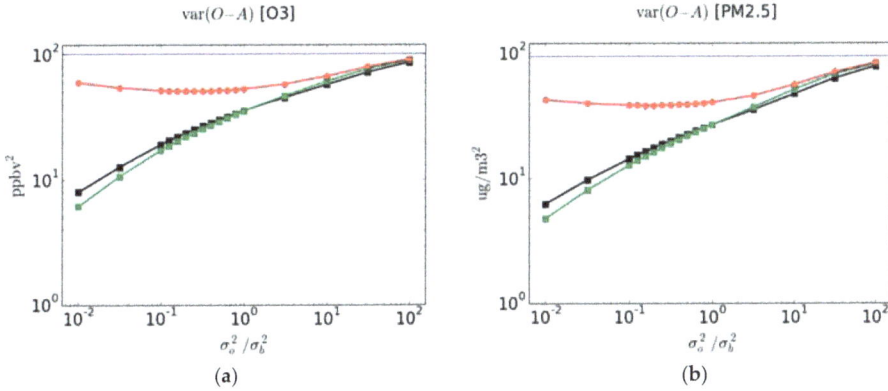

Figure 2. Variance of observation-minus-analysis residuals of O_3 and $PM_{2.5}$ for both active and cross-validation passive observations as a function of $\gamma = \sigma_o^2/\sigma_b^2$. (**a**) is for O_3 with ordinates in $ppbv^2$ units, and (**b**) is for $PM_{2.5}$ with ordinates in $(\mu g/m^3)^2$. Red curve results from the evaluation at the passive observation sites (average of the 3-fold subsets). Black curve results from evaluation at the active observation sites with analyses using all observations. Green curve results from the evaluation at the active observation sites in the cross-validation experiment (i.e., using 2/3 of the observations; average of the three subsets). Blue curve is the variance of observation-minus-model.

The difference between the verification against passive observations in cross-validation analyses (red curve) and the verification against active observations using full analyses (black curve) can be attributed to two effects: (1) the analysis used in the cross-validation uses 2/3rd of the total number of observations and thus the analysis error has larger variance than analyses using all observations, (2) since the analysis error variance has typically a local minimum at the individual active observation sites and increases away from it (see for example Figure 4a in [28] i.e., Part II of this paper), an evaluation of the analysis error at passive sites (i.e., away from the active sites) has larger error variance than those evaluated at the active sites [16]. We may call this the distance effect of passive observation sites. In order to separate these two effects, we also display the 3-fold average of the metric verifying against active observation for the cross-validation analyses as a green curve with squares. Thus in summary we display a;

- red curve: using analysis with $2N_s/3$ observations with an evaluation at passive sites
- green curve: using analysis with $2N_s/3$ observations with an evaluation at active sites
- black curve: using analysis with N_s observations with an evaluation at active sites.

The difference between the red and green curves show the influence of distance between passive and active observation sites, whereas the difference between the green and black curves show the influence of having different number of observations in creating the analysis for verification.

Let us first examine the results of verifying against active observations. As the observation weights get smaller (i.e., $\gamma \gg 1$), the analysis draws closer to the background, so that $var(O − A)$ increases toward $var(O − B)$. On the other end, when $\gamma \ll 1$, the $var(O − A)$ continuously decreases as γ diminishes to ultimately reach zero. This is in effect an expected result from the inner working of an analysis scheme that the analysis error variance goes when the observation error variance goes to zero. This effect does not depend on the observed values or the model values. For this reason, the var$(O - A)$ using active observations cannot provide a true measure of the quality of an analysis.

Now let us examine the results of verifying against passive observations with cross-validation analyses. As the observation weights get smaller (i.e., $\gamma \gg 1$), as for active observations the analysis draws closer to the background, so that var$(O - A)$ increases toward var$(O - B)$. On the other end when $\gamma \ll 1$, the var$(O - A)$ using passive observation increases as γ diminishes, whereas the var$(O - A)$ evaluated at the active observation sites (green and black curves) decreases, indicating that the analysis tries to overfit active observations which results in a spatially noisy analysis between the active observation sites. Somewhere in between these two extreme values of γ lies a minimum of var$(O - A)$ where there is neither an overfitting nor an underfitting to the active observations. This "optimal" ratio, that is found by inspection, actually corresponds the optimal analysis. It is also where the analysis error variance with respect to the truth is minimum, but to show this last statement requires an extensive analysis of the problem that we will discuss in part two of this study.

We also computed the verification of the subset of active observations used in the cross-validation experiments with green curves. The difference between the black and green curves indicate the effect of having more observations in the analysis. One would expect that having a larger number of observations in the full analysis active var$(O - A)$ compared to the active var$(O - A)$ for the cross-validation analyses would result in slightly smaller var$(O - A)$. This is indeed observed between the black and green curves when the observation weight is small (i.e., $\gamma \gg 1$). However, surprisingly, when the observation weight is large, $\gamma \ll 1$, we observe the opposite. This intriguing behavior may indicate an inconsistency between the assumption of uniform error variances for σ_o^2 and σ_b^2 (assumed in the input error statistics) and the real spatial distribution of error variances. This discrepancy being simply amplified when the observation weight is large and when there are less observations to produce the analysis.

The difference between var$(O - A)$ at passive sites and active sites (with the same number of observations to construct the analyses) is substantial. For O_3 and for an optimal ratio, the var$(O - A)$ at passive sites is 51.02 ppbv^2 (red curve) while at active sites is 22.77 ppbv^2 (green curve). For PM2.5 and for an optimal ratio, the var$(O - A)$ at passive sites is 38.09 $(\mu g/m^3)^2$ (red curve) while at active sites is 15.41 $(\mu g/m^3)^2$ (green curve). For both species, the error variance at active sites gives a significant overestimation of the error variance by more than a factor of 2.

In Figure 3, we present the correlation metric between the observations and the analysis using, as in Figure 2, the verification against passive observations in cross-validation analyses (red curve), the verification against active observations using full analyses (black curve) and the verification against active observations in the cross-validation analyses (green curve). The blue curve depicts the correlation between the model and the observations, that is the prior correlation.

The evaluation against passive observations with cross-validation analyses (red curve) shows a maximum at the same values of $\gamma = \sigma_o^2/\sigma_b^2$ than for the var$(O - A)$. We argue that the same arguments of underfitting and overfitting are responsible for this maximum. The correlation between the active observations and the analysis (black and green curves) increases as the observation weight increases (γ decreases), theoretically reaching a value 1 for $\sigma_o^2 = 0$, which is again unrealistic and simply shows the impact of ill-prescribed error statistics in an analysis scheme. The gain in correlation between independent observations and analysis is significant. For O_3, it increases from a value of 0.55 with respect to the model to a value of 0.74 with respect to an optimal analysis (when $\gamma = \sigma_o^2/\sigma_b^2$ is optimal). For PM2.5, the correlation against the model has a value of 0.3 which basically has no skill, to a value of 0.54 for optimal analysis, which represent a modest but useable skill. The correlation evaluated at the active sites for an optimal ratio, is 0.85 for O_3 (green curve) and 0.74 for PM2.5 (green curve), being a substantial overestimation with respect to values obtained at passive sites.

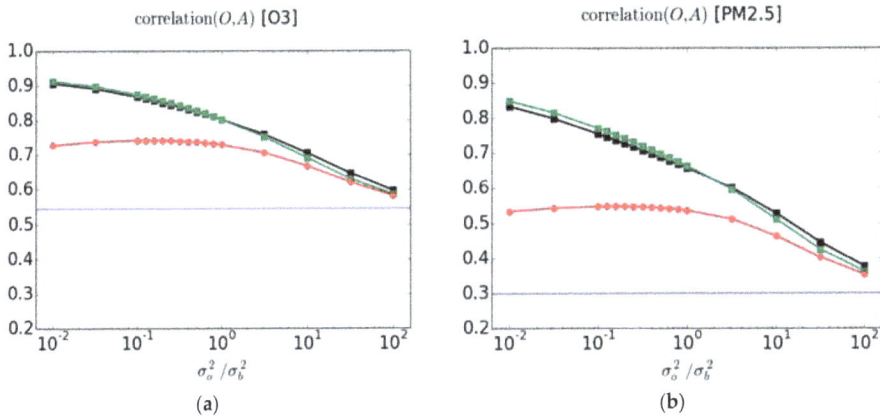

Figure 3. Correlation between observations and analysis for (**a**) O_3 and (**b**) $PM_{2.5}$ for both active and cross-validation passive observations as a function of $\gamma = \sigma_o^2/\sigma_b^2$. The red, black and green curves are as in Figure 2.

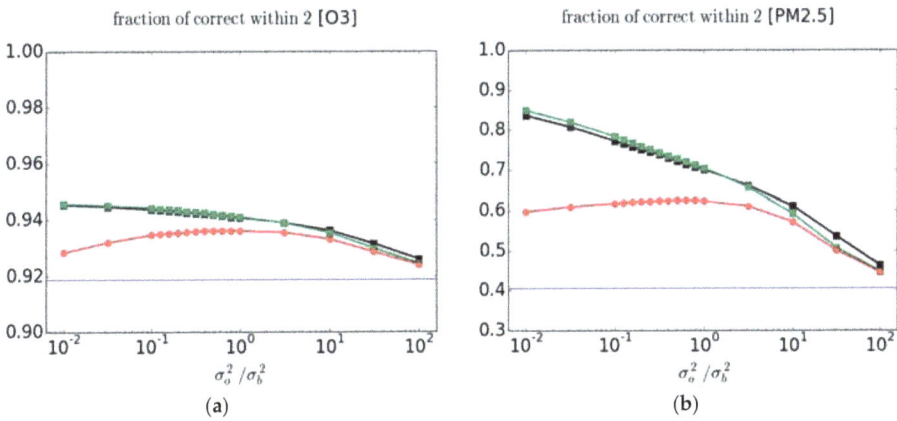

Figure 4. FC2 for (**a**) O_3 and (**b**) $PM_{2.5}$ for both active and cross-validation passive observations as a function of $\gamma = \sigma_o^2/\sigma_b^2$. The red, black and green curves are as in Figure 2.

Another metric that we have considered is the FC2, Equation (5) [3]. The evaluation of this metric against passive and active observations is presented in Figure 4 for O_3 (left panel) and $PM_{2.5}$ (right panel). Note that the scale in the ordinate is quite different between the left and right panels. Although the results bear similarity with the correlation between O and A presented in Figure 3, the maximum with passive observations is reached at larger γ values than those obtained for $var(O - A)$ or $cor(O, A)$, which are identical. Individual fold results are presented in the supplementary materials Figure S3.

The interpretation of this metric is, however, not clear. Although the ratio $z = A/O$ is a dimensionless quantity the spread of z is generally not independent of the variance of A or O and there are cases where it is. So to count the number of occurrence of z between the dimensionless values 0.5 and 2 is confusing. As a simplified illustration, suppose that A is normally distributed as $N(0, \sigma_a^2)$ and similarly with $O \sim N(0, \sigma_o^2)$. The ratio of these two random variables is then a Cauchy distribution whose probability density function (pdf) is $\sigma_o\sigma_a/[\pi(\sigma_o^2 z^2 + \sigma_a^2)]$. The mean, variance and higher moments of Cauchy probability distributions are not defined since the integral of the pdf is not bounded; only the mode is defined. Cauchy distributions also have a spread parameter, which in this case is equal to

σ_a/σ_o. If the variance of A and O are equal, then the number count between the dimensionless bounds 0.5 and 2 depends only on the shape of the probability distribution function, not on the variance. If the variance of A and O are different, then it also depends on the ratio of variances. Furthermore, in principle this metric also depends on the bias (which is not the case here for these analyses). It may be a difficult metric to interpret but if used as a quality control, the FC2 have the unique ability of rejecting too low as well as too high values of z.

In Figure 5 we present the bias between observations and analyses, and where the verification is made against passive and active observations as done with the other metrics. Bias is not a dimensionless quantity; note that the range and scale presented for O_3 and $PM_{2.5}$ in Figure 5 are different. The blue curve is the mean $(O - B)$ and thus indicates that for O_3 in average over all observation stations (for the time and dates considered) the model overpredicts, and that for $PM_{2.5}$ the model underpredicts.

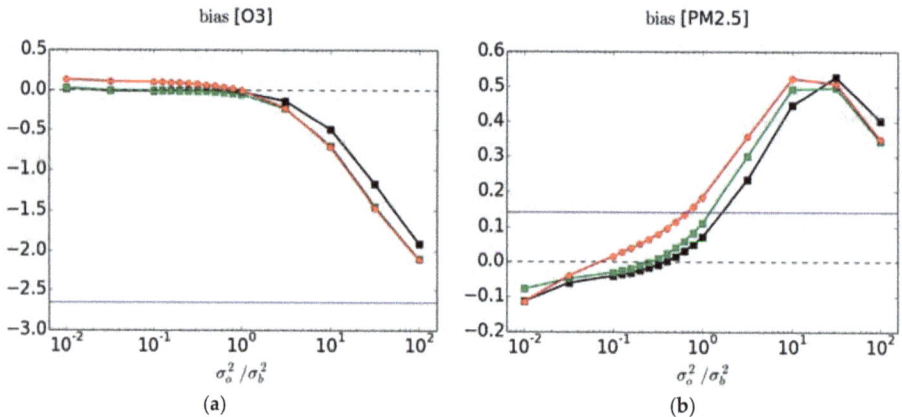

Figure 5. Bias between observation and analysis for (**a**) O_3 and (**b**) $PM_{2.5}$ for both active and cross-validation passive observations as a function of $\gamma = \sigma_o^2/\sigma_b^2$. The red, black and green curves are as in Figure 2.

Contrary to all metric results seen so far, the 3-fold variability of the bias is substantial: it is of the order of ± 0.5 ppbv (in average) for O_3 at passive sites and of the order of ± 0.1 µg/m³ (in average) for the $PM_{2.5}$ at passive sites (results shown in the supplementary material Figure S4). The distinction between the red, black and green curves may not be statistically significant for both O_3 and $PM_{2.5}$. However, the difference between the analysis bias and model bias is large and statistically significant (see supplementary material). For O_3, the model bias is eliminated at the passive observation sites (red curve) as long as the observation weight $\gamma \leq 1$. The situation is not so clear for $PM_{2.5}$. In fact, when the observation weight is small, we get the intriguing result that the bias of the analysis is larger bias than the model. How can that be when the observation weight is small (i.e., $\gamma > 1$); should the analysis not be close to the model values? This apparent contradiction reveals a more complex issue underlying the bias metric.

To explore the possible causes, we have calculated the bias per bin of model values, displayed in Figure 6. In order to have a decent sample size per bin, we collect all the $(O - A)$ and $(O - B)$ over time and observation sites, create bins of model values and calculate the statistic per bin (and not per station as before). The result shows that the model bias is nearly linearly dependent on the model values (black boxes in the bias panel). Both O_3 and $PM_{2.5}$ show an underprediction for low model values and an overprediction for large model values. The origin of this bias is not known but one would argue that it is not directly related to chemistry as such since both constituents, O_3 and $PM_{2.5}$, present the same feature. Possible explanations could be related to the model boundary layer, the emissions being too low for low polluted areas and too large for polluted areas, insufficient transport away from polluted areas to unpolluted areas, species destruction/scavenging could be too low in low polluted areas and too high in polluted areas. The lower panels of Figure 6a,b show the count of stations per model bin size. We observe that the majority of stations have O_3 model values in the range of 40 to 55 ppbv, where the bias is negative. Over all the stations, this gives rise to a negative mean $(O - B)$, and this is how we make the claim that the model overpredicts. However, for $PM_{2.5}$ the situation is different: the majority of stations lie in the low model value range, and there are gradually less stations for increasingly larger model values. Although the $(O - B)$ have large negative values in the high model value bin while small model value bins have positives $(O - B)$'s, the effect over all stations is to yield a modestly positive mean $(O - B)$ and thus the model underestimates the $PM_{2.5}$. The results of the analysis evaluated at the passive observation sites are presented with the yellow and grey histogram boxes. In yellow, near optimal analyses with optimal observation weight, as determined by the minimum of $\mathrm{var}(O - A)$ are used, and in grey non-optimal analyses with $\gamma = 10$.

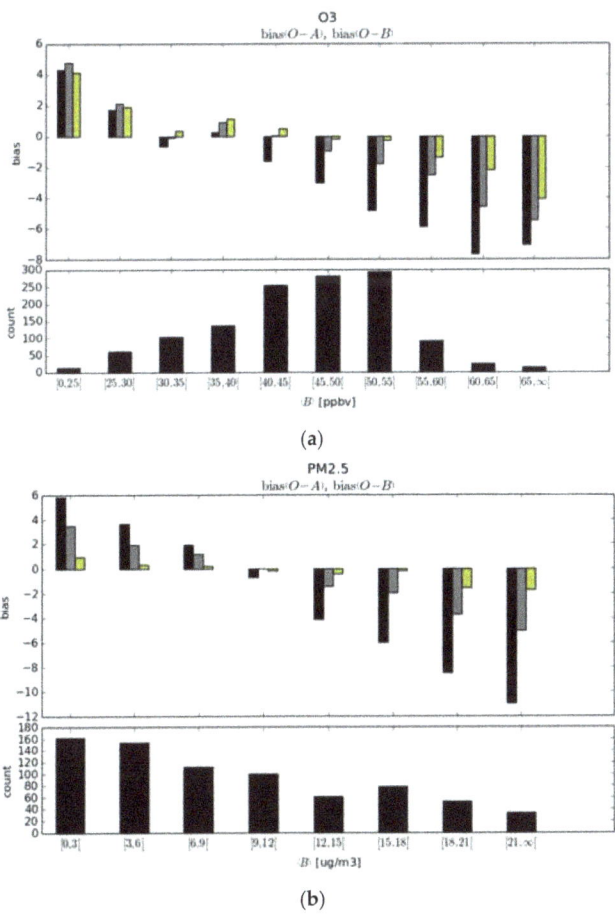

(a)

(b)

Figure 6. Biases per bin of model values. Figure (**a**), presents the statistics for O_3 and in (**b**), for $PM_{2.5}$. In the upper portion, (**a**,**b**) are the residual statistics per bin; in black, the $(O - B)$, in grey, the $(O - A)$ at passive observation sites (mean of the 3-fold subsets) for a non-optimal analysis with $\gamma = 10$, and in yellow, the $(O - A)$ at passive observation sites (mean of the 3-fold subsets) using the optimal observation weight. In the lower portion, (**a**,**b**) are the station number count per model values.

We observe that the effect of the optimal analysis is nearly insentive to model bin values, where near zero biases are obtained in most of the range except for very small and very large model values. The fact that we are not able to capture the full benefit of analysis on all model values may be an artefact of the assumption that we are using uniform observation and background error variances whereas the model values varies considerably. In grey, we used the non-optimal analyses with a small observation weight were we set $\gamma = 10$. In the non-optimal case, the state-dependent bias is still present but appears to be nearly perfectly anti-symmetric, positive in the low model value bins and nearly the exact opposite in high model value bins. Since for O_3 the majority of observations lie in the range 40 to 55 ppbv, $(O - A)$ for the optimal analyses at passive observation sites is nearly zero. However, for the non-optimal analysis with $\gamma = 10$, the $(O - A)$ at passive sites is negative, i.e., the analysis is overpredicting, as shown in Figure 5.

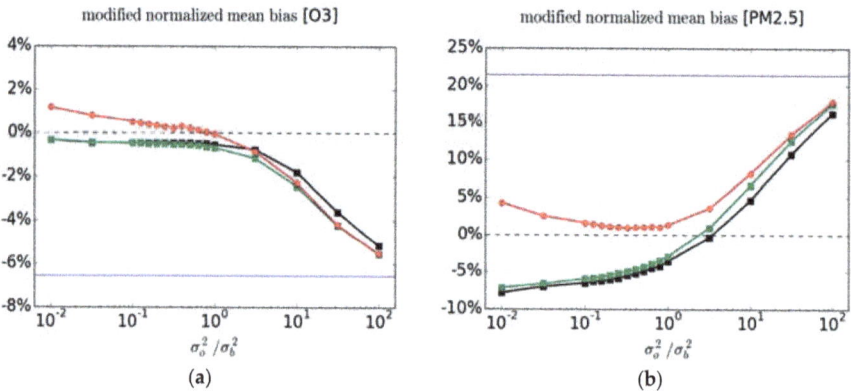

Figure 7. Modified normalized mean bias (MNMB) between observation and analysis for (**a**) O_3 and (**b**) PM$_{2.5}$ for both active and cross-validation passive observations as a function of $\gamma = \sigma_o^2/\sigma_b^2$. The red, black and green curves are as in Figure 2.

For PM$_{2.5}$, the weighted sum of the $(O - A)$ bins is such that over all stations the bias for an optimal analysis is nearly zero. In the case of the non-optimal analysis with $\gamma = 10$, the weighted sum of the nearly anti-symmetric $(O - A)$ bias per bin gives more weight to the positive bias at smaller model values, so that overall there is a positive $(O - A)$, as in Figure 5.

To circumvent the state-dependency of the $(O - A)$ biases it is useful to consider instead a fractional bias metric, such as the modified normalized mean bias, MNMB Equation (4). The MNMB metric is a dimensionless measure and as defined with a factor of 2, Equation (4), represents a % error. The MNMB metric is a relative measure with respect to the mean observed-analysis value and is thus less sensitive to spatially varying distribution of the concentrations, revealing instead the intrinsic difference between the fields. The MNMB for O_3 and PM$_{2.5}$ for passive and active observations are displayed in Figure 7 using the same color as in Figure 2. We note immediately that the MNMB analysis bias does not exceed the MNMB model bias as we observed for the bias metric of PM$_{2.5}$ (Figure 5 right panel). The MNMB bias also varies smoothly as a function of γ (at variance with the bias metric for PM$_{2.5}$—Figure 5).

Furthermore, examining the 3-fold variability of the cross-validation analysis MNMB at the passive sites and the variability of the MNMB at the active sites (see Figure S5 in supplementary materials), we infer that for PM$_{2.5}$, where we can actually deduce that the difference between the cross-validation and the validation against active observations is statistically significant when $\gamma < 1$. There is also another important point to make; although analyses are designed to reduce the error variance, it so happens that for a near optimal analysis the fractional bias MNMB is very small, around 1% for O_3 and about 1–2% for $PM_{2.5}$. We argue that it results from an optimal use of observations.

There is also some information to gain from the variance of observed-minus-analysis per bin size, as illustrated in Figure 8, using the same color histograms as in Figure 6. We note that for O_3, the model error variance against observations increases gradually with larger model values. However, the fraction of analysis variance vs. model variance is roughly uniform across all bins. This can be explained by the fact that the observation and background error variances are uniform, and thus the reduction of variance across all bins is uniform as well. However, the situation is different for $PM_{2.5}$. We note a relatively poor performance of the model at low model values, with standard deviation of 7 $\mu g/m^3$. For slightly larger model values (3–6 $\mu g/m^3$), the error variance is smaller to 5.5 $\mu g/m^3$ and then increases almost linearly with model values. The fraction of analysis variance vs. model variance decreases steadily with larger model values. These results thus indicate that the assumption that observation and background error variances are uniform and independent of the model value may have to be revisited.

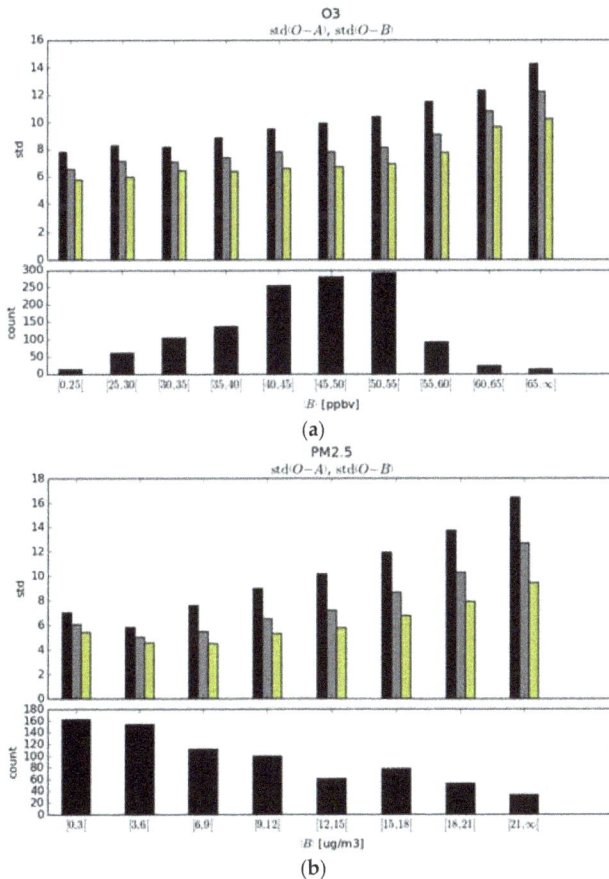

Figure 8. Same as Figure 6 except that we display the variance of analysis-minus-passive observations per bin of model values.

4. Conclusions

We have developed an approach by which analyses can be evaluated and optimized without using a model forecast but rather by partitioning the original observation data set into a training set, to create the analysis, and an independent (or passive) set, used to evaluate the analysis. This kind of evaluation by partitioning is called cross-validation.

The need for such a technique came about from our desire to evaluate our operational surface air quality analyses that are created off-line with no assimilation cycling. Evaluating a surface air quality analysis based on its chemical forecast does in fact require additional information or assumptions, such as vertical correlation, aerosol speciation and bin distribution (while surface measurement is primarily about mass) or unobserved chemical variables correlations, and so on. So that the quality of the chemical forecast is not solely dependent on the quality of the analysis and, if there are compensating errors, can actually be a misleading assessment of the quality of the analysis.

We have applied this cross-validation procedure to the operational analyses of surface O_3 and $PM_{2.5}$ over North America for a period of 60 days and present an evaluation using different metrics; bias, modified normalized mean bias, variance of observation-minus-analysis residuals, correlation between observation and analysis, and fraction of correction within a factor of 2.

Our results show that, in terms of variance and correlation, the verification of analyses against active observations always yield an overestimation of the accuracy of the analysis. This overestimation also increases as the observation weight increases. On the other hand for biases, the distinction between the verification against active observations and passive observations is unclear and drowned in the sample variability. However, using a fractional bias metric, in particular the MNMB, shows that the verification against passive observations can be close to one percent for an optimal analysis while the verification against active observations is much larger.

Results also show the importance of having an optimal analysis for verification. The variance of the analysis with respect to independent observations is minimum and the correlation between the analysis and independent observations is maximum for an optimal analysis. By being a compromise between an overfit to the active observations (which produce noisy analysis field) and an underfit, the optimal analysis offers the best use of observations throughout. At optimality, the analysis fractional bias (MNMB) at the passive observation sites has only one or two percent error whereas the fractional bias of the model is 6.5% for O_3 and 21% for $PM_{2.5}$. The correlation between the analysis and independent observations is also significantly improved with an optimal analysis: the correlation between the model and independent observations is 0.55 for O_3 and increases to 0.74 with the analysis, while for $PM_{2.5}$ the correlation between the model and independent observations is only 0.3 (which is basically no skill) but rises to 0.54 for the analysis.

We also argue that the fraction of correct within a factor of 2, is a metric whose interpretation is unclear as it mixes information about bias, variance and probability distribution in a non-uniform way and does not seem to add anything new to other metrics. The bias is also very sensitive to sample variability and can lead to wrong conclusions. For example, we have seen that the mean analysis bias can be larger than the mean model bias, whether verifying against active or passive observations. However, since an analysis is always closer to the truth than its prior (i.e., the model), it results in an apparent contradiction. This implies that the bias metric cannot be used to faithfully compare model states accurately. Such wrongful conclusions do not arise, however, with the MNMB. We thus recommend avoiding using bias as a measure of truthfulness, and use instead a fractional bias measure such as the MNMB.

We also found that errors in the GEM-MACH model grow almost linearly with the model value. This is particularly evident for the bias where the model underestimates at small model values and overestimates at large model values. Furthermore, this occurs in equal ways for O_3 and $PM_{2.5}$, thus indicating that the source of this bias is not related to chemistry. The fact that, over the entire domain, the model overestimates O_3, and underestimates $PM_{2.5}$ is simply a result of the concentrations. We have not conducted a systematic study of model error for other times of the day and other periods of the year, but it would be very interesting to look at this, to see whether or not changes of biases are due primarily to changes in the distribution of values rather than a fundamental change in the bias per model value bin.

Finally, we have also examined the variance against independent observations per model value bin, and concluded that the error variance is not quite uniform with model values but increases slowly with model values for O_3 and in a more pronounced way for $PM_{2.5}$.

In part two, we will focus on the estimation of the analysis error variance and develop a mathematical formalism that permits the comparison of different diagnostics of variance under different assumptions, optimizes the analysis parameters and gains confidence on the estimate of analysis error as we obtain coherent estimated values across different diagnostics.

Supplementary Materials: The following are available online at www.mdpi.com/xxx/s1, Figure S1: Verification of variance for O_3 and $PM_{2.5}$ for the individual sets. Figure S2: Same as Figure S1 but for the correlation between observations and analysis. Figure S3: Same as Figure S1 but for the fraction of correct within a factor of 2. Figure S4: Same as Figure S1 but for bias. Figure S5: Same as Figure S1 but for modified normalized mean bias.

Acknowledgments: We are grateful to the US/EPA for the use of the AIRNow database for surface pollutants and to all provincial governments and territories of Canada for kindly transmitting their data to the Canadian Meteorological Centre to produce the surface analysis of atmospheric pollutants. We are also thankful for the proof read by Kerill Semeniuk, and for three anonymous reviewers for their comments and help in improving the manuscript.

Author Contributions: This research was conducted as a joint effort by both authors. R.M. contributed to the theoretical development and wrote the paper, and M.D.-J. conducted all experiments design and execution, proof reading and introduced a new diagnostic of optimal analysis error that was furher extended to passive observations space.

Conflicts of Interest: The authors declare no conflict of interest. The founding sponsors, which is the government of Canada, had no role in the design of the study; in the collection, analyses, or interpretation of data; in the writing of the manuscript, and in the decision to publish the results.

References

1. Ménard, R.; Robichaud, A. The chemistry-forecast system at the Meteorological Service of Canada. In Proceedings of the ECMWF Seminar Proceedings on Global Earth-System Monitoring, Reading, UK, 5–9 September 2005; pp. 297–308.
2. Robichaud, A.; Ménard, R. Multi-year objective analysis of warm season ground-level ozone and $PM_{2.5}$ over North-America using real-time observations and Canadian operational air quality models. *Atmos. Chem. Phys.* **2014**, *14*, 1769–1800. [CrossRef]
3. Robichaud, A.; Ménard, R.; Zaïtseva, Y.; Anselmo, D. Multi-pollutant surface objective analyses and mapping of air quality health index over North America. *Air Qual. Atmos. Health* **2016**, *9*, 743–759. [CrossRef] [PubMed]
4. Moran, M.D.; Ménard, S.; Pavlovic, R.; Anselmo, D.; Antonopoulus, S.; Robichaud, A.; Gravel, S.; Makar, P.A.; Gong, W.; Stroud, C.; et al. *Recent Advances in Canada's National Operational Air Quality Forecasting System*, 32nd ed.; Springer: Dordrecht, The Netherlands, 2014.
5. Pudykiewicz, J.A.; Kallaur, A.; Smolarkiewcz, P.K. Semi-lagrangian modelling of tropospheric ozone. *Tellus B* **1997**, *49*, 231–248. [CrossRef]
6. Cressie, N.; Wikle, C.K. *Statistics for Spatio-Temporal Data*; Wiley: Hoboken, NJ, USA, 2011.
7. Schneider, P.; Castell, N.; Vogt, M.; Dauge, F.R.; Lahoz, W.A.; Bartonova, A. Mapping urban air quality in near real-time using observations from low-cost sensors and model information. *Environ. Int.* **2017**, *106*, 234–247. [CrossRef] [PubMed]
8. Lindström, J.; Szpiro, A.A.; Oron, P.D.; Richards, M.; Larson, T.V.; Sheppard, L. A flexible spatio-temporal model for air pollution and spatio-temporal covariates. *Environ. Ecol. Stat.* **2014**, *21*, 411–433. [CrossRef] [PubMed]
9. Carmichael, G.R.; Sandu, A.; Chai, T.; Daescu, D.N.; Constantinescu, E.M.; Tang, Y. Predicting air quality: Improvements through advanced methods to integrate models and measurements. *J. Comput. Phys.* **2008**, *227*, 3540–3571. [CrossRef]

10. Dabberdt, W.F.; Carroll, M.A.; Baumgardner, D.; Carmichael, G.; Cohen, R.; Dye, T.; Ellis, J.; Grell, G.; Grimmond, S.; Hanna, S.; et al. Meteorological research needs for improved air quality forecasting: Report of the 11th prospectus development team of the US weather research program. *Bull. Am. Meteorol. Soc.* **2004**, *85*, 563–586. [CrossRef]

11. Sportisse, B. A review of current issues in air pollution modeling and simulation. *Comput. Geosci.* **2007**, *11*, 159–181. [CrossRef]

12. Elbern, H.; Strunk, A.; Nieradzik, L. Inverse modelling and combined state-source estimation for chemical weather. In *Data Assimilation*; Lahoz, W., Khattatov, B., Ménard, R., Eds.; Springer: Berlin/Heidelberg, Germany, 2010; pp. 491–513.

13. Bocquet, M.; Elbern, H.; Eskes, H.; Hirtl, M.; Žabkar, R.; Carmicheal, G.R.; Flemming, J.; Inness, A.; Pagaoski, M.; Pérez Camaño, J.L.; et al. Data assimilation in atmospheric chemistry models; current status and future prospects for coupled chemistry meteorology models. *Atmos. Chem. Phys.* **2015**, *15*, 5325–5358. [CrossRef]

14. Chai, T.; Carmichael, G.R.; Sandu, A.; Tang, Y.H.; Daescu, D.N. Chemical data assimilation of transport and chemical evolution over the pacific (TRACE-P) aircraft measurements. *J. Geophys. Res.* **2006**, *111*, D02301. [CrossRef]

15. Sandu, A.; Chai, T. Chemical data assimilation—An overview. *Atmosphere* **2011**, *2*, 426–463. [CrossRef]

16. Marseille, G.J.; Barkmeijer, J.; De Haan, S.; Verkle, W. Assessment and tuning of data assimilation systems using passive observations. *Q. J. R. Meteorol. Soc.* **2016**, *142*, 3001–3014. [CrossRef]

17. Ménard, R.; Deshaies-Jacques, M.; Gasset, N. A comparison of correlation-length estimation methods for the objective analysis of surface pollutants at Environment and Climate Change Canada. *J. Air Waste Manag. Assoc.* **2016**, *66*, 874–895. [CrossRef] [PubMed]

18. Cohn, S.E.; Da Silva, A.; Guo, J.; Sienkiewicz, M.; Lamich, D. Assessing the effects of data selection with the DAO physical-space statistical analysis system. *Mon. Weather Rev.* **1998**, *126*, 2913–2926. [CrossRef]

19. Houtekamer, P.L.; Mitchell, H.L. A sequential ensemble Kalman filter for atmospheric data assimilation. *Mon. Weather Rev.* **2001**, *129*, 123–137. [CrossRef]

20. Efron, B.; Tibshirani, R.J. *An Introduction to Boostrap*; Chapman & Hall: New York, NY, USA, 1993.

21. Seigneur, C.; Pun, B.; Pai, P.; Louis, J.F.; Solomon, P.; Emery, C.; Morris, R.; Zahniser, M.; Worsnop, D.; Koutrakis, P.; et al. Guidance for the performance evaluation of three-dimensional air quality modeling systems for particulate matter and visibility. *J. Air Waste Manag. Assoc.* **2000**, *50*, 588–599. [CrossRef] [PubMed]

22. Chang, J.C.; Hanna, S.R. Air quality model performance evaluation. *Meteorol. Atmos. Phys.* **2004**, *87*, 167–196. [CrossRef]

23. Savage, N.H.; Agnew, P.; Davis, L.S.; Ordóñez, C.; Thorpe, R.; Johnson, C.E.; O'Connor, F.M.; Dalvi, M. Air quality modelling using the Met Office Unified Model (AQUM OS24-26): Model description and initial evaluation. *Geosci. Model Dev.* **2013**, *6*, 353–372. [CrossRef]

24. Katragkou, E.; Zanis, P.; Tsikerdekis, A.; Kapsomenakis, J.; Melas, D.; Eskes, H.; Flemming, J.; Huijnen, V.; Inness, A.; Schultz, M.G.; et al. Evaluation of near surface ozone over Europe from the MACC reanalysis. *Geosci. Model Dev.* **2015**, *8*, 2299–2314. [CrossRef]

25. Ménard, R. Error covariance estimation methods based on analysis residuals: Theoretical foundation and convergence properties derived from simplified observation networks. *Q. J. R. Meteorol. Soc.* **2016**, *142*, 257–273. [CrossRef]

26. Desroziers, G.; Berre, L.; Chapnik, B.; Poli, P. Diagnosis of observation, background, and analysis-error statistics in observation space. *Q. J. R. Meteorol. Soc.* **2005**, *131*, 3385–3396. [CrossRef]

27. Daley, R. *Atmospheric Data Analysis*; Cambridge University Press: New York, NY, USA, 1991; p. 457.

28. Ménard, R.; Deshaies-Jacques, M. Evaluation of analysis by cross-validation, Part II: Diagnostic and optimization of analysis error covariance. *Atmosphere* **2018**, *9*, 70. [CrossRef]

atmosphere

MDPI

Article

Evaluation of Analysis by Cross-Validation, Part II: Diagnostic and Optimization of Analysis Error Covariance

Richard Ménard *and Martin Deshaies-Jacques

Air Quality Research Division, Environment and Climate Change Canada, 2121 Transcanada Highway, Dorval, QC H9P 1J3, Canada; martin.deshaies-jacques@canada.ca
* Correspondence: richard.menard@canada.ca; Tel.: +1-514-421-4613

Received: 7 November 2017; Accepted: 13 February 2018; Published: 15 February 2018

Abstract: We present a general theory of estimation of analysis error covariances based on cross-validation as well as a geometric interpretation of the method. In particular, we use the variance of passive observation-minus-analysis residuals and show that the true analysis error variance can be estimated, without relying on the optimality assumption. This approach is used to obtain near optimal analyses that are then used to evaluate the air quality analysis error using several different methods at active and passive observation sites. We compare the estimates according to the method of Hollingsworth-Lönnberg, Desroziers et al., a new diagnostic we developed, and the perceived analysis error computed from the analysis scheme, to conclude that, as long as the analysis is near optimal, all estimates agree within a certain error margin.

Keywords: data assimilation; statistical diagnostics of analysis residuals; estimation of analysis error; air quality model diagnostics; Desroziers et al. method; cross-validation

1. Introduction

At Environment and Climate Change Canada (ECCC) we have been producing hourly surface pollutants analyses covering North America [1–3] using an optimum interpolation scheme which combines the operational air quality forecast model GEM-MACH output [4] with real-time hourly observations of O_3, $PM_{2.5}$, PM_{10}, NO_2, and SO_2 from the AirNow gateway with additional observations from Canada. These analyses are not used to initialize the air quality model and we wish to evaluate them by cross-validation, that is by leaving out a subset of observations from the analysis to use them for verification. Observations used to produce the analysis are called active observations while those used for verification are called passive observations.

In a first-part paper of this study, i.e., Ménard and Deshaies-Jacques [5], we have examined different verification metrics using either active or passive observations. As we changed the ratio of observation error to background error variances $\gamma = \sigma_o^2/\sigma_b^2$, while keeping the sum $\sigma_o^2 + \sigma_b^2$ equal to var$(O - B)$, we found a minimum in var$(O - A)$ in the passive observation space. In this second-part paper, we formalize this result, develop the principles of estimation of the analysis error covariance by cross-validation, and apply it to estimate and optimize the analysis error covariance of ECCC's surface analyses of O_3 and $PM_{2.5}$.

When we refer to analysis error, or analysis error covariance, it is important to distinguish the perceived analysis error with the true analysis error [6]. The perceived analysis error is the analysis error that results from the analysis algorithm itself, whereas the true analysis error is the difference between the analysis and the true state. Analysis schemes are usually derived from an optimization of some sort. In a variational analysis scheme for example, the analysis is obtained by minimizing a cost function with some given or prescribed observation error and background error covariances, $\tilde{R}$ and $\tilde{B}$

respectively. In a linear unbiased analysis scheme, the gain matrix $\tilde{\mathbf{K}}$ is obtained by minimum variance estimation, yielding an expression of the form, $\tilde{\mathbf{K}} = \tilde{\mathbf{B}}\mathbf{H}(\mathbf{H}\tilde{\mathbf{B}}\mathbf{H} + \tilde{\mathbf{R}})^{-1}$, where $\mathbf{H}$ is the observation operator. The perceived analysis error covariance is then derived as $\tilde{\mathbf{A}} = (\mathbf{I} - \tilde{\mathbf{K}}\mathbf{H})\tilde{\mathbf{B}}$. In order to derive an expression for the perceived analysis error covariance we in fact assume that given error covariances, $\tilde{\mathbf{R}}$ and $\tilde{\mathbf{B}}$, are error covariances with respect to the true state, i.e., the true error covariances. We also assume that the observation operator is not an approximation with some error, but is the true error-free observation operator. Of course, in real applications neither of $\tilde{\mathbf{R}}$ and $\tilde{\mathbf{B}}$ are covariance measures with respect to the true state, but only a more or less accurate estimate of those. Daley [6] have argued that in principle for an arbitrary gain matrix $\tilde{\mathbf{K}}$, the true analysis error covariance A can be computed as $\mathbf{A} = (\mathbf{I} - \tilde{\mathbf{K}}\mathbf{H})\mathbf{B}(\mathbf{I} - \tilde{\mathbf{K}}\mathbf{H})^T + \tilde{\mathbf{K}}\mathbf{R}\tilde{\mathbf{K}}^T$ provided that we know the true observation and background error covariances, R and B. This expression is a quadratic matrix equation, and has the property that the true analysis error variance, $tr(\mathbf{A})$ is minimum when $\tilde{\mathbf{K}} = \mathbf{B}\mathbf{H}(\mathbf{H}\mathbf{B}\mathbf{H} + \mathbf{R})^{-1} = \mathbf{K}$. In that sense, the analysis is truly optimal. The optimal gain matrix K is called the Kalman gain. It thus illustrates that although an analysis is obtained through some minimization principle, the resulting analysis error is not necessarily the true analysis error.

One of the main sources of information to obtain the true R and B is from the $var(O - B)$ statistic. However, it has always been argued that this is not possible without making some assumptions [7–9], the most useful one being that background errors are spatially correlated while the observation errors are spatially uncorrelated, or at least on a much shorter length-scale. Even under those assumptions, different estimation methods such as the Hollingsworth-Lönnberg method [10], the maximum likelihood gives different error variances and different correlation lengths [11]. Other methods use $var(O - B)$ for rescaling but assume that the observation error is known. The assumption that the observation error is known is also debated as they contain representativeness errors [12] that include observation operator errors. How to obtain an optimal analysis is thus unclear.

The evaluation of the true or perceived analysis error covariance using its own active observations is also a misleading problem unless the analysis is already optimal. Hollingsworth and Lönnberg [13] addressed this issue for the first time where they noted that in the case of an optimal gain (i.e., optimal analysis), the statistics of observation-minus-analysis residuals $O - \hat{A}$ are related to the analysis error by $\mathrm{E}[(O - \hat{A})(O - \hat{A})^T] = \mathbf{R} - \mathbf{H}\hat{\mathbf{A}}\mathbf{H}^T$, where $\hat{\mathbf{A}}$ is the optimal analysis error covariance and H and R are the observation operator and observation error covariance respectively. The caret (ˆ) over A indicates that the analysis uses an optimal gain. In the context of spatially uncorrelated observation errors, the off-diagonal elements of $\mathrm{E}[(O - \hat{A})(O - \hat{A})^T]$ would then give the analysis error covariance in observation space. Hollingsworth and Lönnberg [13] argued that for most practical purposes, the negative intercept of $\mathrm{E}[(O - \hat{A})(O - \hat{A})^T]$ at zero distance and the prescribed observation weight should be nearly equal, and thus could be used as an assessment of optimality of an analysis. However, in case where such agreement does not exist, an estimate of the actual analysis error is not possible. Another method, proposed by Desroziers et al. [14], argued that the diagnostic $\mathrm{E}[(O - \hat{A})(\hat{A} - B)^T]$ should be equal to the analysis error covariance in observation space but, again, only if the gain is optimal and the innovation covariance consistency is respected [15].

The impasse of the estimation of the true analysis error seems to be tied with using active observations, i.e., using the same observations as those used to create the analysis. A robust approach that does not require an optimal analysis is to use observations whose errors are uncorrelated with the analysis error. For example, if we assume that observation errors are temporarily (serially) uncorrelated, an estimation of the analysis error can be made with the help of a forecast model initialized by the analysis by verifying the forecast against these observations. This is the essential assumption used traditionally in meteorological data assimilation to assess indirectly the analysis error by comparing the resulting forecast with observations. As forecast error grows with time, the observation-minus-forecast can be used to assess whether an analysis is better than another. In a somewhat different method but making the same assumption, Daley [6] used the temporal (serial) correlation of the innovations to diagnose the optimality of the gain matrix. This property was first established in the context of Kalman filter estimation theory by Kailath [16]. However, both the traditional meteorological forecast approach and the Daley method [6] are subject to limitations: they assume that the model forecast has no bias and the analysis corrections are made correctly on all the variables needed to initialize the model. In practice, improper initialization of unobserved meteorological variables gives rise to spin-up problems or imbalances. Furthermore, with the traditional meteorological approach, compensation due to model error can occur, so that an optimal analysis does not necessarily yield an optimal forecast [7].

An alternative approach introduced by Marseille et al. [17], which we will follow here, is to use independent observation or passive observations to assess the analysis error. The essential assumption of this method is that the observations have spatially uncorrelated errors, so that the observations used for verification, i.e., the passive observations, have uncorrelated errors with the analysis. The advantage of this approach is that it does not involve any model to propagate the analysis information to a later time. Marseille et al. [17] then showed that by multiplying the Kalman gain with an appropriate scalar value, one can reduce the analysis error. In this paper, we go further by using principles of error covariance estimation to obtain a near optimal Kalman gain. In addition we impose the innovation covariance consistency [15] and show that all diagnostics of analysis error variance nearly agree with one another. These include the Hollingsworth and Lönnberg [13], the Desroziers et al. [14] and new diagnostics that we will introduce.

The paper is organized as follows. First we present in Section 2 the theory and diagnostics of analysis error covariance in both passive and active observation spaces, as well as a geometrical representation. This leads us to a method to minimize the true analysis error variance. In Section 3, we present the experimental setup on how we obtain near optimal analyses and presents the results of several diagnostics in active and passive observation spaces, and compare with the analysis error variance obtained from the optimum interpolation scheme itself. In Section 4, we discuss the statistical assumptions being used, how and if they can be extended and how this formalism can be used in other applications such as the estimation of correlated observation errors with satellite observations. Finally, we draw some conclusions in Section 5.

2. Theoretical Framework

This section is composed of mainly three parts. In Sections 2.1 and 2.2, we first describe the diagnostics to obtain the true error covariances using passive observations whether the analysis is optimal or not. We then give a geometric interpretation in Section 2.3 that indicate the way to obtain the optimal analysis, that is, one that minimizes the true analysis error. Lastly, in Sections 2.4 and 2.5, we formulate the diagnostics of analysis error for optimal analyses.

2.1. Diagnostic of Analysis Error Covariance in Passive Observation Space

Let us decompose the observation space in two disjoint sets; the active observation set or training set $\{\mathbf{y}\}$ used to create the analysis, and the independent or passive observation set $\{\mathbf{y}_c\}$ used to evaluate the analysis. An analysis built from prescribed background and observation error covariances, $\tilde{\mathbf{B}}$ and $\tilde{\mathbf{R}}$ respectively, is given by

$$\mathbf{x}^a = \mathbf{x}^f + \tilde{\mathbf{B}}\mathbf{H}^T(\mathbf{H}\tilde{\mathbf{B}}\mathbf{H}^T + \tilde{\mathbf{R}})^{-1}\mathbf{d} = \mathbf{x}^f + \tilde{\mathbf{K}}\mathbf{d} \tag{1}$$

where $\tilde{\mathbf{K}}$ is the gain matrix built from the prescribed error covariances, $\mathbf{H}$ is the observation operator for the active observation set, $\mathbf{d}$ is the active innovation vector, $\mathbf{d} = \mathbf{y} - \mathbf{H}\mathbf{x}^f = \varepsilon^o - \mathbf{H}\varepsilon^f$, $\mathbf{x}^f$ is the background state or model forecast, ε^o is the active observation error and ε^f the background error. The observation-minus-analysis residual $(O - A)$ for the active set is given by [14,15,18].

$$
\begin{aligned}
(O - A) \ &= \mathbf{y} - \mathbf{H}\mathbf{x}^a = \varepsilon^o - \mathbf{H}\varepsilon^a \\
&= \mathbf{d} - \mathbf{H}\tilde{\mathbf{B}}\mathbf{H}^T (\mathbf{H}\tilde{\mathbf{B}}\mathbf{H}^T + \tilde{\mathbf{R}})^{-1}\mathbf{d} \\
&= \tilde{\mathbf{R}}(\mathbf{H}\tilde{\mathbf{B}}\mathbf{H}^T + \tilde{\mathbf{R}})^{-1}\mathbf{d}
\end{aligned}
\tag{2}
$$

where ε^a is the analysis error. The analysis interpolated at the passive observation sites can be denoted by $\mathbf{H}_c\mathbf{x}^a$, where $\mathbf{H}_c$ is the observation operator at passive observation sites. The observation-minus-analysis residual at the passive observation sites $(O - A)_c$ is then given by

$$
\begin{aligned}
(O - A)_c \ &= \mathbf{y}_c - \mathbf{H}_c\mathbf{x}^a = \varepsilon_c^o - \mathbf{H}_c\varepsilon^a \\
&= \mathbf{d}_c - \mathbf{H}_c\tilde{\mathbf{B}}\mathbf{H}^T (\mathbf{H}\tilde{\mathbf{B}}\mathbf{H}^T + \tilde{\mathbf{R}})^{-1}\mathbf{d}
\end{aligned} \ '
\tag{3}
$$

where $\mathbf{d}_c = \mathbf{y}_c - \mathbf{H}_c\mathbf{x}^f = \varepsilon_c^o - \mathbf{H}_c\varepsilon^f$ is the innovation at the passive observation sites. Note that the formalism introduced here is general, and can be used with any set of independent observations such as different instruments or observation networks as long as the $\mathbf{H}_c$ operator is properly defined. Consequently, for generality, we distinguish the passive observation errors or independent observation error, ε_c^o, from the active observation error ε^o.

There are two important statistical assumptions from which we derive cross-validation diagnostics. Assuming that the observation errors are spatially uncorrelated, it follows that

$$
\mathrm{E}[\varepsilon^o(\varepsilon_c^o)^T] = \mathbf{0}
\tag{4}
$$

where E[ ] is the mathematical expectation that represents the mean over an ensemble of realizations. It has been argued by Marseille et al. [17] that representativeness error can violate this assumption for a close pair of active-passive observations, but we will neglect this effect. Also, assuming that observation errors are uncorrelated with background error, we have

$$
\mathrm{E}[\varepsilon^o(\mathbf{H}\varepsilon^f)^T] = \mathbf{0}, \ \mathrm{E}[\varepsilon_c^o(\mathbf{H}_c\varepsilon^f)^T] = \mathbf{0}
\tag{5}
$$

We come now to the most important property: since the analysis is a linear combination of the active observations and the background state, the analysis error is then uncorrelated with the passive observation errors,

$$
\mathrm{E}[(_c\varepsilon^a)(\varepsilon_c^o)^T] = \mathbf{0}
\tag{6}
$$

and thus we get the following cross-validation diagnostic in passive observation space,

$$
\mathrm{E}[(O - A)_c(O - A)_c^T] = \mathbf{R}_c + \mathbf{H}_c\mathbf{A}\mathbf{H}_c^T
\tag{7}
$$

similarly to Marseille et al. [17]. A very important point to note is that $\mathbf{A}$ is the true analysis error covariance—it does not assume that the gain is optimal. The matrices in Equation (7) are of the dimension of the passive observation space and $\mathbf{R}_c$ is the observation error covariance matrix for the passive or independent observations.

2.2. A Complete Set of Diagnostics of Error Covariances in Passive Observation Space

It is also possible to define a set of diagnostics that would determine, in principle, the true error covariances, $\mathbf{R}$, $\mathbf{B}$ and $\mathbf{A}$. From Equations (4) and (5) we get another cross-validation diagnostic,

$$
\mathrm{E}[(O - B)_c(O - B)^T] = \mathbf{H}_c\mathbf{B}\mathbf{H}^T
\tag{8}
$$

This diagnostic is related to the Hollingsworth-Lönnberg [10] estimation of the spatially correlated part of the innovation, and in practice can be dominated by sampling error. Note that it is not a square matrix and an estimation of parameters of **B** may not be trivial.

We can also obtain the innovation covariance matrix in passive observation space (identical in fact to the one in active observations space) as

$$E[(O-B)_c(O-B)_c^T] = \mathbf{R}_c + \mathbf{H}_c\mathbf{B}\mathbf{H}_c^T \tag{9}$$

The system of Equations (7)–(9) gives a complete set of equations to determine the true **R**, **B** and **A** at the passive observation sites provided that by interpolation/extrapolation we can obtain $\mathbf{H}_c\mathbf{B}\mathbf{H}_c^T$ from $\mathbf{H}_c\mathbf{B}\mathbf{H}^T$.

Nevertheless, and for sake of completeness, we also investigated the meaning of the statistics $E[(O-A)_c(O-B)^T]$ and came to the conclusion that it can interpreted as a misfit to the Desroziers et al. estimate of **B** [14] with the true value **B**. We recall that the first iterate of the Desroziers et al. estimate for **B** in active observation space is given by $\mathbf{HB}^D\mathbf{H}^T = E[(\mathbf{Hx}^a - \mathbf{Hx}^f)(\mathbf{y} - \mathbf{Hx}^f)^T]$ [15], where we used the superscript D indicate the Desroziers et al. first iterate estimate. We can actually generalize this diagnostic to be a cross-covariance between state space and active observation space as

$$\mathbf{B}^D\mathbf{H}^T = E[(\mathbf{x}^a - \mathbf{x}^f)(\mathbf{y} - \mathbf{Hx}^f)^T] \tag{10}$$

By applying $\mathbf{H}_c$ to Equation (10) we can then introduce a generalized Desroziers et al. estimate of the background error covariance **B** between passive and active observation spaces as,

$$E[(A-B)_c(O-B)^T] = \mathbf{H}_c\mathbf{B}^D\mathbf{H}^T \tag{11}$$

Since $(O-A)_c = (O-B)_c - (A-B)_c$, we get with Equations (8) and (10),

$$E[(O-A)_c(O-B)^T] = \mathbf{H}_c(\mathbf{B} - \mathbf{B}^D)\mathbf{H}^T \tag{12}$$

That is the difference between the true **B** and the Desroziers et al. first estimate $\mathbf{B}^D$ in the cross passive-active observation spaces. Similarly to Equation (8) this diagnostics requires spatial interpolation of error covariances from active sites to passive sites of basically noisy statistics. The estimation of **B** from this diagnostics is further complicated by the fact that what is being interpolated, that is $\mathbf{B} - \mathbf{B}^D$, may not even be positive definite, but the augmented matrix,

$$\mathrm{cov}\begin{pmatrix} (O-B) \\ (O-A)_c \end{pmatrix} = \begin{bmatrix} \mathbf{R} + \mathbf{HBH}^T & \mathbf{H}(\mathbf{B} - \mathbf{B}^D)\mathbf{H}_c^T \\ \mathbf{H}_c(\mathbf{B} - \mathbf{B}^D)\mathbf{H}^T & \mathbf{R}_c + \mathbf{H}_c\mathbf{A}\mathbf{H}_c^T \end{bmatrix} \tag{13}$$

is positive definite. Except for the diagonal of Equation (13), we have not attempted in this study to conduct this complete estimation of **R**, **B** and **A**, but rather focused on getting a reliable estimate of the analysis error covariance **A**.

2.3. Geometrical Interpretation

A geometrical illustration of some of the relationships obtained above can be made by using a Hilbert space representation of random variables in observation space. A 2D representation for the analysis of a scalar quantity was used in Desroziers et al. [14] to illustrate their a posteriori diagnostics. We will generalize this approach to include passive observations by considering a 3D representation.

As in Desroziers et al. [14] let's consider the analysis of a scalar quantity. Several variables are to be considered in this observation space: y^o the active observation (or measurement) of the scalar quantity, y^b the background (or prior) value equivalent in observation space (i.e., $y^b = H\ x^b$), y^a the analysis in observation space (i.e., $y^a = H\ x^a$), and for verification y^c an independent observation (or passive observation) that is not used to compute the analysis. Each of these quantities is a random variable as they each contain random errors, and any linear combination of random variables in observation space also belong to observation space. For example, $y^o - y^b$ is the innovation (commonly denoted by O-B) and that belongs to observation space, $y^a - y^b$ is the analysis increment in observation space (commonly denoted by A-B), and $y^o - y^a$ is the analysis residual in observation space (commonly denoted by O-A). We can also define an inner product of any random variables in observation space. y_1, y_2, as

$$\langle y_1, y_2 \rangle := E[(y_1 - E(y_1))(y_2 - E(y_2))] \tag{14}$$

The squared norm then represents the variance,

$$\|y\|^2 := \langle y, y \rangle = \sigma_y^2 \tag{15}$$

so the inner product has the following geometric interpretation

$$\langle y_1, y_2 \rangle = \|y_1\| \|y_2\| \cos \theta \tag{16}$$

where $\cos \theta$ is the correlation coefficient. Uncorrelated random variables are thus statistically orthogonal. With this inner product, the observation space forms a Hilbert space of random variables.

Figure 1 illustrates the statistical relationship in observation space between: the active observation y^o (illustrated as O in the figure), the prior or background y^b (i.e., B), the analysis y^a (i.e., A), and the independent observation y^c (i.e., O_c). The origin T corresponds to the truth of the scalar quantity, and also corresponds to the zero of the central moment of each random variables, e.g., $y - E[y]$, since each variables are assumed to be unbiased. We also assume that the background, active and passive observations errors are uncorrelated to one another, so the three axes; ε^o for the active observation error, ε^b for the background error, and ε_c^o for the passive observation error are orthogonal. The plane defined by ε^o and ε^b axes is the space where the analysis takes place, and is called the analysis plane. However, since we define the analysis to be linear and unbiased, only linear combinations of the form $y^a = k y^o + (1 - k) y^b$ where k is a constant are allowed. The analysis A then lies on the line (B, O). The thick lines in Figure 1 represent the norm of the associated error. For example, the thick line along the ε^o axis depict the (active) observation standard deviation σ_o, and similarly for the other axes and other random variables. Since the active observation error is uncorrelated with the background error, the triangle ΔOTB is a right triangle, and by Pythagoras theorem we have, $\overline{(y^o - y^b)^2} := \langle (O - B), (O - B) \rangle = \sigma_o^2 + \sigma_b^2$. This is the usual statement that the innovation variance is the sum of background and observation error variances. The analysis is optimum when the analysis error $\|\varepsilon^a\|_2 = \sigma_a^2$ is minimum, in which case the line (T, A) is perpendicular to line (O, B).

Now let's consider the passive observation O_c. The passive observation error is perpendicular to the analysis plane, thus the triangle ΔO_cTA is a right triangle,

$$\overline{(y^c - y^a)^2} := \langle (O - A)_c, (O - A)_c \rangle = \sigma_c^2 + \sigma_a^2 \tag{17}$$

where σ_c^2 is the passive observation error variance. The most important fact to stress here is that the orthogonality expressed in Equation (17) is true whether or not the analysis is optimal. Furthermore, as the distance $\overline{(y^c - y^a)^2}$ varies with the position of A along the line (O, B), the distance $\overline{(y^c - y^a)^2}$ reaches a minimum value when σ_a^2 is minimum that is when the analysis is optimal. We thus also argue from this representation that there is always a minimum, and the minimum is unique. Finally, we note that ΔBTO$_c$ is also a right triangle so that $\overline{(y^c - y^b)^2} := \langle (O - B)_c, (O - B)_c \rangle = \sigma_c^2 + \sigma_b^2$, which is the scalar version of Equation (9).

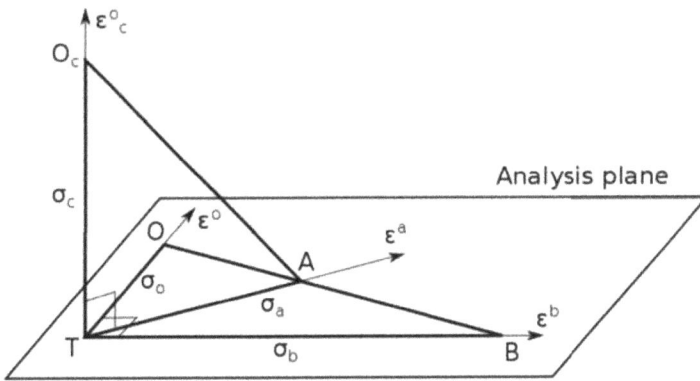

Figure 1. Hilbert space representation of a scalar analysis and cross-validation problem. The arrows indicate the directions of variability of the random variables, and the plane defined by the background and observation errors ε^b, ε^o defines the analysis plane. The thick lines represent the norm associated with the different random variables. T indicate the truth, O the active observation, B the background, A the analysis and O_c the passive observation.

To extend this formalism to random vectors requires to define a proper matricial inner product as briefly described in Section 1.2 of Caines [19]. It is important here to distinguish the stochastic metric space from the observation vector space. In the stochastic metric space a "scalar" needs not to be a number, but only a non-random quantity invariant with respect to $E[\ ]$. Hence, we can define a Hilbert space with the following (stochastic scalar) matricial inner product $\langle \mathbf{y}, \mathbf{w} \rangle = E[(\mathbf{y} - E(\mathbf{y}))(\mathbf{w} - E(\mathbf{w})^T)] = \text{cov}(\mathbf{y}, \mathbf{w})$. The matrix nature of $\text{cov}(\mathbf{y}, \mathbf{w})$ pertains to the observation vector space, but it remains a scalar with respect to the stochastic Hilbert space herein defined. In order to obtain a true scalar ($\in \mathbb{R}$), one would need to define a metric on the observation space matrices as well, such as the trace. Also to be able to compare active and passive observations, projectors need to be introduced on the observation space, implying yet another metric structure on the observation space. We do not carry out this formalism here as it would represent a rather lengthy development that would distract us from the main purpose of this paper; this will be considered in a future manuscript. Finally, we remark that Hilbert space representation of random variables in infinite dimensional space (i.e., continuous space) can also be defined, see Appendix 1 of Cohn [20].

2.4. Error Covariance Diagnostics in Active Observation Space for Optimal Analysis

An analysis is optimal if the analysis error $E[(\varepsilon^a)^T \varepsilon^a]$ is minimum. This implies that the gain matrix using the prescribed error covariances, $\tilde{\mathbf{K}}$ in Equation (1), must be identical to the gain using the true error covariances, i.e., $\tilde{\mathbf{K}} = \mathbf{B}\mathbf{H}^T(\mathbf{H}\mathbf{B}\mathbf{H}^T + \mathbf{R})^{-1}$ [14,15]. It is important to mention that necessary and sufficient conditions to obtain the true error covariances $\mathbf{H}\mathbf{B}\mathbf{H}^T$ and $\mathbf{R}$ in observation space, are: 1—the Kalman gain condition, $\mathbf{H}\tilde{\mathbf{K}} = \mathbf{H}\mathbf{K}^{true}$ and 2—the innovation covariance consistency, $E[(O - B)(O - B)^T] = \mathbf{H}\tilde{\mathbf{B}}\mathbf{H}^T + \tilde{\mathbf{R}}$. For a proof see the Theorem on error covariance estimates in Ménard [15].

From the optimality of the analysis (or Kalman gain) alone, we derive that $E[(\hat{A} - T)(O - B)^T] = 0$ or $E[(\hat{A} - T)(O - \hat{A})^T] = 0$. Indeed, from Equation (2), we get $(O - \hat{A}) = \mathbf{R}(\mathbf{H}\mathbf{B}\mathbf{H} + \mathbf{R})^{-1}d$, and for the analysis error in observation space we get, $(\hat{A} - T) = \mathbf{R}(\mathbf{H}\mathbf{B}\mathbf{H}^T + \mathbf{R})^{-1}\mathbf{H}\varepsilon^f + \mathbf{H}\mathbf{B}\mathbf{H}^T(\mathbf{H}\mathbf{B}\mathbf{H}^T + \mathbf{R})^{-1}\varepsilon^o$, from which we derive the expectations above. Using the geometrical representation in Section 2.3 the distance between A and T is minimum, when ΔTAO and ΔTAB (Figure 1) are right triangles. We should also note that for the scalar problem, the Kalman gain depends only on the ratio of observation to background error variances and thus the scalar Kalman gain is optimal if the ratio of the prescribed error variances is equal to the ratio of the true error variances.

If in addition to the optimality of the analysis or Kalman gain we add the innovation covariance consistency then we get three different statistical diagnostics of the (optimal) analysis error covariance. Hollingsworth and Lönnberg [13] was the first to introduce a statistical diagnostic of analysis error in the active observation space, as:

$$E[(O - \hat{A})(O - \hat{A})^T] = \mathbf{R} - \mathbf{H}\hat{\mathbf{A}}_{HL}\mathbf{H}^T \tag{18}$$

Here we use a subscript, *HL* to indicate that this is the Hollingsworth-Lönnberg estimate. Equation (18) can obtained from the covariance of $(O - \hat{A})$ and that, for an optimal gain matrix, $\mathbf{H}\hat{\mathbf{A}}\mathbf{H}^T = \mathbf{R} - \mathbf{R}(\mathbf{HBH} + \mathbf{R})^{-1}\mathbf{R}$ which derives from the usual formula, $\hat{\mathbf{A}} = \mathbf{B} - \mathbf{BH}^T(\mathbf{HBH} + \mathbf{R})^{-1}\mathbf{HB}$. Geometrically it derives from the fact that the triangle $\Delta T\hat{A}O$ is a right triangle, and from the innovation covariance consistency that implies that the triangle ΔOTB is a right triangle. Using data to construct $E[(O - \hat{A})(O - \hat{A})^T]$, an estimated analysis error covariance obtained from Equation (18) is symmetric but could be non-positive definite as it is obtained by subtracting two positive definite matrices. The effect of misspecification in the prescribed error covariances resulting from a lack of innovation covariance consistency will be discussed in the result Section 3 and in Appendix B.

Inspired from the geometrical interpretation that ΔTAB is also be a right triangle we derived the following diagnostic,

$$E[(\hat{A} - B)(\hat{A} - B)^T] = \mathbf{HBH}^T - \mathbf{H}\hat{\mathbf{A}}_{MDJ}\mathbf{H}^T = \mathbf{H}(\mathbf{B} - \hat{\mathbf{A}}_{MDJ})\mathbf{H}^T \tag{19}$$

where *MDJ* stands for Ménard-Deshaies-Jacques. This relationship is obtained by using the expression $(\hat{A} - B) = \mathbf{HBH}^T(\mathbf{HBH} + \mathbf{R})^{-1}\mathbf{d}$, the innovation covariance consistency and the formula for the optimal analysis error covariance $\hat{\mathbf{A}} = \mathbf{B} - \mathbf{BH}^T(\mathbf{HBH} + \mathbf{R})^{-1}\mathbf{HB}$. As for the *HL* diagnostics, the estimated error covariance obtained from this diagnostic is symmetric by construction but may not be positive definite. Another way of looking at Equation (19) is that it expresses, in observation space, the error reduction due to the use of observations.

Another diagnostic of analysis error covariance was proposed by Desroziers et al. [14]. By combining $(O - \hat{A})$ and $(\hat{A} - B)$ we get

$$E[(O - \hat{A})(\hat{A} - B)^T] = \mathbf{H}\hat{\mathbf{A}}_D\mathbf{H}^T \tag{20}$$

where the subscript *D* denotes the Desroziers et al. estimate. By construction, the estimated analysis error covariance is not necessarily symmetric. A geometrical derivation is provided in Appendix A. We also provide in Appendix B a sensitivity analysis on the departure from innovation covariance consistency for each diagnostics in both active and passive observation spaces.

2.5. Error Covariance Diagnostics in Passive Observation Space for Optimal Analysis

We can also derive optimal analysis diagnostics in the passive observation space. Considering the 3D geometric interpretation, and in particular the tetrahedron (O_c, T, A, B), we notice that since ΔTAB is a right triangle, so is $\Delta O_c AB$, which is a projection of the triangle ΔTAB on the plane passing through O_c, O and B. We thus have $E[(\hat{A} - B)_c(\hat{A} - B)_c^T] + E[(O - \hat{A})_c(O - \hat{A})_c^T] = E[(O - B)_c(O - B)_c^T]$. Combining this result with Equation (7) and using, $E[(O - B)_c(O - B)_c^T] = \mathbf{H}_c\mathbf{BH}_c^T + \mathbf{R}_c$, we then get

$$E[(\hat{A} - B)_c(\hat{A} - B)_c^T] = \mathbf{H}_c\mathbf{BH}_c^T - \mathbf{H}_c\hat{\mathbf{A}}_{MDJ}\mathbf{H}_c^T \tag{21}$$

Note that our analysis diagnostic is the only diagnostic that is valid in both active and passive observation spaces, i.e., Equation (21) is similar to Equation (19).

The other, less direct, diagnostic for optimal analysis is simply based on Equation (7) that is

$$\underset{\gamma,\, L_c}{\operatorname{argmin}}\{E[(O-A)_c(O-A)_c^T]\} = \mathbf{R}_c + \mathbf{H}_c\hat{\mathbf{A}}(\gamma, L_c)\mathbf{H}_c^T \tag{22}$$

These five diagnostics will be used later in the results section.

3. Results with Near Optimal Analyses

3.1. Experimental Setup

We will just give here a short summary of the experimental setup we are using in this study. More details can be found in the Part I paper (Ménard and Deshaies-Jacques [5]). A series of hourly analyses of O_3 and $PM_{2.5}$ at 21 UTC for a period of 60 days (14 June to 12 August 2014) were performed using an optimum interpolation scheme combining the operational air quality model GEM-MACH forecast and the real-time AirNow observations (see Section 2 of [5] for further details). The analyses are made off-line so they are not used to initialize the model. As input error covariances, we use uniform observation and background error variances, with $\tilde{\mathbf{R}} = \sigma_o^2\mathbf{I}$ and $\tilde{\mathbf{B}} = \sigma_b^2\mathbf{C}$, where $\mathbf{C}$ is a homogeneous isotropic error correlation based on a second-order autoregressive model. The correlation length is estimated by using a maximum likelihood method using at first, error variances obtained from a local Hollingworth-Lönnberg fit [11] (and only for the purpose of obtaining a first estimate of the correlation length). We then conduct a series of analyses by changing error variance ratio $\gamma = \sigma_o^2/\sigma_b^2$ while at the same time respecting the innovation variance consistency condition, $\sigma_o^2 + \sigma_b^2 = \mathrm{var}(O-B)$. This corresponds basically in searching for the minimum of the *tr* (trace) of Equation (7) while the trace of the innovation covariance consistency, $tr[\mathbf{R}+\mathbf{HBH}^T] = tr\{E[(O-B)(O-B)^T]\}$, is respected.

First the observations are separated into 3 sets of observations of equal number and distributed randomly in space. By leaving out one set of observations for verification and constructing analyses with the remaining 2 other sets, we construct a cross-validation setup from which we can evaluate the diagnostic $\mathrm{var}(O-A)_c = tr\{E[(O-A)_c(O-A)_c^T]\}$ in passive observation space. This constitutes our first guess experiment that we will refer to as **iter 0**. No observation or model bias correction was applied, nor were the mean of the innovation at the stations were removed prior to performing the analysis. The variance statistics are first computed at the station using the 60-day members, and then averaged over the domain to give the equivalent of $tr\{E[(O-A)_c(O-A)_c^T]\}$. We repeat the procedure by exhausting all permutations possible, in this case 3. The mean value statistic for the three verifying subsets are then averaged. More details can be found in Section 2 and beginning of Section 3 of Part I [5].

The red curve on Figure 2 illustrates how this diagnostic varies with and exhibits a minimum. This minimum can easily be understood by referring to Figure 1: as the analysis point A changes position along the line (O, B) the distance $\|O_c - A\|$ reaches a minimum, and this is what we observe in Figure 2.

In our next step, **iter 1**, we first re-estimate the correlation length by applying a maximum likelihood method, as in Ménard [15], using the **iter 0** error variances that are consistent with the optimal ratio $\hat{\gamma}$ obtained in **iter 0** and the innovation variance consistency. Then, with this new correlation length, we estimate a new optimal ratio $\hat{\gamma}$ (**iter 1**), which turn out to be very close to the value obtained in **iter 0**. We recall that we use uniform error variances, both to keep things simple but also because the optimal ratio is obtained by minimizing a domain-averaged variance $\mathrm{var}(O_c - A)$. A summary of the error covariance parameters obtained for **iter 0** and **iter 1** are presented in Table 1.

var(*O*−*A*) [O3]

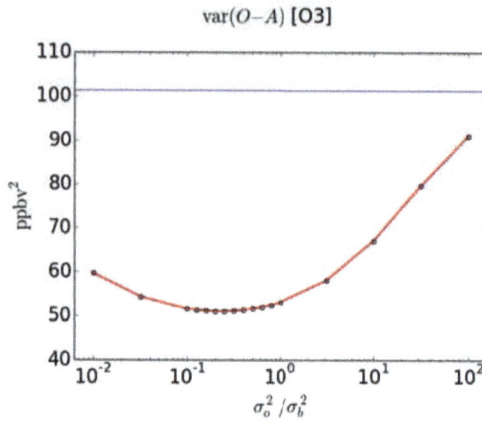

Figure 2. Red line, variance of observation-minus-analysis of O_3 in passive observation space. Blue line, variance of observation-minus-model.

Table 1. Input error statistics for the first experiment and optimized variance ratio experiment.

Experiment		L_c (km)	$\left\langle (O-B)^2 \right\rangle$	$\hat{\gamma} = \hat{\sigma}_o^2/\hat{\sigma}_b^2$	$\hat{\sigma}_o^2$	$\hat{\sigma}_b^2$	χ^2/N_s
O_3	iter 0	124	101.25	0.22	18.3	83	2.23
O_3	iter 1	45	101.25	0.25	20.2	81	1.36
$PM_{2.5}$	iter 0	196	93.93	0.17	13.6	80.3	2.04
$PM_{2.5}$	iter 1	86	93.93	0.22	16.9	77	1.25

We have also added in Table 1, the χ^2/N_s diagnostic values [15] which is the 60-day mean value of $\chi_k^2/N_s(k)$ where $N_s(k)$ is the total number of observations available at time t_k and $\chi_k^2 = d_k^T \left(H\tilde{B}H + \tilde{R} \right)^{-1} d_k$. The χ^2/N_s diagnostics [15] is closely related to the J_{min} diagnostic used in variational methods, and should have a value of 1 in the case where the innovations are consistent with the prescribed error statistics. When there is innovation covariance consistency then $\chi^2/N_s = 1$ but the reverse is not true. We observe that there was a significant improvement from **iter 0** to **iter 1** in terms of χ^2/N_s but is still not equal to one. We thus refer the analysis of **iter 1** as near optimal.

The repeated application of this sequence of estimation methods, i.e., find the correlation length by maximum likelihood and optimize the variance ratio, converges really fast and in practice there is no need to go beyond **iter 1**. Figure 3 displays iterates 0 to 4 with our estimation procedure for O_3. With one iteration update we nearly converge. A similar procedure was used in Ménard [15], where the variance and correlation length (estimated by maximum likelihood) were estimated in sequence, which taught us that a slow and fictitious drift in estimated variances and correlation length can occur when the correlation model is not the true correlation. So in regard of similar considerations that may occur here, we do not extend our iteration procedure beyond the first iterate.

3.2. Statistical Diagnostics of Analysis Error Variance

For each of these experiments, statistics related diagnostics for analysis error variance, discussed in Section 2, are computed and the results are presented in Table 2 for the verification made against active observations, and in Table 3 to the verification made against the passive observations. If the analysis was truly optimal the different diagnostics would all agree.

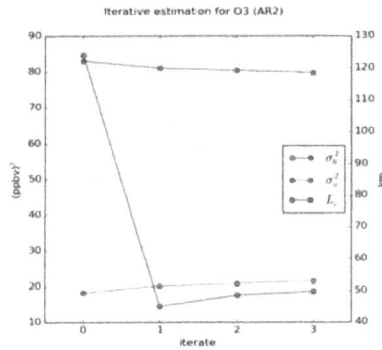

Figure 3. Optimal estimates of σ_o^2, σ_b^2 and maximum likelihood estimate of correlation length L_c for the first four iterates. Blue, is the optimal background error variance, green, the optimal observation error variance and in red the correlation length (in km, with labels on the right side of the figure).

Table 2. Analysis statistics against active observations.

Experiment		Active var$(\hat{A} - B)$	Active $tr(\mathbf{H}\hat{A}_{MDJ}\mathbf{H}^T)/N_s$	Active $tr(\mathbf{H}\hat{A}_D\mathbf{H}^T)/N_s$	Active var$(O - \hat{A})$	Active $tr(\mathbf{H}\hat{A}_{HL}\mathbf{H}^T)/N_s$
O_3	iter 0	60.29	22.69	9.61	24.33	−6.03
O_3	iter 1	67.66	13.32	13.68	11.26	8.94
$PM_{2.5}$	iter 0	62.29	17.98	7.71	16.78	−3.18
$PM_{2.5}$	iter 1	66.3	10.68	9.51	9.57	7.33

In the second, third and last column of Table 2 are tabulated estimates of the analysis error variance at the active location sites, i.e., $tr(\mathbf{HAH}^T)/N_s$, obtained by three different methods. The second column is an estimate given with our method $\sigma_b^2 - \text{var}(\hat{A} - B) = tr(\mathbf{H}\hat{A}_{MDJ}\mathbf{H}^T)/N_s$. The third column is the Desroziers et al. estimate of analysis error [14], Equation (20), and the last column is the estimate using the method proposed by Hollingsworth and Lönnberg [13], Equation (18). We note that the analysis error variance estimate provided by the first two methods is fairly consistent for an updated correlation length estimate, i.e., **iter 1** (but not **iter 0**). We also note that χ^2/N_s is closer to one for **iter 1**. These two facts indicate that the updated correlation length (**iter 1**) with uniform error variances is closer to the innovation covariance consistency. The Hollingsworth and Lönnberg [13] method however, is very sensitive and negatively biased in the lack of innovation covariance consistency.

Estimate of the analysis error variance at the passive observation locations, i.e., $tr(\mathbf{H}_c\mathbf{A}\mathbf{H}_c^T)/N_s$, provided by two different methods are given by Equation (21) in column 3 and by Equation (22) in column 5 of Table 3. As for the estimate at the active locations (Table 2), there is a general agreement on the analysis error estimates with the updated correlation length (**iter 1**), although this distinction is not that clear for $PM_{2.5}$.

Table 3. Analysis statistics against passive observations.

Experiment		Passive var$[(\hat{A} - B)_c]$	Passive $tr(\mathbf{H}_c\hat{A}_{MDJ}\mathbf{H}_c^T)/N_s$	Passive var$[(O - \hat{A})_c]$	Passive var$[(O - \hat{A})_c] - \sigma_{oc}^2$
O_3	iter 0	56.95	26.03	51.02	32.72
O_3	iter 1	52.04	28.95	48.95	28.75
$PM_{2.5}$	iter 0	62.29	22.65	38.09	24.49
$PM_{2.5}$	iter 1	66.3	24.62	38.28	21.38

We note also that the analysis error variance at the active sites is smaller than the analysis error variance at the passive observation sites. This involves in particular the fact that since the passive observation are away from the active observation sites, the reduction of variance at the passive observation sites is smaller than at the active observation sites.

3.3. Comparison with the Perceived Analysis Error Variance

We computed the analysis error covariance **A** resulting from the analysis scheme, the so-called perceived analysis covariance [6], using the expression,

$$\mathbf{A} = \tilde{\mathbf{B}} - \tilde{\mathbf{B}}\mathbf{H}^T (\mathbf{H}\tilde{\mathbf{B}}\mathbf{H} + \tilde{\mathbf{R}})^{-1}\mathbf{H}\tilde{\mathbf{B}} = \tilde{\mathbf{B}} - \mathbf{G}\mathbf{G}^T \tag{23}$$

Contrary to the statistical diagnostics above, the perceived analysis error variance is obtained at each model grid point. We then compared the perceived analysis error variance at the active observation sites with the estimated active analysis error variance obtained from previous diagnostics. Since we showed that in both active and passive observations sites the different diagnostics agrees when the analysis is optimal, this would indicate that if the perceived analysis error variance agrees with the analysis error variance diagnostics, that the whole map of analysis error variance is credible.

In order to calculate the perceived analysis error covariance Equation (23) we first perform a Choleski decomposition of $\mathbf{H}\tilde{\mathbf{B}}\mathbf{H}^T + \tilde{\mathbf{R}} = \mathbf{L}\mathbf{L}^T$, where **L** is a lower triangular matrix. Then with a forward substitution we obtain $\mathbf{L}^{-1}$, from which we compute $\mathbf{G} = \tilde{\mathbf{B}}\mathbf{H}^T\mathbf{L}^{-T}$. The perceived analysis error variance for the ozone optimal analysis (i.e., O_3 **iter 1**) is displayed in Figure 4 (A similar figure but for $PM_{2.5}$ is given in supplementary material). We note that although the input statistics used for the analysis are uniform (i.e., uniform background and observation error variances, and homogeneous correlation model), the computed analysis error variance at the active observation location displays large variations, which is attributed to the non-uniform spatial distribution of the active observations.

In Figure 5 we display a histogram of those variances for the ozone optimal analysis O_3 **iter 1** (panel **b**) and for the first experiment O_3 **iter 0** (panel **a**) without optimization (A similar figure is but for $PM_{2.5}$ is given in supplementary material).

Note that median or mean values of variances are significantly different between the optimal and non-optimal analysis cases. Although the observation and background errors are uniform in both optimal and non-optimal analyses, in the optimal analysis case the perceived analysis uncertainty at the observation location is distributed more equally across all values, indicating that there is a better propagation of information, measured by analysis error variance, across the different observation sites. The exception being the isolated observation sites, for which we observe a maxima on the high end of the histograms. At those sites the analysis error variance is simply obtained by the scalar equation $1/\sigma_a^2 = 1/\sigma_o^2 + 1/\sigma_b^2$. For O_3 **iter 1** the scalar analysis error variance gives 16.2, and for O_3 **iter 0** we get 15.0, thus explaining the secondary maxima on the high end of the histogram.

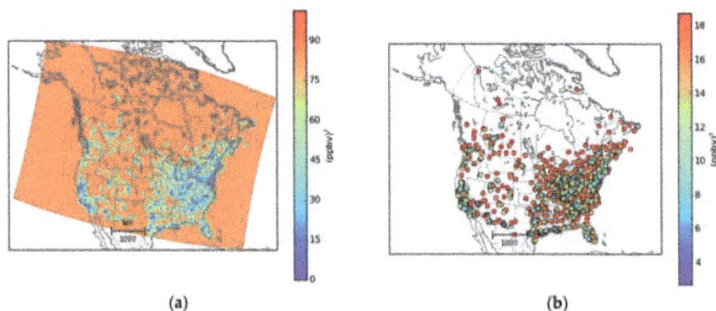

(a)   (b)

Figure 4. Analysis error variance for ozone optimal analysis case O_3 **iter 1**. (**a**) is the analysis error on the model grid and (**b**) at the active observation sites. Note that the color bar of the left and right panels are different. The maximum of the color bar for the left panel correspond to $\sigma_o^2 + \sigma_b^2$.

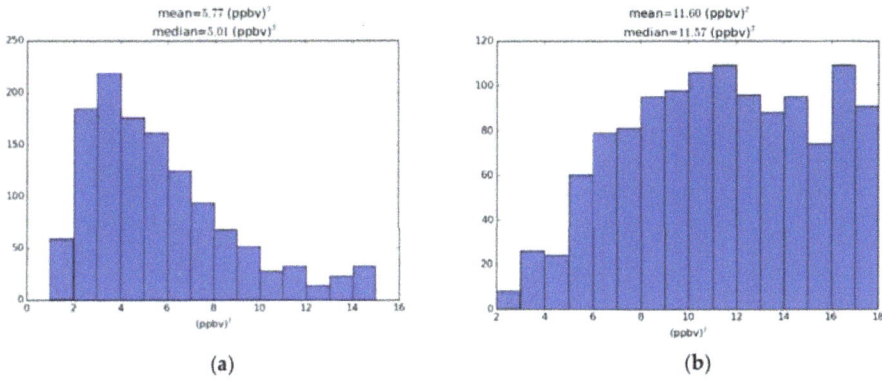

Figure 5. Distribution (histogram) of the ozone analysis error variance at the active observation locations. (a) First analysis experiment O$_3$ **iter 0** (no optimization) on the left panel; (b) Optimal analysis case O$_3$ **iter 1**. Note that the scales are different between the left and right panels.

The mean perceived analysis error variance for all experiments is presented in Table 4. Comparing these values with the estimated values of analysis error variance based on diagnostics in Table 2 we note that for both optimal experiments, O$_3$ **iter 1** and PM$_{2.5}$ **iter 1**, the perceived analysis error variance roughly agrees with all analysis error variances estimated with diagnostics (Table 2).

Table 4. Perceived analysis error variance. Mean over active observation sites.

Experiment		Perceived $tr(\mathbf{H}\hat{\mathbf{A}}_P\mathbf{H}^T)/N_s$
O$_3$	iter 0	5.77
O$_3$	iter 1	11.60
PM$_{2.5}$	iter 0	4.37
PM$_{2.5}$	iter 1	8.21

However, for the non-optimal analyses, O$_3$ **iter 0** and PM$_{2.5}$ **iter 0**, there is a general disagreement between all estimated values. Looking more closely, however, we note that the agreement in the optimal case is not perfect. The perceived analysis error variance is about 20% lower than the best estimates $tr(\mathbf{H}\hat{\mathbf{A}}_{MDJ}\mathbf{H}^T)/N_s$ and $tr(\mathbf{H}\hat{\mathbf{A}}_D\mathbf{H}^T)/N_s$. The optimal χ^2/N_s values in the "optimal" cases are slightly above one, thus indicating that the innovation covariance consistency is not exact and some further tuning of the error statistics could be done. More on that matter will be presented in Section 4.5.

4. Discussion on the Statistical Assumptions and Practical Applications

4.1. Representativeness Error with In situ Observations

The statistical diagnostics presented in Section 2 derive from the assumption that the observation errors are horizontally uncorrelated and uncorrelated with the background error. Although this assumption is never entirely observed in reality, there are ways to work around it. In the case of in situ observations, and assuming that any systematic error have been removed, random errors are still present, due to the difference between the observation and the model's equivalent of the observation—called representativeness error (see Janjic et al. [12] for a review). Representativeness error is due to unresolved scales and processes in the model and interpolation or forward observation model errors. These errors are typically roughly at the scale of the model grid [21,22], so typically a few tens of kilometers for air quality models. This should not be confused with the representativeness of an observation, where, for example, remote stations are representative of large area (e.g., several hundreds of kilometers), whereas urban and suburban stations are at the scale of human activity in the cities, traffic and industries, etc. and are, depending on the chemical specie, of a few kilometers and less.

Representativeness error of in situ measurements can be discarded altogether by simply filtering any pair of observations that are in the range of a few model grid sizes, both in assimilation and estimation of error statistics [1] or in pairs of passive-active observations for cross-validation [17]. Once this filtering is done, the assumption on observation errors being spatially uncorrelated and uncorrelated with the background error then applies.

4.2. Correlated Observation-Background Errors

In any case, it is interesting to show how the different diagnostics, introduced in Section 2, depends on the statistical assumptions of the observation error. One way to get an understanding of the effect of these assumptions is to look at it from a geometrical point of view, using the representation introduced in Section 2.2. Note that the same results can be obtained analytically, but the geometrical interpretation gives a simple and appealing way of looking at the problem.

Let us consider the effect on the analysis of observation error correlated with background error. The case where the observation error is uncorrelated with background error is represented in Figure 6 on the panel **a** and when they are correlated on the panel **b**.

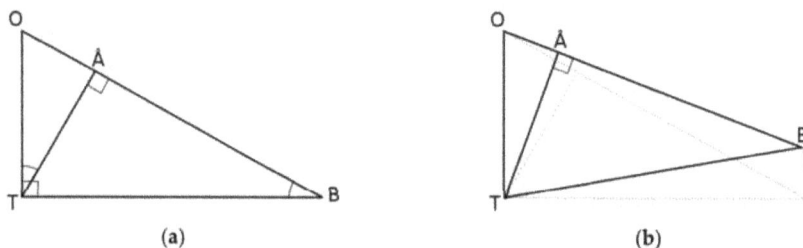

Figure 6. Geometrical representation of the analysis. (**a**) for observation errors uncorrelated with the background error. (**b**) with correlated errors. T indicate the truth, O the observation, B the background and Â the optimal analysis.

The observation and background error variances are kept unchanged, with the same (O, T) length and (B, T) length in both panels. In the case of correlated errors the angle $\angle BTO$ is no longer a right angle. Yet, it is still possible to obtain an optimal analysis, Â, as a linear combination of the observation and the background, on the line (O, B), for which the distance Â to T (i.e., the analysis error variance) is minimum. In this case, $(\hat{A}, T) \perp (O, B)$. Note that for strongly correlated errors and when $\sigma_b^2 > \sigma_o^2$, although Â is still on the line (O, B), it may actually lie outside the segment $[O, B]$. Yet, the principles and theory still hold in that case.

When the observation error is uncorrelated with the background error, $(O, T) \perp (B, T)$, the triangles $\triangle OT\hat{A}$ and $\triangle TB\hat{A}$ are similar and it follows that $\langle (O - \hat{A})(\hat{A} - B) \rangle = \|(\hat{A} - T)\|^2$, which is the Desroziers et al. [14] diagnostic for analysis error variance. However, when the observation error is correlated with the background error (right panel of Figure 6), the triangles $\triangle OT\hat{A}$ and $\triangle TB\hat{A}$ are no longer similar triangles and the Desroziers et al. [14] diagnostics for analysis error does not hold (see derivation in Appendix A). However, the *HL*, Equation (18), and *MDJ* diagnostic, Equation (19), depend only on having right triangles $\triangle OT\hat{A}$ and $\triangle TB\hat{A}$, and not on the orthogonality of (B, T) with (O, T). Therefore, the *HL* and *MDJ* diagnostics are valid with or without correlated observation-background errors.

4.3. Estimation of Satellite Observation Errors with In situ Observation Cross-Validation

One of the important problems in satellite assimilation is the estimation of the satellite observation error, which could be addressed with a simple modification of our cross-validation procedure. Let us assume that we have in situ observations that we assume to have uncorrelated errors between themselves (or use a filter with a minimum distance as discussed in Section 4.1), with the background errors and the satellite observation errors. Yet, the satellite observation errors could be correlated with the background error. Satellite observations could come from a multi-channel instrument with channel-correlated observation errors, as found with many instruments, and yet our validation procedure can still be used. Let us consider that the analyses comprise of satellite and in situ observations but, for the purpose of cross-validation, we use only 2/3rd of the in situ observations in the analysis, and keep the remaining 1/3rd as passive to carry out the cross-validation procedure.

The first thing to note is that the passive in situ observations have uncorrelated errors with the analysis error (the analysis is composed of satellite observations and 2/3rd of the in situ observations). We then use Equation (7) where the interpolation of the analysis is made only at the in situ active observations. Minimizing $E[(O - A)_c^T (O - A)]$ (i.e., the trace of the left hand side of Equation (7)) results in finding the optimal in situ observation weight. Then, computing the analysis error covariance in the satellite observation space from the analysis scheme (either from a Hessian of a variational cost function, or with an explicit gain as in Equation (23)), i.e., $\mathbf{H}_{sat}\hat{\mathbf{A}}\mathbf{H}_{sat}^T$, we use the HL formulation Equation (18) to obtain the satellite observation error covariance,

$$\mathbf{H}_{sat}\hat{\mathbf{A}}\mathbf{H}_{sat}^T + E[(O - A)_{sat}(O - A)_{sat}^T] = \mathbf{R}_{sat} \tag{24}$$

The Equation (24) has the important properties that the estimated observation error covariance is symmetric and positive definite by construction. Then, a new analysis could be carried out to obtain a more realistic $\mathbf{H}_{sat}\hat{\mathbf{A}}\mathbf{H}_{sat}^T$, with a resulting updated $\mathbf{R}_{sat}$, and so forth until convergence.

4.4. Remark on Cross-Validation of Satellite Retrievals

As a last remark, it appears that cross-validation of satellite retrieval observations using a k-fold approach where the observations are used as passive observations to validate the analysis can be a difficult problem. Retrievals from passive remote sensing at nadir generally involve a prior or climatology or a model assumption over different regions, and is thus likely to have spatially correlated errors and errors correlated with the background error. It does not mean, however, that nothing can be done in that case. For example, for certain sensors, such as infrared sensors, it is possible to disentangle the prior from the retrieval, so that by an appropriate transformation of the measurements, observations can be practically decorrelated from the background [23,24]. However to the authors' knowledge, such an approach have never been undertaken for visible measurements such as for NO_2 or AOD's.

4.5. Lack of Innovation Covariance Consistency and Its Relevance to the Statistical Diagnostics

The error covariance diagnostics for optimal analysis, presented in Sections 2.4 and 2.5, depends on the innovation covariance consistency, $E[(O - B)(O - B)^T] = \mathbf{H}\tilde{\mathbf{B}}\mathbf{H}^T + \tilde{\mathbf{R}}$, and our results presented in Section 3 have shown that the different estimates for the optimal analysis error variance are close, but do not strictly agree to each other. This disagreement is related to the lack of innovation consistency as follows.

Let us introduce a departure matrix $\boldsymbol{\Delta}$ from innovation covariance consistency as,

$$E[\mathbf{dd}^T](\mathbf{H}\tilde{\mathbf{B}}\mathbf{H}^T + \tilde{\mathbf{R}})^{-1} = \mathbf{I} + \boldsymbol{\Delta} \tag{25}$$

The trace of Equation (25), which is related to χ^2, is given by

$$E[\chi^2] = tr\{E[\mathbf{dd}^T](\mathbf{H\tilde{B}H}^T + \mathbf{\tilde{R}})^{-1}\} = N_s + tr(\mathbf{\Delta}) \tag{26}$$

We recall that in the experiment **iter 1** we got χ^2/N_s values of 1.36 for O_3 and 1.25 for $PM_{2.5}$ (see Table 1), indicating that the innovation covariance consistency is deficient, although less serious than with the experiment **iter 0** where values of 2 and higher have been obtained.

If we take into account the fact that there can be a difference between $E[\mathbf{dd}^T]$ and $(\mathbf{H\tilde{B}H}^T + \mathbf{\tilde{R}})$ and we rederive the (active) analysis error covariance for *HL*, *MDJ* and *D* schemes, we get (see Appendix B)

$$tr\{\mathbf{H\hat{A}}_{HL}\mathbf{H}^T\} = tr\{\mathbf{H\hat{A}}^{true}\mathbf{H}^T\} + tr\{\mathbf{\tilde{R}\Delta}\} - tr\{\mathbf{H\tilde{B}H}^T\mathbf{\Delta}\} + tr\{\mathbf{error}_{MDJ}\} \tag{27}$$

$$tr\{\mathbf{H\hat{A}}_{MDJ}\mathbf{H}^T\} = tr\{\mathbf{H\hat{A}}^{true}\mathbf{H}^T\} - tr\{\mathbf{error}_{MDJ}\} \tag{28}$$

$$tr\{\mathbf{H\hat{A}}_D\mathbf{H}^T\} = tr\{\mathbf{H\hat{A}}^{true}\mathbf{H}^T\} - tr\{\mathbf{error}_D\} \tag{29}$$

where $tr\{\mathbf{error}_{MDJ}\}$ $=$ $tr\{(\mathbf{H\tilde{B}H}^T)(\mathbf{H\tilde{B}H}^T + \mathbf{\tilde{R}})^{-1}\mathbf{\Delta}(\mathbf{H\tilde{B}H}^T)\}$, $tr\{\mathbf{error}_D\}$ $=$ $tr\{\mathbf{\tilde{R}}(\mathbf{H\tilde{B}H}^T + \mathbf{\tilde{R}})^{-1}\mathbf{\Delta}(\mathbf{H\tilde{B}H}^T)\}$. We note that although the error terms are complex expressions, they all depend linearly on $\mathbf{\Delta}$. Thus, the disagreement between the *HL*, *MDJ* and *D* analysis error variance estimates is due to lack of innovation covariance consistency.

5. Conclusions

We showed that analysis error variance can be estimated and optimized, without using a model forecast, by partitioning the original observation data set into a training set, to create the analysis, and an independent (or passive) set, used to evaluate the analysis. This kind of evaluation by partitioning is called cross-validation. The method derives from assuming that the observations have spatially uncorrelated errors or, minimally, that the independent (or passive) observations have uncorrelated errors with the active observation, and are uncorrelated the background error. This leads to the important property that passive observations are uncorrelated with the analysis error and can then be used to evaluate the analysis [17].

We have developed a theoretical framework and a geometric interpretation that has allowed us to derive a number of statistical estimation formulas of analysis error covariance that can be used in both passive and active observation spaces. It is shown that by minimizing the variance of observation-minus-analysis residuals in passive observation space we actually identify the optimal analysis. This has been done with respect to a single parameter, namely the ratio of observation to background error variances, to obtain a near optimal Kalman gain. The optimization is also done under the constraint of the innovation covariance consistency [14,15]. This optimization could have been done with more than one error covariance parameter but this has not been attempted here. The theory does suggest, however, that the minimum is unique.

Once an optimal analysis is identified we conduct an evaluation of the analysis error covariance using several different formulas; Desroziers et al. [14], Hollingsworth Lönnberg [13], and one that we develop in this paper which works in either active or passive observation spaces. As a way to validate the analysis error variance computed by the analysis scheme itself, the so-called perceived analysis error variance [6], we compare it with the values obtained from the different statistical diagnostics of analysis error variance.

This methodology arises from a need to assess and improve ECCC's surface air quality analyses using our operational air quality model GEM-MACH and real-time surface observations of O_3 and $PM_{2.5}$. Our method applied the theory in a simplified way. First by considering the averaged observation and background error variances and finding an optimal ratio $\gamma = \sigma_o^2/\sigma_b^2$ using as a constraint the trace of the innovation covariance consistency [15]. Second, using a single parameter correlation model, its correlation length, we used the maximum likelihood estimation [11] to obtain near optimal analyses. Also we did not attempt to account for representativeness error in the observations by, for example, filtering observations that are close. Despite all these limitations, our results show that with near optimal analyses, all estimates of analysis error variance roughly agree with each other, while disagreeing strongly when the input error statistics are not optimal. This check on estimating the analysis error variance gives us confidence that the method we propose is reliable, and provides us an objective method to evaluate different analysis components configurations, such as the type of background error correlation model, the spatial distribution of error variances and possibly the use of thinning observations to circumvent effects of representativeness errors.

The methodology introduced here for estimating analysis error variances is general and not restricted to the case of surface pollutant analysis. It would be desirable to investigate other areas of applications, such as surface analysis in meteorology and oceanography. The method could, in principle, provide guidance for any assimilation system. By considering the observation space subdomain [25], proper scaling, local averaging [26], or other methods discussed in Janjic et al. [12] it may also be possible to extend this methodology to spatially varying error statistics. Based on our verification results in Part I [5], we found that there is a dependence between model values and error variances, which we will investigate further in view of our next operational implementation of the Canadian surface air quality analysis and assimilation.

One strong limitation of the optimum interpolation scheme we are using (i.e., homogeneous isotropic error correlation and uniform error variances), which is also the case for most 3D-Var implementations, is the lack of innovation covariance consistency. Ensemble Kalman filters seem, however, much better in that regard although they have their own issues with localization and inflation. Experiments with chemical data assimilation using an ensemble Kalman filter does gives χ^2/N_s values very close to unity after simple adjustments for observation and model error variances [27]. We thus argue that ensemble methods, such as the ensemble Kalman filter, would produce analysis error variance estimates that are much more consistent between the different diagnostics.

Estimates of analysis uncertainties can also be obtained by resampling techniques, such as the jackknife method and bootstrapping [28]. In bootstrapping with replacement, the distribution of the analysis error is obtained by creating new analyses by replacing and duplicating observations from an existing set of observations [28]. This technique relies on the assumption that each member of the dataset is independent and identically distributed. For surface ozone analyses where there is persistence to next day and the statistics is spatially inhomogeneous, the assumption of statistical independence may not be adequate. The comparison of these resampling estimates of analysis uncertainties could be compared with our analysis error variance estimates to help us identify limitations and areas of improvement.

Supplementary Materials: The following are available online at http://www.mdpi.com/2073-4433/9/2/70/s1, Figure S1: Analysis error variance for ozone optimal analysis case PM$_{2.5}$ **iter 1**. Left panel is the analysis error on the model grid and on the right panel at the active observation sites. Note that the color bar of the left and right panels are different. The maximum of the color bar for the left panel correspond to $\sigma_o^2 + \sigma_b^2$, Figure S2: Distribution (histogram) of the ozone analysis error variance at the active observation locations. First analysis experiment PM$_{2.5}$ **iter 0** (no optimization) on the left panel, and optimal analysis case PM$_{2.5}$ **iter 1** on the right panel.

Acknowledgments: We are grateful to the US/EPA for the use of the AIRNow database for surface pollutants and to all provincial governments and territories of Canada for kindly transmitting their data to the Canadian Meteorological Centre to produce the surface analysis of atmospheric pollutants. We are also thankful for the proof read by Kerill Semeniuk, and for two anonymous reviewers for their comments and help in improving the manuscript.

Author Contributions: This research was conducted as a joint effort by both authors. R.M. contributed to the theoretical development and wrote the paper, and M.D.-J. conducted all experiments design and execution, proof reading and introduced a new diagnostic of optimal analysis error that was further extended to passive observations space.

Conflicts of Interest: The authors declare no conflict of interest. The founding sponsors, which is the government of Canada, had no role in the design of the study; in the collection, analyses, or interpretation of data; in the writing of the manuscript, and in the decision to publish the results.

Appendix A. A Geometrical Derivation of the Desroziers et al. Diagnostic

Let u and v be two random variables of a real Hilbert space as defined in Section 2.3. Two properties of Hilbert spaces are: the polarization identity

$$\langle u, v \rangle = \frac{1}{4}\left\{ \|u + v\|^2 - \|u - v\|^2 \right\} \tag{A1}$$

and the parallelogram identity [19]

$$\|u + v\|^2 + \|u - v\|^2 = 2\left(\|u\|^2 + \|v\|^2 \right) \tag{A2}$$

Combining these two equations, we get:

$$\langle u, v \rangle = \frac{1}{2}\left\{ \|u + v\|^2 - \|u\|^2 - \|v\|^2 \right\} \tag{A3}$$

Let $u = O - A$ and $v = A - B$, then $u + v = O - B$, and so we have

$$\langle (O - A), (A - B) \rangle = \frac{1}{2}\left\{ \|O - B\|^2 - \|O - A\|^2 - \|A - B\|^2 \right\} \tag{A4}$$

If we assume that the analysis is optimal, so that $\angle OAT$ is a right triangle (see Figure 1), then

$$\|\hat{A} - T\|^2 + \|O - \hat{A}\|^2 = \|O - T\|^2 \tag{A5}$$

and similarly that $\angle TAB$ is a right triangle,

$$\|\hat{A} - T\|^2 + \|\hat{A} - B\|^2 = \|B - T\|^2 \tag{A6}$$

and substitute these expression into Equation (A4) we get

$$\langle (O - \hat{A}), (\hat{A} - B) \rangle = \frac{1}{2}\left\{ \|O - B\|^2 - \|O - T\|^2 - \|B - T\|^2 - 2\|\hat{A} - T\|^2 \right\} \tag{A7}$$

If in addition we assume that we have uncorrelated observation-background errors, that is

$$\|O - B\|^2 = \|O - T\|^2 + \|B - T\|^2 \tag{A8}$$

we then get

$$\langle (O - \hat{A}), (\hat{A} - B) \rangle = \|\hat{A} - T\|^2 \tag{A9}$$

Note that the difference with the result obtained in Appendix A (property 6) of Ménard [15] where it is shown that the necessary and sufficient condition for the perceived analysis error covariance in active observation space to meet the Desroziers et al. diagnostics for analysis error, is to have innovation covariance consistency and uncorrelated observation-background errors. The result obtained here is different; it concerns the optimal analysis error covariance.

Appendix B. Diagnostics of Analysis error Covariance and the Innovation Covariance Consistency

Let us introduce a departure matrix Δ from innovation covariance consistency as,

$$E[dd^T](H\tilde{B}H^T + \tilde{R})^{-1} = I + \Delta \tag{A10}$$

The *HL* diagnostic, Equation (18), can be expanded as

$$
\begin{aligned}
E[(O-\hat{A})(O-\hat{A})^T] &= E[\mathbf{dd}^T] + \mathbf{H}\tilde{\mathbf{B}}\mathbf{H}^T(\mathbf{H}\tilde{\mathbf{B}}\mathbf{H}^T + \tilde{\mathbf{R}})^{-1}E[\mathbf{dd}^T](\mathbf{H}\tilde{\mathbf{B}}\mathbf{H}^T + \tilde{\mathbf{R}})^{-1} \\
&\quad -E[\mathbf{dd}^T](\mathbf{H}\tilde{\mathbf{B}}\mathbf{H}^T + \tilde{\mathbf{R}})^{-1}\mathbf{H}\tilde{\mathbf{B}}\mathbf{H}^T - \mathbf{H}\tilde{\mathbf{B}}\mathbf{H}^T(\mathbf{H}\tilde{\mathbf{B}}\mathbf{H}^T + \tilde{\mathbf{R}})^{-1}E[\mathbf{dd}^T] \\
&= \tilde{\mathbf{R}} - \mathbf{H}(\tilde{\mathbf{B}} - \tilde{\mathbf{B}}\mathbf{H}(\mathbf{H}\tilde{\mathbf{B}}\mathbf{H}^T + \tilde{\mathbf{R}})^{-1}\mathbf{H}\tilde{\mathbf{B}})\mathbf{H}^T + \mathbf{error}_{HL} \\
&= \tilde{\mathbf{R}} - \mathbf{H}\hat{\mathbf{A}}\mathbf{H}^T + \mathbf{error}_{HL}
\end{aligned}
\tag{A11}
$$

where the $\mathbf{error}_{HL}$ is given as

$$
\mathbf{error}_{HL} = \Delta\tilde{\mathbf{R}} - (\mathbf{H}\tilde{\mathbf{B}}\mathbf{H}^T)\Delta^T + \mathbf{H}\tilde{\mathbf{B}}\mathbf{H}^T(\mathbf{H}\tilde{\mathbf{B}}\mathbf{H} + \tilde{\mathbf{R}})^{-1}\Delta(\mathbf{H}\tilde{\mathbf{B}}\mathbf{H}^T)
\tag{A12}
$$

which includes three error terms. A scalar version of Equation (A11) is given as

$$
error_{HL} = \frac{\sigma_b^2 \Delta}{1+\gamma} - \sigma_b^2 \Delta + \sigma_o^2 \Delta
\tag{A13}
$$

The error analysis for the *MDJ* diagnostic, Equation (19) is

$$
\begin{aligned}
E[(\hat{A}-B)(\hat{A}-B)] &= \mathbf{H}\tilde{\mathbf{K}}E[\mathbf{dd}^T]\tilde{\mathbf{K}}\mathbf{H}^T \\
&= \mathbf{H}\tilde{\mathbf{B}}\mathbf{H}^T(\mathbf{H}\tilde{\mathbf{B}}\mathbf{H}^T + \tilde{\mathbf{R}})^{-1}\mathbf{H}\tilde{\mathbf{B}}\mathbf{H}^T + \mathbf{error}_{MDJ}
\end{aligned}
\tag{A14}
$$

where the $\mathbf{error}_{MDJ}$ is given as

$$
\mathbf{error}_{MDJ} = \mathbf{H}\tilde{\mathbf{B}}\mathbf{H}^T(\mathbf{H}\tilde{\mathbf{B}}\mathbf{H}^T + \tilde{\mathbf{R}})^{-1}\Delta(\mathbf{H}\tilde{\mathbf{B}}\mathbf{H}^T)
\tag{A15}
$$

There is only one term, and its scalar version is given as,

$$
error_{MDJ} = \frac{\sigma_b^2 \Delta}{1+\gamma}
\tag{A16}
$$

The error analysis for the Desroziers et al. diagnostic, Equation (20) is

$$
\begin{aligned}
E[(O-\hat{A})(\hat{A}-B)] &= E[\mathbf{dd}^T](\mathbf{H}\tilde{\mathbf{B}}\mathbf{H}^T + \tilde{\mathbf{R}})^{-1}\mathbf{H}\tilde{\mathbf{B}}\mathbf{H}^T - \mathbf{H}\tilde{\mathbf{B}}\mathbf{H}^T(\mathbf{H}\tilde{\mathbf{B}}\mathbf{H}^T + \tilde{\mathbf{R}})E[\mathbf{dd}^T](\mathbf{H}\tilde{\mathbf{B}}\mathbf{H}^T + \tilde{\mathbf{R}})^{-1}\mathbf{H}\tilde{\mathbf{B}}\mathbf{H}^T \\
&= \mathbf{H}\tilde{\mathbf{B}}\mathbf{H}^T - \mathbf{H}\tilde{\mathbf{B}}\mathbf{H}^T(\mathbf{H}\tilde{\mathbf{B}}\mathbf{H}^T + \tilde{\mathbf{R}})^{-1}\mathbf{H}\tilde{\mathbf{B}}\mathbf{H}^T + \mathbf{error}_D
\end{aligned}
\tag{A17}
$$

where the $\mathbf{error}_D$ is given as

$$
\mathbf{error}_D = \{\mathbf{I} - \mathbf{H}\tilde{\mathbf{B}}\mathbf{H}^T(\mathbf{H}\tilde{\mathbf{B}}\mathbf{H}^T + \tilde{\mathbf{R}})^{-1}\}\Delta(\mathbf{H}\tilde{\mathbf{B}}\mathbf{H}^T) = \tilde{\mathbf{R}}(\mathbf{H}\tilde{\mathbf{B}}\mathbf{H}^T + \tilde{\mathbf{R}})^{-1}\Delta(\mathbf{H}\tilde{\mathbf{B}}\mathbf{H}^T)
\tag{A18}
$$

Again only one error term that is similar to the *MDJ* diagnostic, and its scalar version is given as,

$$
error_D = \frac{\sigma_o^2 \Delta}{1+\gamma}
\tag{A19}
$$

Error analysis for the diagnostics using passive observations can also be derived. For the passive *MDJ* diagnostics, Equation (21), we have similarly to (A14),

$$
\begin{aligned}
E[(\hat{A}-B)_c(\hat{A}-B)_c] &= \mathbf{H}_c\tilde{\mathbf{K}}E[\mathbf{dd}^T]\tilde{\mathbf{K}}\mathbf{H}_c^T \\
&= \mathbf{H}_c\tilde{\mathbf{B}}\mathbf{H}^T(\mathbf{H}\tilde{\mathbf{B}}\mathbf{H}^T + \tilde{\mathbf{R}})^{-1}\mathbf{H}\tilde{\mathbf{B}}\mathbf{H}_c^T + \mathbf{error}_{MDJ\_passive}
\end{aligned}
\tag{A20}
$$

with a single error term given as,

$$
\mathbf{error}_{MDJ\_passive} = \mathbf{H}_c\tilde{\mathbf{B}}\mathbf{H}^T(\mathbf{H}\tilde{\mathbf{B}}\mathbf{H}^T + \tilde{\mathbf{R}})^{-1}\Delta(\mathbf{H}\tilde{\mathbf{B}}\mathbf{H}_c^T)
\tag{A21}
$$

To express a scalar version of this equation we need to account for the background error correlation ρ between the active observation location and the passive observation location, and thus expressed as,

$$error_{MDJ\_passive} = \frac{\rho^2 \sigma_b^2 \Delta}{1 + \gamma} \tag{A22}$$

Finally, the fundamental diagnostic of cross-validation Equation (7) does not depend explicitly on the innovation covariance consistency. However, attaining its true minimum by tuning only γ and L_c as would suggest Equation (22), does introduce some innovation in-consistency, which all other optimal diagnostics Equations (18)–(21) has to account for.

References

1. Ménard, R.; Robichaud, A. The chemistry-forecast system at the Meteorological Service of Canada. In Proceedings of the ECMWF Seminar Proceedings on Global Earth-System Monitoring, Reading, UK, 5–9 September 2005; pp. 297–308.
2. Robichaud, A.; Ménard, R. Multi-year objective analysis of warm season ground-level ozone and PM$_{2.5}$ over North-America using real-time observations and Canadian operational air quality models. *Atmos. Chem. Phys.* **2014**, *14*, 1769–1800. [CrossRef]
3. Robichaud, A.; Ménard, R.; Zaïtseva, Y.; Anselmo, D. Multi-pollutant surface objective analyses and mapping of air quality health index over North America. *Air Qual. Atmos. Health* **2016**, *9*, 743–759. [CrossRef] [PubMed]
4. Moran, M.D.; Ménard, S.; Pavlovic, R.; Anselmo, D.; Antonopoulus, S.; Robichaud, A.; Gravel, S.; Makar, P.A.; Gong, W.; Stroud, C.; et al. Recent advances in Canada's national operational air quality forecasting system. In Proceedings of the 32nd NATO-SPS ITM, Utrecht, The Netherlands, 7–11 May 2012.
5. Ménard, R.; Deshaies-Jacques, M. Evaluation of analysis by cross-validation. Part I: Using verification metrics. *Atmosphere* **2018**, in press.
6. Daley, R. The lagged-innovation covariance: A performance diagnostic for atmospheric data assimilation. *Mon. Weather Rev.* **1992**, *120*, 178–196. [CrossRef]
7. Daley, R. *Atmospheric Data Analysis*; Cambridge University Press: New York, NY, USA, 1991; p. 457.
8. Talagrand, O. A posteriori evaluation and verification of analysis and assimilation algorithms. In *Proceedings of Workshop on Diagnosis of Data Assimilation Systems, November 1998*; European Centre for Medium-Range Weather Forecasts: Reading, UK, 1999; pp. 17–28.
9. Todling, R. Notes and Correspondence: A complementary note to "A lag-1 smoother approach to system-error estimaton": The intrinsic limitations of residuals diagnostics. *Q. J. R. Meteorol. Soc.* **2015**, *141*, 2917–2922. [CrossRef]
10. Hollingsworth, A.; Lönnberg, P. The statistical structure of short-range forecast errors as determined from radiosonde data. Part I: The wind field. *Tellus* **1986**, *38A*, 111–136. [CrossRef]
11. Ménard, R.; Deshaies-Jacques, M.; Gasset, N. A comparison of correlation-length estimation methods for the objective analysis of surface pollutants at Environment and Climate Change Canada. *J. Air Waste Manag. Assoc.* **2016**, *66*, 874–895. [CrossRef] [PubMed]
12. Janjic, T.; Bormann, N.; Bocquet, M.; Carton, J.A.; Cohn, S.E.; Dance, S.L.; Losa, S.N.; Nichols, N.K.; Potthast, R.; Waller, J.A.; et al. On the representation error in data assimilation. *Q. J. R. Meteorol. Soc.* **2017**. [CrossRef]
13. Hollingsworth, A.; Lönnberg, P. The verification of objective analyses: Diagnostics of analysis system performance. *Meteorol. Atmos. Phys.* **1989**, *40*, 3–27. [CrossRef]
14. Desroziers, G.; Berre, L.; Chapnik, B.; Poli, P. Diagnosis of observation-, background-, and analysis-error statistics in observation space. *Q. J. R. Meteorol. Soc.* **2005**, *131*, 3385–3396. [CrossRef]
15. Ménard, R. Error covariance estimation methods based on analysis residuals: Theoretical foundation and convergence properties derived from simplified observation networks. *Q. J. R. Meteorol. Soc.* **2016**, *142*, 257–273. [CrossRef]
16. Kailath, T. An innovation approach to least-squares estimation. Part I: Linear filtering in additive white noise. *IEEE Trans. Autom. Control* **1968**, *13*, 646–655.

17. Marseille, G.-J.; Barkmeijer, J.; de Haan, S.; Verkley, W. Assessment and tuning of data assimilation systems using passive observations. *Q. J. R. Meteorol. Soc.* **2016**, *142*, 3001–3014. [CrossRef]

18. Waller, J.A.; Dance, S.L.; Nichols, N.K. Theoretical insight into diagnosing observation error correlations using observation-minus-background and observation-minus-analysis statistics. *Q. J. R. Meteorol. Soc.* **2016**, *142*, 418–431. [CrossRef]

19. Caines, P.E. *Linear Stochastic Systems*; John Wiley and Sons: New York, NY, USA, 1988; p. 874.

20. Cohn, S.E. The principle of energetic consistency in data assimilation. In *Data Assimilation*; Lahoz, W., Boris, K., Richard, M., Eds.; Springer: Berlin/Heidelberg, Germany, 2010.

21. Mitchell, H.L.; Daley, R. Discretization error and signal/error correlation in atmospheric data assimilation: (I). All scales resolved. *Tellus* **1997**, *49A*, 32–53. [CrossRef]

22. Mitchell, H.L.; Daley, R. Discretization error and signal/error correlation in atmospheric data assimilation: (II). The effect of unresolved scales. *Tellus* **1997**, *49A*, 54–73. [CrossRef]

23. Joiner, J.; da Silva, A. Efficient methods to assimilate remotely sensed data based on information content. *Q. J. R. Meteorol. Soc.* **1998**, *124*, 1669–1694. [CrossRef]

24. Migliorini, S. On the quivalence between radiance and retrieval assimilation. *Mon. Weather Rev.* **2012**, *140*, 258–265. [CrossRef]

25. Chapnik, B.; Desroziers, G.; Rabier, F.; Talagrand, O. Properties and first application of an error-statistics tunning method in variational assimilation. *Q. J. R. Meteorol. Soc.* **2005**, *130*, 2253–2275. [CrossRef]

26. Ménard, R.; Deshiaes-Jacques, M. Error covariance estimation methods based on analysis residuals and its application to air quality surface observation networks. In *Air Pollution and Its Application XXV*; Mensink, C., Kallos, G., Eds.; Springer International AG: Cham, Switzerland, 2017.

27. Skachko, S.; Errera, Q.; Ménard, R.; Christophe, Y.; Chabrillat, S. Comparison of the ensemlbe Kalman filter and 4D-Var assimilation methods using a stratospheric tracer transport model. *Geosci. Model Dev.* **2014**, *7*, 1451–1465. [CrossRef]

28. Efron, B. *An Introduction to Boostrap*; Chapman & Hall: New York, NY, USA, 1993; p. 436.

atmosphere

MDPI

Article

Overview of the Model and Observation Evaluation Toolkit (MONET) Version 1.0 for Evaluating Atmospheric Transport Models

Barry Baker [1,2,*] and Li Pan [1,2]

1   NOAA Air Resources Laboratory, College Park, MD 20740, USA; li.pan@noaa.gov
2   Cooperative Institute for Climate and Satellites, University of Maryland at College Park,
    College Park, MD 20740, USA
*   Correspondence: Barry.Baker@noaa.gov; Tel.: +1-301-683-1373

Received: 16 June 2017; Accepted: 21 October 2017; Published: 31 October 2017

Abstract: This paper describes the development and initial applications of the Model and Observation Evaluation Tool (MONET) v1.0. MONET was developed to evaluate the Community Multiscale Air Quality Model (CMAQ) for the NOAA National Air Quality Forecast Capability (NAQFC) modeling system. MONET is designed to be a modularized Python package for (1) pairing model output to observational data in space and time; (2) leveraging the pandas Python package for easy searching and grouping; and (3) analyzing and visualizing data. This process introduces a convenient method for evaluating model output. MONET processes data that is easily searchable and that can be grouped using meta-data found within the observational datasets. Common statistical metrics (e.g., bias, correlation, and skill scores), plotting routines such as scatter plots, timeseries, spatial plots, and more are included in the package. MONET is well modularized and can add further observational datasets and different models.

Keywords: CMAQ; evaluation; air quality; software; visualization; statistics

1. Introduction

Ozone (O_3) and particulate matter smaller than 2.5 μm in diameter ($PM_{2.5}$) are among a handful of criteria air pollutants—pollutants the Clean Air Act requires to be monitored and regulated—that are primarily responsible for adverse impacts on human health [1]. Breathing these pollutants is recognized as major causes of acute and chronic respiratory and cardiovascular diseases, and premature mortalities associated with air pollution [2].

In relation to health hazards caused by air pollutants, air quality also has a direct impact on the economy. Trasande et al. [3] found the cost of medical care for preterm births attributable to $PM_{2.5}$ exposure to be between $2.43 and $9.66 billion. Ghude et al. [4] demonstrates the economic cost of poor air quality in India to crop yields is estimated to be around $1.26 billion annually. Tong et al. [5] reveals a similar influence to crop yields in the United States (U.S.).

Evaluating model simulations is critical for the development and implementation for forecasting [6]. Evaluation of meteorological and air quality parameters are essential in validating and improving model simulations.

Recently, air quality simulations increased from running for days or weeks to months or years, modeling domains increased resolution, and the use of ensembles greatly increased the amount of model output analyzed. The Community Multiscale Air Quality (CMAQ) modeling simulations, used for air quality and air composition modeling, currently run spanning modeling domains of regional to hemispheric scale and modeling times of days to years resulting in terabytes of model output available for analysis. Conventional methods of data analysis, such as spreadsheets, are not suited for a task this large. There are several model evaluation tools available for evaluating meteorological model simulations but few for evaluating air quality models [7]. Although there are many software packages, such as AMET, or visualization tools, such as Verdi, pycmbs, or Panoply, MONET is built using a single open-source language that retrieves, downloads, and analyses both model and observations on a regional scale. MONET is available on all major computing platforms, i.e., Mac, Linux, and PC, and provides the power of a dedicated programming language such as IDL, MATLAB, or ferret if needed.

The National Oceanic and Atmospheric Administration (NOAA) Air Resources Laboratory (ARL) developed the Model and Observation Evaluation Tool (MONET) to aid in the assessment of the National Air Quality Forecasting Capability (NAQFC) [8,9]. MONET reads, interpolates, and organizes model results to observation sites in both space and time resulting in a fast and flexible method to evaluate air quality modeling simulations. Although MONET was originally created to only evaluate CMAQ simulations, it can easily be expanded to include different model outputs (e.g., the Weather Research and Forecasting model, Next Generation Global Prediction System, and Comprehensive Air Quality Model with Extensions), along with adding additional observational sources (e.g., different ground based networks, satellite observations, and other in-situ observations).

This paper describes the structure and functionality of the MONET Python package (version 2.7). A broad description of MONET will be provided, followed by detailed description of how MONET works. Examples of the analysis products available from MONET will be presented and described. Finally, a discussion and future directions of MONET will be provided.

2. Tool Description

MONET pairs observations and gridded model prediction in space and time. This evaluates the model's performance for a set of predicted or diagnosed fields such as aggregating nitrous oxide (NO) and nitrogen dioxide (NO_2) into nitrogen oxides (NO_x). MONET is built in Python v2.7+ and is intended to follow object oriented concepts. Each model and observation has a specific object intended to be used for its specific cases. Stemming from such object structures, a verification object that inherits observation and model objects can be created allowing for targeted verification between sets of models and observations. MONET runs interactively or with a simple adaptable Python script.

Most of the dependencies can be obtained using commonly available Python packages, such as the Anaconda Python Package or the Enthought Canopy distribution. MONET uses the pandas Python package [10,11] to enable fast and efficient data manipulation using meta-data available in the observational datasets. Pairing of model and observations is done using the pyresample package, (available online: https://pyresample.readthedocs.io/en/latest/), which uses pykdTree to interpolate. Several interpolation methods are available; nearest neighbor, Gaussian, elliptical weighted averaging, or a user defined method, such as inverse distance weighting. Analysis is done using scipy, numpy, and scikit-learn Python functions, while plotting is achieved with Basemap (for spatial plotting), matplotlib, and seaborn [12,13]. Figure 1 presents a flowchart of the MONET software.

MONET runs in four phases; (1) creation of a model object that determines the model grid, variable names, and run duration; (2) creation of an observational object that parses observational data; (3) combines and interpolates model results to observations; and (4) plotting and statistical comparisons.

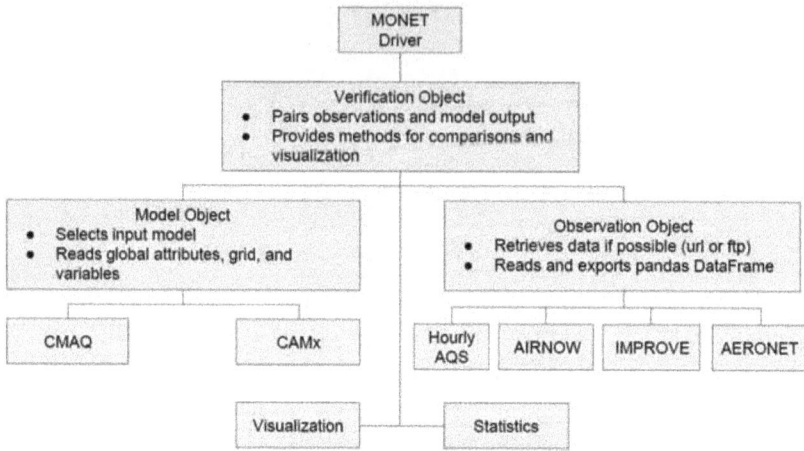

Figure 1. Flow chart of Model and ObservatioN Evaluation Tool (MONET).

2.1. Creation of Model and Observation Objects

MONET is designed to be modular in that each observational network or model used has its own set of specialized functions to handle the differences in each dataset. Currently, MONET can handle the hourly Environmental Protection Agency AirData (EPA AQS; available online: https://aqsdr1.epa.gov/aqsweb/aqstmp/airdata/download_files.html#Raw), including the Chemical Speciation Network (CSN) and AirNow (available online: https://www.airnow.gov) datasets, along with the Interagency Monitoring of Protected Visual Environments (IMPROVE; available online: http://vista.cira.colstate.edu/improve) and Aerosol Robotic Network (AERONET; available online: http://aeronet.gsfc.nasa.gov) datasets. Datasets can be readily downloaded through an FTP or http by using an array of datetime objects given to the observation object. Other networks, such as IMPROVE, require that the data be downloaded manually.

Each observation network is designed to monitor specific sets of air quality parameters, from individual pollutants to derived physical parameters (e.g., AOD). MONET makes it easy to compare individual pollutants or to aggregate model results to pair with the monitor dataset. Examples of this are NO_x (NO + NO_2) in AQS or AIRNOW or particle sulfate, an aggregate of the Aitken, accumulation, and coarse mode contribution to the sulfate particle concentration within CMAQ, from AQS and IMPROVE. MONET includes a set of standard aggregations for different pollutants but is versatile enough to allow user defined aggregations.

The observation objects require that a time range is given as an array of datetime objects to the object instance attribute. This is used to tailor data extraction from a pre-downloaded file covering the prescribed time range or otherwise expand the data retrieval. Then, observation objects read the raw datafile and process it into a pandas DataFrame, thus assigning it to an object instance with common columns of observational data (Obs), date (datetime), local date (datetime_local), latitude (Latitude), and longitude (Longitude). Depending on the network, more meta-data may be available, meaning that an observational object in MONET can be used as a standalone method to analyze observational data without the need for pairing. If adding a new network, only the latitude, longitude, observations, and date or timestamp is needed for the interpolation.

For model objects, the only information needed is the output file(s). In the case of overlapping simulations, the model results with the latest creation time stamp will be used. For instance, if two 48 h simulations with 24 h of overlap are given to MONET, then the first 24 h from the first simulation and the 48 h of the second day are used. MONET assigns the model object file to a class instance attribute and creates corresponding helper functions, such as that defining map projections for spatial plotting and retrieving and aggregating data. In the future, the model object will be done with the xarray Python package to keep a consistent view with observations. The xarray package is an N-Dimensional implementation of the pandas library that is highly suitable for scientific data that allows for larger then memory computations and efficient data extraction and resampling.

2.2. Pairing Observations and Model Results

MONET has a built-in driver for each model and observation pair, meaning that the model and observation are separately imported objects. This creates a simple and efficient method for adding new model results—once a model and observational dataset has its own Python object it can easily follow the same pairing algorithms. The driver creates the model object, reads the time range, and creates the model instance. The model time range is then provided to the observational object, where the data is downloaded and processed into a pandas DataFrame or loaded from a preprocessed file.

At this point, each species available in observations and model prediction are paired. MONET has the ability to calculate or aggregate data, such as the calculation of $PM_{2.5}$ (particulate matter with diameter less than 2.5 µm) mass concentration, from several modeled chemical species. Herein, a helper function retrieves the data in the model object to derive the needed data array. MONET then interpolates the model results for each timestamp that is available in the observational network to the observational site and merges them into a common pandas DataFrame containing pairings of model results and observations for all specified observation locations.

The Pyresample library (available online: https://pyresample.readthedocs.io/en/latest/) interpolates the model to observations. Pyresample uses (k-dimensional) KDTrees to transform data defined in one geometrical grid to another. Resampling uses the nearest neighbor, Gaussian, or a customized method, such as inverse distance weighting. MONET allows the user to define the radius of influence and number of neighbors to be used for data re-gridding.

After re-gridding, depending on the dataset, MONET resamples in time and then assigns these processed data to a separate pandas DataFrame for deriving certain species, such as the 8 h max ozone or the daily $PM_{2.5}$ concentration. Timing for opening model results, processing of observation data, and pairing takes less than 3 min with AIRNOW for a 48 h simulation. At this point, MONET proceeds to analyze the dataset with the observation and model result pairs.

3. Example of Tool Applications

MONET is a new software package developed for verifying the NAQFC and more generally the CMAQ model. It has been newly developed as described above and is available on GitHub. Detailed examples and installation instructions are available at https://github.com/noaa-oar-arl/MONET. Once MONET makes the verification object, all of the non-spatial plots, such as timeseries, scatter plots, and histograms, are executed through as a single line-command with flags specifying the user's selection of geographical extent and variables to be plotted. For instance, a user can specify that the geographical extent of their plot be within the U.S. EPA defined Conterminous United States (CONUS) region. MONET performs verification within the entire domain, a region, a state, a county, a metropolitan area, or a specific site. MONET also compares multiple simulations on a single figure simply by creating two verification objects and passing the figure handle. Figures 2–6 showcases some of the plotting routines found within MONET.

As an example of how easily MONET does a quick pairing and analysis, the following example shows how to use MONET to pair AirNow data and a CMAQ simulation. Begin by entering an interactive python session.

$$ipython\ -pylab \tag{1}$$

Import the MONET object. This is the python object interface to the model and observation objects.

$$import\ monet \tag{2}$$

Set the concentration file or files, in this case concfiles, and the gridcro2d file, gridcro.

$$concfiles = '/path/to/concentration/files/ *.ncf' \tag{3}$$

$$gridcro = '/path/to/gridcro2dfile.ncf' \tag{4}$$

Use the MONET object to pair the AirNow data and CMAQ simulation by creating an instance of the verified airnow object, which imports both the CMAQ and AIRNOW objects.

$$m = monet.vairnow\ (concobj = concfiles,\ gridobj = gridcro) \tag{5}$$

MONET retrieves the observations if not in the current directory and the model is interpolated in space to the observational data points. The paired data can be accessed through the pandas DataFrame contained within the monet object, m.

$$m.df.head() \tag{6}$$

Then simple commands can be used to create different plots. For example, MONET creates a timeseries plot averaged over all of the observations found within the modeling domain with a simple one line execution as follows:

$$m.compare\_param\ (param = 'OZONE',\ timeseries = True) \tag{7}$$

Figure 2a–d shows a time series of the average concentration over the northeastern United States during the summer of 2016. The average value is shown in solid lines, observations in black, and model results in blue and purple. Shaded areas display one standard deviation from the mean. The footer illustates forecasting performance statistical measures specified by the user. The footer also shows the start date, end date, number of sites, and the number of measurements. Figure 2b presents an example of time series root mean square error plot comparing how two simulations performed for the EPA AQS sites in the U.S. EPA CONUS domain. Likewise, Figure 2c displays a mean bias time series of the two simulations. Figure 2d provides an example of speciated $PM_{2.5}$ sulfate using the IMPROVE network over the Pacific region. With the IMPROVE network daily average measurement every three days, MONET interpolates the hourly CMAQ results to each individual site and then averaged to create a daily concentration as if were an IMPROVE measurement. Then, MONET merges the daily averaged CMAQ results into the IMPROVE DataFrame using a mysql like merge found within Pandas.

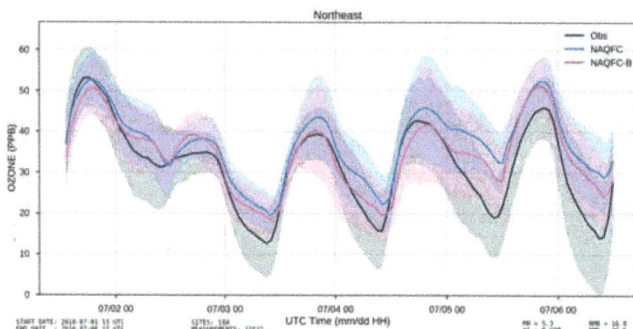

(a)

Figure 2. *Cont.*

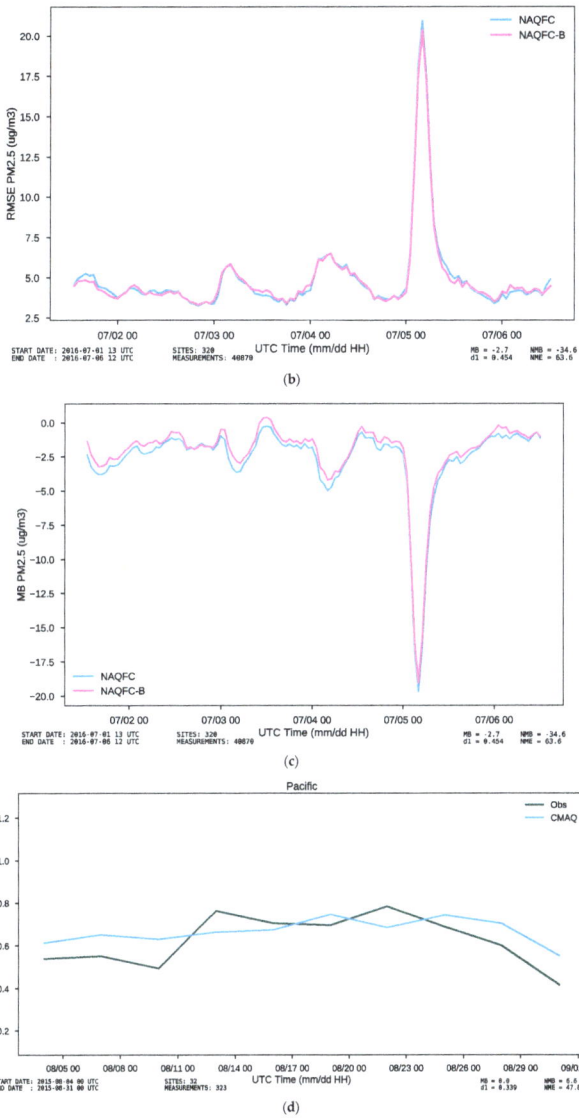

Figure 2. Examples of time-series (**a**) Hourly averaged values are shown in solid lines for EPA AQS observations (black), and two different simulations, National Air Quality Forecast Capability (NAQFC) (blue) and NAQFC-Beta (purple). Shaded regions represent one standard deviation for each hourly averaged time series; (**b**) Hourly averaged values of the root mean square error (RMSE) from EPA AQS observations are shown for two different simulations of PM$_{2.5}$, NAQFC (blue) and NAQFC-Beta (purple); (**c**) Hourly averaged values of the RMSE from EPA AQS observations are shown for two different simulations of PM$_{2.5}$, NAQFC (blue) and NAQFC-Beta (purple); (**d**) Daily averaged values are shown in solid lines for IMPROVE observations (black), and two different simulations, NAQFC (blue) and NAQFC-Beta (purple) in the Pacific region. Shaded regions represent one standard deviation for each daily averaged time series.

Figure 3a presents an example of a kernel density estimation (KDE) comparison. The black line shows an observational KDE and the blue represents the CMAQ model results. This type of plot (Figure 3a) is useful for inspection for distributional discrepancy between the model and observations. Using conditional distributions, such as conditioning on time of day or other meteorological factors like boundary layer height or wind speed, one can provide a more in depth analysis using MONET. Figure 3b displays the difference in the KDE between model and observation results. Difference KDE are important to show model biases in the most probable value.

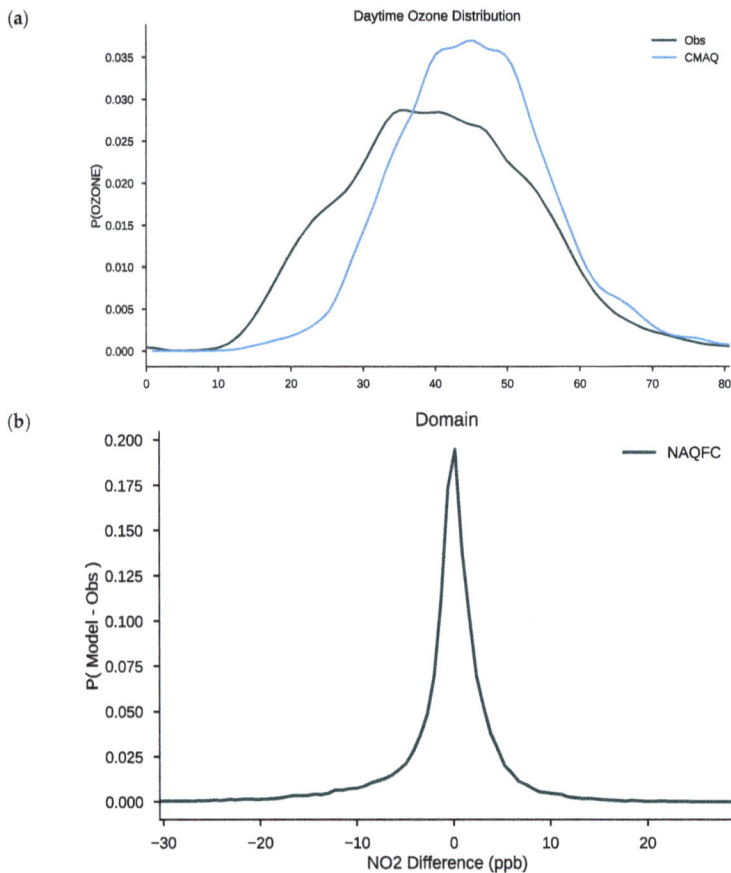

Figure 3. Examples of kernel density estimations (KDEs) (**a**) Daytime kernel density estimations for O_3 concentrations over the continental United States (U.S.) for both observations (black) and a Community Multiscale Air Quality Model (CMAQ) simulation (blue); (**b**) A kernel density estimation of the difference between the simulated NO_2 concentrations and observations.

Figure 4a shows an example of a spatial plot of model results against overlaid observations. The colorbar is customizable by passing colormaps that can be user defined or predefined in matplotlib. The colorbar levels can be tailored individually to any single observation. By default, the colorbar is scaled automatically based on the model data and a discrete colorbar with 15 value bins using the 'viridis' colormap. The 'viridis' colormap is perceptually uniform and sensitive for colorblind individuals. Keyword arguments for the matplotlib.pyplots.imshow can be passed to the MONET function to give additional control over the spatial plot. Spatial plots are useful to give quick glimpses of spatial patterns found within the model. Figure 4b shows an example of a spatial difference plot which is useful for providing spatial assessments for absolute discrepancy. The same type of plot can be created for different metrics such as for RMSE or correlation. In this particular example, it is easy to pertain that over large cities such as Los Angeles or Las Vegas there is a significant under prediction of ozone.

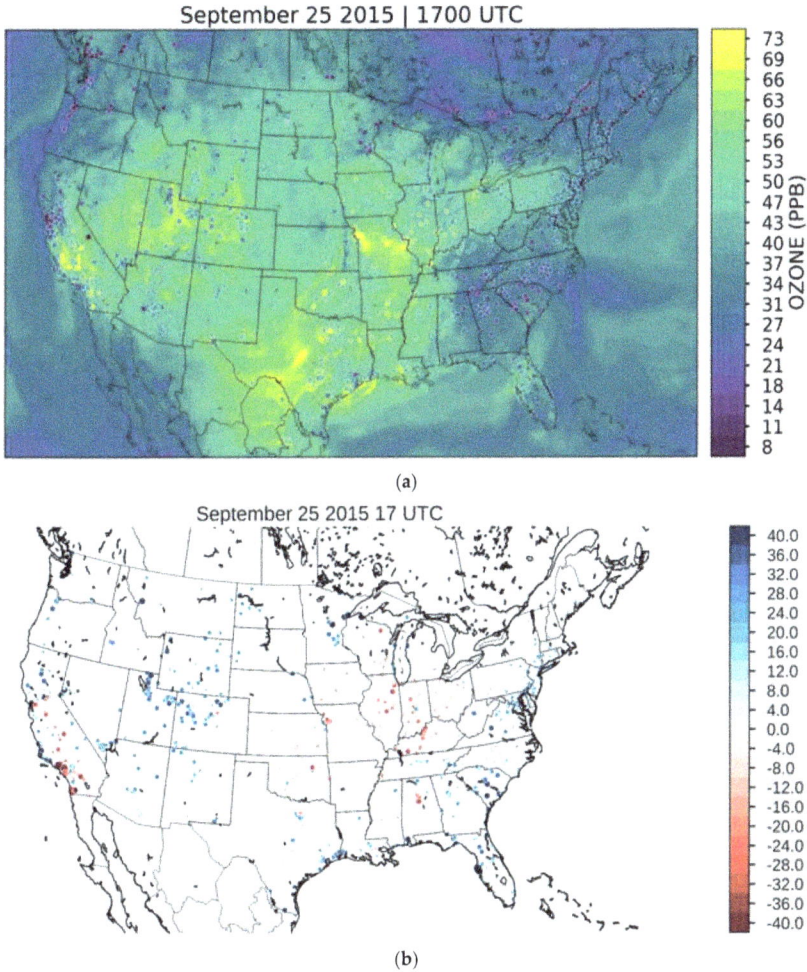

(a)

(b)

Figure 4. Examples of spatial plots. (a) Example of a spatial contour map with a discrete colorbar and observations overlaid; (b) Displayed is the normalized spatial O_3 bias (model–observation) at all of the monitors within the domain. The marker size is dependent on the absolute value of the normalized bias.

Figure 5a presents a scatter plot, while Figure 5b shows a difference scatter plot. Observations are displayed on the x-axis and model results on the y-axis. A one to one line is automatically created along with a line of best fit. The scipy linregress function creates the line of best fit. Currently, there is not a more complicated fitting algorithm available in MONET. The full power of scipy and scikit-learn is available however if a more in-depth analysis is needed and hopefully will be included in a future version of MONET. This type of plot is useful for determining the correlation of simulation to observational results.

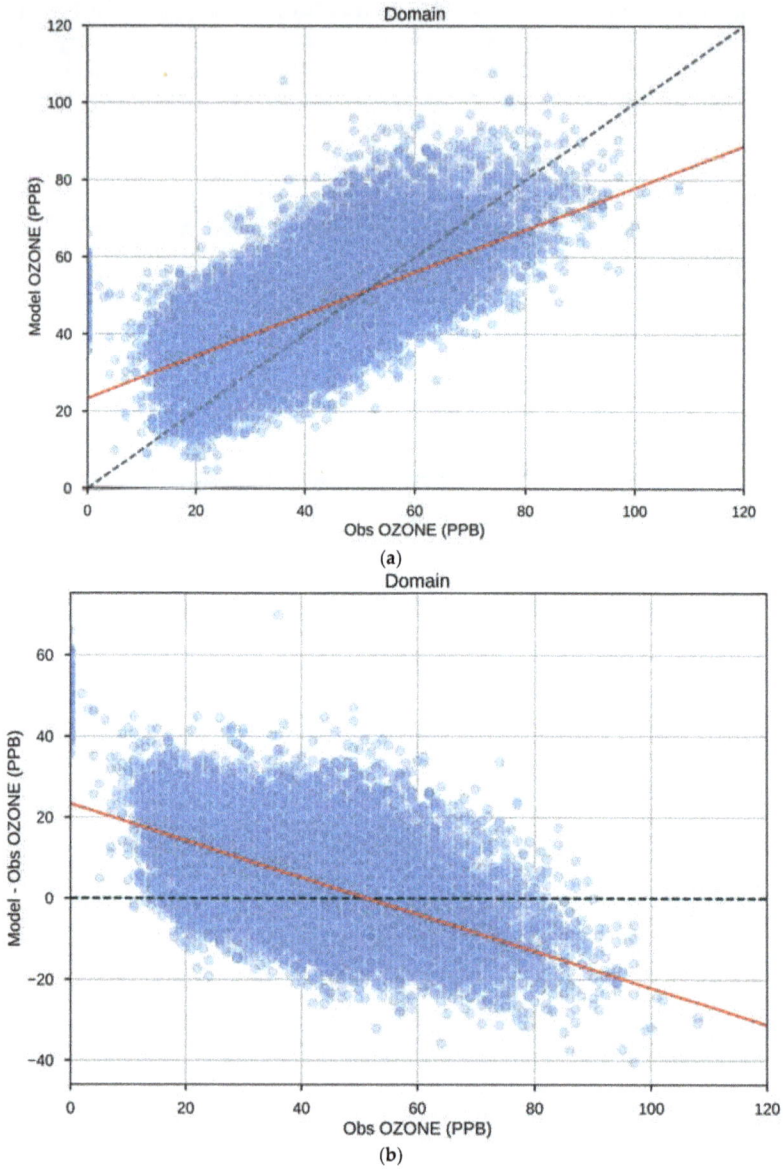

Figure 5. Examples of scatter plots in MONET (**a**) Scatter plot example of model versus ozone; (**b**) Scatter plot example of the difference in model and observation versus observations.

MONET uses the correlation, centered root mean square error, and the standard deviation measures to graphically display on its Taylor diagrams [14]. Taylor diagrams are an efficient way to graphically summarize model performance with observations. Currently, it is up to the user to ensure that data is close to a Gaussian distribution.

The relative merits of various model simulations can be inferred from Figure 6. Simulations that best compare to observations lie closest to the x-axis and have a low RMSE, while simulations that lie above the *x*-axis have a greater RMSE and lower correlation. The concentric arcs around the observation point show RMSE (in this graph at approximately 5.9 on the *x*-axis). Dispersion or variation within the simulations are graphically displayed by the distance from the concentric arcs. Simulations closest to the dotted arc have a standard deviation similar to observations.

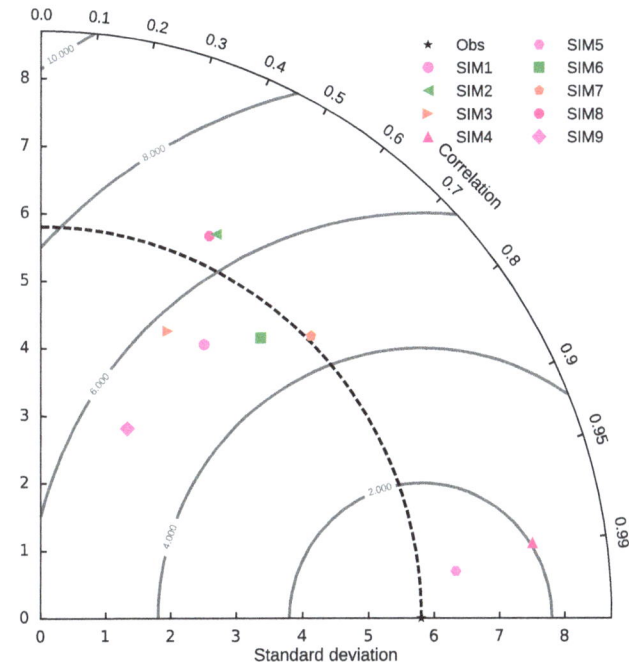

Figure 6. Taylor diagram example available in MONET. Taylor diagrams provide a concise visualization of multiple statistics on a single graph. In this example, the closer the simulation to the *x*-axis the better the correlation. The closer to the dotted line the better the model captured the variability of the model. The RMSE values follow the contours emanating from the *x*-axis.

In this example, SIM5 performed the best as it has the highest RMSE, correlation, and a respectable standard deviation as compared to observations. SIM9 has the lowest standard deviation and SIM2 and SIM8 have the lowest RMSE.

Different visualizations are being developed for future use. Specifically, cumulative distributions, spectral density analysis, and principle component analysis will be added. The addition of different observational data is also a high priority for use with MONET, including data from radiosondes, aircraft, ceilometers, and satellite data. MONET is a simple, easy to use object oriented approach for analyzing model results initially from NAQFC and more generally the CMAQ model.

4. Software and Data Availability

MONET is available on GitHub (https://github.com/noaa-oar-arl/MONET) as a free download. The verification package is written in Python v2.7+. The authors recommend the Anaconda Python package, available for free at https://www.continuum.io/downloads, to install supporting packages described in the readme file included at GitHub. As mentioned previously, MONET uses the Pandas python package to manipulate and store intermediate files. For one day of paired data, the file size is approximately 100 mb for all of the variables and meta-variables in AirNow within the NAQFC domain. MONET uses freely available observational data from the U.S. EPA, U.S. NOAA, U.S. NASA, and U.S. NPS.

5. Conclusions

MONET is a comprehensive software package used to analyze observational data alone or pair observations with gridded model data for chemical transport models. Currently, the observational datasets included are focused within North America, except for AERONET, which is global. MONET performs statistical calculations and creates visualizations to enable researchers to evaluate and improve the scientific understanding of model outputs and observational data. MONET is built entirely on open-source software and while the only model available for use is CMAQ used in the NAQFC, the software is easily able to include different models. To include a new model into MONET, a new object needs to be written that can: (1) read the output and (2) map of the output variables to observation variables needed to pair data; (3) provide an output latitude and longitude for interpolation. Work is ongoing to add the Comprehensive Air Quality Model with Extensions (CAMx) [15] model output to the MONET software package through the use of PseudoNetCDF (https://github.com/barronh/pseudonetcdf).

The MONET software package will continue to grow and improve through internal development at NOAA ARL, and external development through collaborating partners and contributions made through the community. The authors encourage collaborations with intergovernmental agencies, private industry, and academia to improve and further develop MONET. Major software developments are planned for the next release of MONET. One improvement is the inclusion of different models, such as the Weather Research and Forecasting (WRF), CAMx, and the Next Generation Global Prediction System (NGGPS). The next major development planned for MONET is to add different observational sources such as radiosonde, profiler, and radar observations from the Meteorological Assimilation Data Ingest System (MADIS; available online: http://madis.noaa.gov) and AWIPS II (available online: http://www.unidata.ucar.edu/software/awips2/). Another improvement planned is the use of pyresample to re-grid satellite swath data to model grids, as well as calculate comparable variables (e.g., column aggregated aerosol optical depth values from model output).

Analysis enhancements to the MONET software will include implementing non-parametric analysis techniques, such as cumulative distribution frequency plots, and Q-Q plots; adding temporal decomposition capabilities such as Kolmogorov-Zurbenko filtering (Rao and Zurbenko [16]; Wise and Comrie [17]) and spectral density plots; improving the overall speed through the use of parallel processing in pyresample for spatial re-gridding, xarray for temporal averaging, and dask for larger than memory computations. Many improvements to the MONET software are in progress and will be included in the next release.

Supplementary Materials: Initial application of the MONET software is available at https://github.com/noaa-oar-arl/MONET. Please look to the README.md file and the Wiki (https://github.com/noaa-oar-arl/MONET/wiki) for further tutorials. The software can be install with conda package manager: conda install –c bbakernoaa MONET. Included is a sample script along with a fortran style namelist that can be used with the MONET package.

Acknowledgments: The authors would like to thank the reviewers and editors for their dedication and valuable input in improving this manuscript.

Author Contributions: Barry Baker developed the infrastructure and implementation of MONET while Li Pan assisted in testing and evaluation of the NAQFC simulations.

Conflicts of Interest: The authors declare no conflict of interest.

References

1. George, B.J.; Schultz, B.D.; Palma, T.; Vette, A.F.; Whitaker, D.A.; Williams, R.W. An evaluation of EPA's National-Scale Air Toxics Assessment (NATA): Comparison with benzene measurements in Detroit, Michigan. *Atmos. Environ.* **2011**, *45*, 3301–3308. [CrossRef]
2. Fann, N.; Lamson, A.D.; Anenberg, S.C.; Wesson, K.; Risley, D.; Hubbell, B.J. Estimating the National Public Health Burden Associated with Exposure to Ambient $PM_{2.5}$ and Ozone. *Risk Anal.* **2011**, *32*, 81–95. [CrossRef] [PubMed]
3. Trasande, L.; Malecha, P.; Attina, T.M. Particulate Matter Exposure and Preterm Birth: Estimates of U.S. Attributable Burden and Economic Costs. *Environ. Health Perspect.* **2016**, *124*, 1913–1918. [CrossRef] [PubMed]
4. Ghude, S.D.; Jena, C.; Chate, D.M.; Beig, G.; Pfister, G.G.; Kumar, R.; Ramanathan, V. Reductions in India's crop yield due to ozone. *Geophys. Res. Lett.* **2014**, *41*, 5685–5691. [CrossRef]
5. Tong, D.; Mathur, R.; Schere, K.; Kang, D.; Yu, S. The use of air quality forecasts to assess impacts of air pollution on crops: Methodology and case study. *Atmos. Environ.* **2007**, *41*, 8772–8784. [CrossRef]
6. Dennis, R.; Fox, T.; Fuentes, M.; Gilliland, A.; Hanna, S.; Hogrefe, C.; Irwin, J.; Rao, S.T.; Scheffe, R.; Schere, K.; et al. A framework for evaluating regional-scale numerical photochemical modeling systems. *Environ. Fluid Mech.* **2010**, *10*, 471–489. [CrossRef] [PubMed]
7. Appel, K.W.; Gilliam, R.C.; Davis, N.; Zubrow, A.; Howard, S.C. Overview of the Atmospheric Model Evaluation Tool (AMET) V1.1 for Evaluating Meteorological and Air Quality Models. *Environ. Model. Softw.* **2011**, *26*, 434–443. [CrossRef]
8. Lee, P.; McQueen, J.; Stajner, I.; Huang, J.; Pan, L.; Tong, D.; Kim, H.; Tang, Y.; Kondragunta, S.; Ruminski, M.; et al. NAQFC developmental forecast guidance for fine particulate matter ($PM_{2.5}$). *Weather Forecast.* **2017**. [CrossRef]
9. Stajner, I.; Lee, P.; McQueen, J.; Draxler, R.; Dickerson, P.; Upadhayay, S. Update on NOAA's Operational Air Quality Predictions. In *Air Pollution Modeling and Its Application XXIV*; Steyn, D.G., Chaumerliac, N., Eds.; Springer: Cham, Switzerland, 2016; pp. 593–597.
10. McKinney, W. Data Structures for Statistical Computing in Python. In Proceedings of the 9th Python in Science Conference, Austin, TX, USA, 28 June–3 July 2010; pp. 51–56.
11. McKinney, W. *Python for data analysis: Data wrangling with Pandas, NumPy, and IPython*; O'Reilly Media, Inc.: Sebastopol, CA, USA, 2012.
12. Van der Walt, S.; Colbert, S.C.; Varoquaux, G. The NumPy Array: A Structure for Efficient Numerical Computation. *Comput. Sci. Eng.* **2011**, *13*, 22–30. [CrossRef]
13. Hunter, J.D. Matplotlib: A 2D graphics environment. *Comput. Sci. Eng.* **2007**, *9*, 90–95. [CrossRef]
14. Taylor, K.E. Summarizing multiple aspects of model performance in a single diagram. *J. Geophys. Res. Atmos.* **2001**, *106*, 7183–7192. [CrossRef]
15. ENVIRON. *User's Guide, Comprehensive Air Quality Model with Extensions (CAMx), Version 5.40*; ENVIRON: Arlington, TX, USA, 2011.
16. Rao, S.T.; Zurbenko, I.G. Detecting and Tracking Changes in Ozone Air Quality. *Air Waste* **1994**, *44*, 1089–1092. [CrossRef] [PubMed]
17. Wise, E.K.; Comrie, A.C. Extending the Kolmogorov–Zurbenko Filter: Application to Ozone, Particulate Matter, and Meteorological Trends. *J. Air Waste Manag. Assoc.* **2005**, *55*, 1208–1216. [CrossRef] [PubMed]

MDPI

St. Alban-Anlage 66

4052 Basel, Switzerland

Tel. +41 61 683 77 34

Fax +41 61 302 89 18

http://www.mdpi.com

Atmosphere Editorial Office

E-mail: atmosphere@mdpi.com

http://www.mdpi.com/journal/atmosphere

www.ingramcontent.com/pod-product-compliance
Lightning Source LLC
Chambersburg PA
CDIIW051047210320
41597CB00033B/5810